AF316565

Nanoporous Gold
From an Ancient Technology to a High-Tech Material

Nanoporous Gold
From an Ancient Technology to a High-Tech Material

Edited by

Arne Wittstock
*Nanoscale Synthesis and Characterization Laboratory, Livermore, CA, USA;
Institute of Applied and Physical Chemistry, Bremen, Germany*

Jürgen Biener
Nanoscale Synthesis and Characterization Laboratory, Livermore, CA, USA

Jonah Erlebacher
*Department of Materials Science and Engineering, Johns Hopkins University,
Baltimore, MD, USA*

Marcus Bäumer
Institute of Applied and Physical Chemistry, Bremen, Germany

RSCPublishing

RSC Nanoscience & Nanotechnology No. 22

ISBN: 978-1-84973-374-8
ISSN: 1757-7136

A catalogue record for this book is available from the British Library

Published by The Royal Society of Chemistry,
Thomas Graham House, Science Park, Milton Road,
Cambridge CB4 0WF, UK

Registered Charity Number 207890

For further information see our web site at www.rsc.org

Printed in the United Kingdom by Henry Ling Limited, at the Dorset Press, Dorchester, DT1 1HD

Preface

High surface area nano-materials have recently attracted significant interest due to potential applications in various technological fields such as catalysis, sensors and actuators, energy harvesting and storage, and optics. In contrast to bulk materials containing micrometer and macroscopic features, the properties of high surface area nano-materials are not dominated by atoms within the bulk, but rather by their nanoscale architecture. Nanoporous gold is a prime example of a fascinating class of mesoporous bulk-metals that have been intensively investigated in recent years.

Over centuries, humans coveted gold for its colour, persistent lustre, and relative scarcity. Indeed, the value of gold extends beyond its market price; it is a highly recycled, "green" and sustainable material-resource, generally non-toxic for humans and the environment. Considering gold's rich and intimate role in human history, it is surprising that many of the familiar characteristics of a gold nugget can be transformed just by bringing it into a nanoporous form, in particular making it chemically less noble. Instead of gold's yellow colour, nanoporous gold appears coppery, its macroscopic dimensions may be sensitive to the surrounding gas atmosphere and facilitate (i.e. catalyse) chemical reactions, even at room temperature. Clearly, reforming pure bulk gold material with nanostructure provides a high surface area material with fascinating, non-intuitive properties. In fact, the changes observed in the properties of gold from the macro- to nano-structured regime are perhaps the most dramatic of any known material and, consequently, have import to a variety of possible technical applications.

This book will provide a broad overview of the emerging novel properties of nanoporous gold as the prime example of corrosion derived bulk nano-structured (high surface area) materials from a multidisciplinary perspective. The chapters are organised so as to approach different facets of the material in a logical progression beginning with fundamental aspects ranging all the way to

RSC Nanoscience & Nanotechnology No. 22
Nanoporous Gold: From an Ancient Technology to a High-Tech Material
Edited by Arne Wittstock, Jürgen Biener, Jonah Erlebacher and Marcus Bäumer
© Royal Society of Chemistry 2012
Published by the Royal Society of Chemistry, www.rsc.org

its applications in various technological sectors. The intent of this approach is to provide a comprehensive and up-to-date platform to understand this class of nanomaterial through the prototypical example of nanoporous gold. In this respect, this book is not solely addressing specialists of a particular academic field but rather a multidisciplinary readership ranging from graduate students to professional researchers in academia and industry.

The book is organized as follows. Chapter 1 will provide a comprehensive introduction to the topic: gold as a resource, its origin, use throughout history, and current production will be described as well as the novel properties arising when bringing the material into a nanostructure. Chapter 2 will cover the general background of Au alloy corrosion and the generation of nanostructures followed by Chapter 3 which will report on experimental model studies on the evolution of the Au nanostructure by corrosion. Chapter 4 will discuss the mechanical properties of nanoporous gold. This chapter will emphasize how critical characteristics like the elastic modulus or the hardness of the material can be measured and how this material relates to its bulk counterpart – leading to the conclusion that smaller can indeed be stronger. Chapter 5 will give an overview of state-of-the-art synthesis and preparation techniques of nanoporous gold materials for a variety of specific technical applications. Subsequent chapters (6 to 10) will cover the advanced applications of nanoporous gold, including optics, catalysis, electro-catalysis (e.g. energy harvesting applications), and sensors. The chapters are written by the world's leading scientists in the particular field. Each chapter will cover one technique/application so that the reader can easily target its favoured topic and extract current, state-of-the-art information of interest to their field or industry.

As the technologically and scientifically most mature corrosion derived nanostructured bulk metal, the subject of nanoporous gold can be forecast to impact the material science and development of advanced devices in a variety of technical fields. We particularly hope this book is suitable to give a review and substantial introduction to gold-based, high surface area nano-materials and their technological applications, for a broad readership. We thank all contributors to this book and we hope that you enjoy reading it.

Arne Wittstock
Jürgen Biener
Jonah Erlebacher
Marcus Bäumer
Livermore, Baltimore, and Bremen

Contents

RSC Nanoscience & Nanotechnology No. 22
Nanoporous Gold: From an Ancient Technology to a High-Tech Material
Edited by Arne Wittstock, Jürgen Biener, Jonah Erlebacher and Marcus Bäumer
© Royal Society of Chemistry 2012
Published by the Royal Society of Chemistry, www.rsc.org

Chapter 10   Nanoporous Gold in Sensor Applications   224
I-Wen Sun and Po-Yu Chen

CHAPTER 1

Introduction to Nanoporous Gold

ARNE WITTSTOCK,*[a,b] JÜRGEN BIENER[a] AND
MARCUS BÄUMER[b]

[a] Lawrence Livermore National Laboratory, Nanoscale Synthesis and
Characterization Laboratory, 7000 East Avenue, Livermore, CA 94500,
USA. [b] University of Bremen, Institute of Applied and Physical Chemistry,
Leobener Strasse NW2, 28359 Bremen, Germany
*Email: wittstock1@llnl.gov

1.1   Nanoporous Gold

Nanoporous gold is a corrosion-derived bulk nanostructured material. It is
generated by the corrosion of an alloy of Au and a less noble metal, such as Ag
or Cu. By electrochemical removal (dealloying) of the less noble constituent, the
remaining gold undergoes a self-organization process forming a three-dimensional bicontinuous porous network of interconnected ligaments (Figure 1.1).
Depending on the preparation conditions, the resulting pores and ligaments can
be as small as 5 nm, but are typically around 30 to 40 nm. By IUPAC definition,
the as-prepared material is mesoporous. Due to its high porosity and small
feature size, this material has a specific surface area in the range of $10 \, \mathrm{m^2 \, g^{-1}}$.
The void or pore volume in the resulting material mostly depends on the concentration of the less noble metal (*e.g.* Ag) in the starting compound. Because of
fundamental limitations for bulk dealloying, such as the 'parting limit' (see
Chapter 2) and the stability of the evolving porous network, alloys containing
between 60 at.% and 80 at.% Ag are most viable. Processing (dealloying) of
according alloys results in pore volumes between about 60% and 80%.

RSC Nanoscience & Nanotechnology No. 22
Nanoporous Gold: From an Ancient Technology to a High-Tech Material
Edited by Arne Wittstock, Jürgen Biener, Jonah Erlebacher and Marcus Bäumer

Published by the Royal Society of Chemistry, www.rsc.org

Early experimental work on corrosion-derived nanoporous Au by Pickering and Swann in the 1960s[1] and by Forty in the 1970s[2,3] focused on the corrosion aspect using this material and its starting alloys, respectively, as a model system for studies on the molecular mechanism of alloy corrosion. With the onset of nanotechnology in the late 1990s and the early 2000s, researchers revealed and developed the potential of this material for a variety of technological aspects. As a consequence, the number of publications dealing with nanoporous Au has increased steeply by about 40% per year, from about 11 publication in the year 2001 to more than 150 in the year 2010.[4] One of the reasons for the success of this material is the comparatively simple preparation of this nanomaterial using bench-top corrosion techniques to generate bulk samples several millimeters in size and even larger. By avoiding financially demanding techniques, such as electron beam lithography, this material became available to a variety of research groups working on the optical or mechanical properties, the catalysis or the electrochemistry of the material.

Besides the availability of the material for different research groups, another crucial factor fuelling interest in this material is its structural and chemical flexibility (see Figure 1.1). Microfabrication of the material using fast ion bombardment has been used to generate various micrometer-sized patterns and structures of interest for mechanical tests, for example.[5,6] Temperature-activated ripening of the nanostructures opens the door to materials with pores and ligaments in the size regime between about 30 nm and several micrometers, without losing the typical bicontinuous structure of the material.[5,7,8] By using templating techniques, such as slip casting of alloy-coated polystyrene beads and subsequent removal of the template, hierarchical nanoporous Au can be generated as well with relative densities as low as 2 to 3%.[9] In addition to these structural variations, the materials surface can be chemically modified with metals, organic entities or metal oxides bringing forward its applications in electrochemistry (*e.g.* fuel-cell applications), sensorics, and catalysis.[10–14]

Although the term 'nanotechnology' is rather new, the use of nanomaterials can be dated back several hundreds or even thousands of years. The first reports on the use of corrosion to generate nanoporous gold can be related to pre-Columbian civilizations, such as the Incans (see Chapter 2). Here, the superficial dealloying and subsequent burnishing of a comparably cheaper Au–Cu alloy (removal of the Cu from the alloy surface) was used to generate a shiny gold surface, giving the work piece the allure of pure gold. Undoubtedly, this must have caused severe frustration in the Spanish conquistadores when melting the looted, apparently pure, gold pieces back in Spain. However, artisans throughout the centuries have used this superficial enrichment of Au alloys as a means of gilding. The technique was thus dubbed depletion gilding or 'mis-en-colour', accordingly. For these reasons, when dealing with nanoporous gold, we speak of an ancient material yet with a novel technological impact.

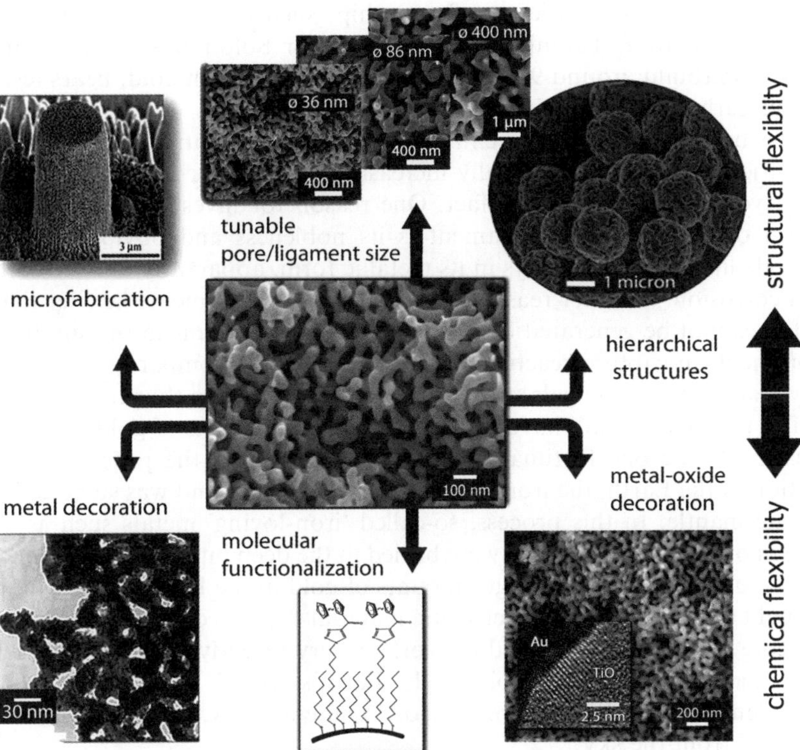

Figure 1.1 Nanoporous Au material: its various structural variations ranging from microfabrication (Chapter 4, adapted from Biener *et al.*[5] Copyright 2006 American Chemical Society), tunable pore and ligament sizes or hierarchical structures based on templating techniques and its chemical variations, ranging from decoration of the surface of the ligaments with metals such as Pt or Pd (Chapter 9, adapted from Ding and Chen[10] with permission of Cambridge University Press), organic entities or even metal oxides, such as titania (Chapter 8).

1.2 Gold—Some Facts

Gold is an element that has inspired mankind at all times. Around 700 BC, the famous Greek poet Hesiod described the five ages of mankind in his poem *Works and Days*.[15] The first age he calls the golden age of mankind, free from later gradual deterioration of moral values. Indeed, gold was the first metal recognized by humans even before bronze and iron.[16] Traces of gold can be found in early human settlements ($\sim$8000 BC) in the Euphrat and Tigris river system, an area that is today part of the Iraq. Archeological findings of gold from later high civilizations such as Egypt and Mesopotamia can be dated back as early as 4000 BC. Back in those days, gold was already used as a means of payment, in the form of rings (about 2700 BC) and later in the form of coins (since 600 BC). The earliest craftsmanship, such as the funeral mask of the Egyptian Pharaoh Tutankhamun (1223 BC) or Solomon's famous temple in Jerusalem (build around 950 BC), allegedly overlaid with gold, bears testimony to this early and lasting fascination with gold.

The belief in gold as the embodiment of value continued throughout the centuries. Today, the drastically increasing demand for gold as a safe investment very much reflects this fact. One reason for investing in gold as a safe haven of treasure and investment is its nobleness and obvious inability to corrode like iron. Gold stays in its metallic form, apparently unaffected by dirt and corrosion. Another reason is that gold is rare. Elements heavier than iron (^{56}Fe) cannot be generated by fusion reactions that occur in the sun but result from neutron-capture reactions as in supernovae, a comparatively rare astrophysical event. In the galaxy, elements such as gold are thus inherently rare.

The fact that we still have unexpectedly high amounts of gold in the Earth's crust might be one fortunate cosmic coincidence. In the process of the formation of the Earth, the iron segregated into the core and was surrounded by a silicate mantle. In this process, so-called 'iron-loving' metals such as gold as well as other precious metals were buried in the deep interior of our planet. The presence of unexpectedly large amounts of gold in the Earth's crust, though, is thought to originate from meteoritic material deposited after the formation of the core mantle. Willbold and co-workers very recently provided experimental data that the presence of gold in the Earth's crust is due to a meteor bombardment about 3.9 billion years ago.[17] Quite literally, one could say that our gold fell from the sky.

Yet, the gold in the Earth's crust is not evenly distributed. About 40% of all gold mined within the last 120 years has come from just one area, the Witwatersrand Basin in South Africa (Transvaal and Orange Free State). Since the average concentration of gold in the Earth's crust is only about 2–5 parts per billion (by weight),[16,18] mining of traces in many areas of the world is not economically viable. Hence, gold such as that in the Witwatersrand Basin is mined from deposits where it is largely enriched. The average gold content of the ore in these so-called *conglomerate deposits* is about 12 ppm.[16] Although these numbers still sound very small, the gold in those deposits is enriched by a factor of more than 1000, and mining becomes economically viable. The explanation for this enrichment is that hot fluids ($\sim$400 °C) in the Earth's crust are able to dissolve gold traces in the soil,[19] yet at shear zone faults, the pressure and temperature drop, and gold precipitates, forming veins (*primary deposits*).[18] Later sedimentation and compaction of sand and shingle deposits can lead to *conglomerate deposits, i.e.* gold deposits finely dispersed in rock.

The mining of gold strongly profits from the development of production processes to remove the gold from deposits and bringing it into a concentrated and pure form. The total production of gold up to the end of the Roman Empire is estimated to be about 10 000 t, slowing down in the medieval ages to a total of about 2000 to 3000 t. With the advent of the Californian Gold Rush in 1848, the production of gold increased steeply to about 40 t per year.[16] However, these numbers are very small when compared to the current mining and production of gold. In 2010, the worldwide annual production (mining) of gold was 2500 t, more than 20 times the annual production from 1848.[20] In recent years, the production of gold from South Africa, mainly from the above-mentioned famous Witwatersrand Basin, has been continuously declining from 464 t in 1998 to only 190 t in 2010, making it still the fourth largest producer of gold. Today, the largest producer of gold is China, with an annual production of 345 t, followed by Australia (255 t) and the USA (230 t). Importantly, the recycling of gold has becomes increasingly relevant. In 2010, the US Department of Geological Survey reported that in the USA, 205 t of gold was generated from recycling of new and old scrap, a number that was even higher than the actual consumption.[21] This reflects the fact that gold is a very sustainable resource.

The mining and production of gold are of considerable economic value. Already, in 1997, about 500 000 people worldwide were employed in the mining and production of gold. The value associated with gold primary production at this time was already in the range of ~$30 billion. Since then, it has increased drastically to about $100 billion in 2010 (based on an average price of gold of $40/g). Most gold is used in the form of jewellery (50%) or as a means of exchange (coins and money) and monetary assets (40%).[22] The remaining 10% of the world's gold production goes into the fields of industry and technology. Here, it is mostly used in the field of electronics, mainly because of its high electrical conductivity and corrosion resistivity. Another typical application of gold is its use as an 'inert' coating. For example, gold alloyed to silver and palladium (Pallacid) forms a very resistant coating against mineral acids, especially at higher temperatures.

Yet, gold, as a nanomaterial, is also playing an emerging and fascinating role in modern technologies such as biomedicine, water purification, fuel cells, exhaust-gas purification, energy-efficient glazing coatings, touch-sensitive screens, and even solar cells, to name just a few.[22] At first glance, some of these applications seem to contradict the apparent inertness of gold, as they involve, for example, chemical activity at the gold surface. However, it is the unique combination of gold and its nanostructure that is opening doors to these high-tech applications. With the anticipated ongoing increase in worldwide research funding on nanotechnology to be over $12 billion per year in 2015 (according to Cientifica Ltd), gold as a nanomaterial certainly is opening up a variety of promising new perspectives also in the applied and industrial sector.

1.3 What Makes 'Nano' Special?

As mentioned above, a crucial factor for various applications of nanoporous gold is its nanostructure. The term 'nano' or 'nanomaterial' implies that the critical features and structures of the material are several nanometers (10^{-9} m) in size, typically below 100 nm. This is less than one-hundredth of the diameter of a human hair. When dealing with 'nano'-sized materials, new material properties and applications emerge.[23,24] But why, and to what extent, should the properties of a material such as gold 'change' when they become nano-sized? One important factor is the reduced characteristic length and the increased surface-to-volume ratio of the materials. When moving from objects with characteristic length-scales (*e.g.* diameter, edge length) in the order of centimeters or millimeters to the nanometer scale, the surface-to-volume ratio is increased by a factor of more than a million. The proportion of surface atoms in ligaments (pillars) with diameters in the range of tens of nanometers is in the order of several percent. These hence constitute a sizable proportion of the total number of atoms in the material (Figure 1.2).

Nanosized objects are to a large extent determined by the physicochemical properties of their surfaces, as distinguished from their bulk properties. In a way, nano-sized materials bridge the realm of isolated atoms on the subnanometer scale ($\text{Å} = 10^{-10}$ m) and bulk materials. Atoms at the surface have a lower number of neighboring atoms. Considering a cut through a bulk crystal, the atoms along this cut have fewer neighbors and accordingly a lower co-ordination number (CN). Since gold crystallizes in a face-centered cubic lattice, the CN of an atom in the bulk is 12, meaning that every gold atom is surrounded by 12 nearest neighbors. An Au atom at the surface has a reduced coordination number; depending on whether it is located on a terrace, edge, or kink site, the CN is reduced to 9, 7, and 6, respectively (Figure 1.2). In the following, effects related to surfaces and low coordination of atoms are exemplarily discussed.

One effect is that the charge redistribution due to low coordination of surface atoms gives rise to surface stress (see Chapter 7).[25–29] The excess charge from unsaturated bonds is redistributed into in-plane bonds, which are strengthened or weakened, depending on whether the additional charge is distributed into bonding or antibonding states. Compared to the bond length of atoms within the bulk, the distance between surface atoms can thus be altered, leading to surface stress. For compressive stress, the surface tends to expand compared to the bulk, whereas for tensile stress, the surface tends to shrink.[27] A typical consequence of surface stress is a reconstruction of the surface. As an example, the Au(111) surface reconstructs in the commonly known Herringbone reconstruction.[30,31] The tensile surface stress of the Au(111) surface is com-pensated by the incorporation of about 4% additional atoms into the surface as compared to the bulk. Recently, several investigations have shown that in the case of nanoporous metals (npAu, npPt), a change in surface stress can lead to macroscopically detectable strain of the entire material, an effect formerly known only for piezo ceramics.[25,26,32,33]

An additional effect is used in catalysis (see Chapters 8–10). The low coor-dination of surface atoms enhances the reactivity of (metal) surfaces. The desorption enthalpy of CO on gold surfaces is (among others) a function of the roughness that is the coordination number of surface atoms (Figure 1.2).[34] Theoretically, there are two ways in which the coordination of a surface atom and the geometry of the surface, respectively, can influence the interaction with molecules and the activation barrier for the reaction; one is electronic, and the other is geometric. The adsorption and bonding of a molecule on a metal surface are determined by the electronic structure of the substrate. A concept describing the interaction (adsorption) of various molecules with transition metals was introduced by Nørskov and co-workers.[35] In transition metals,

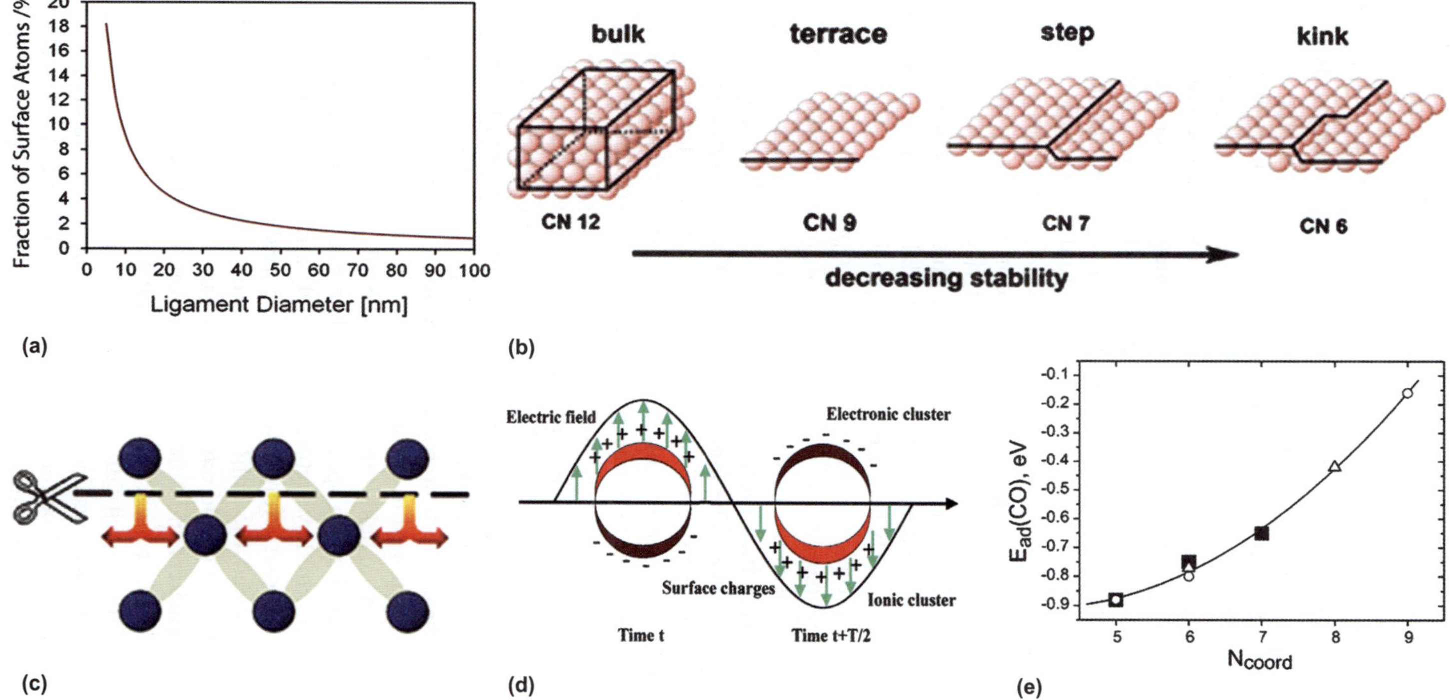

Figure 1.2 Characteristics of nanomaterials. (a) When reducing the length such as the diameter of a ligament or pillar to only nanometers ($nm = 10^{-9}$ m), the proportion of surface atoms reaches several, even tens of, percent. (b) Surface atoms have a lower number of nearest neighbors (coordination number, CN). (c) Imaginary cut through the bulk of a material. The electron density of 'dangling bonds' is redistributed into in-plane bonds, giving rise, for example, to surface stress (see Chapter 7). (d) Coherent excitation of electrons in conduction bands near the surface gives rise to surface plasmons (Chapter 6). In the case of nanomaterials such as nanoporous gold, the optical properties are determined by surface plasmons (image courtesy X. Lang and M. Chen, Chapter 6). (e) Adsorption and bonding of molecules on the surface are a function of the CN and roughness of the surface, respectively (Chapter 8). The energy of adsorption (E_{ads}) decreases (stronger bonding) with decreasing CN (calculated for the adsorption of CO on Au(332) and Au(321); reproduced from Biener *et al.*[25] by permission of the PCCP Owner Societies).

the so-called d-bands determine the reactivity of the substrate, as they are the highest-lying electronic states. The d-band center is defined as the first moment ('average') of the density of d-states. The position (*i.e.* energy) of the d-band center with respect to the Fermi level affects the ability of a metal surface atom to form a bond with an adsorbate. For example, transition metals tend to have higher d-states in case of low coordination numbers (kink- and edge-sites). Accordingly, these atoms interact more strongly with adsorbates than atoms in a close-packed surface.[36] Additionally, the geometry of the substrate provides specific adsorption sites that can be crucial for the activation and reaction of the adsorbed molecules.[37] The impact of the latter geometrical effect depends on the transition state, *i.e.* the optimal local geometry of atoms to form and break bonds. Certainly, the electronic and the geometrical effect are hardly distinguishable, as the local geometry of a surface atom always affects its electronic structure. Yet, both effects contribute to the catalytic performance of a metal. In the case of gold, the presence of low coordinated atoms even determines whether the surface shows any catalytic activity.[38]

A third effect is employed in optics (see Chapter 6). This visual effect of the size of the gold structure can be demonstrated by colloidal solutions of (spherical) gold particles, which vary between red/ruby for particles in the range of several nanometers and the typical gold-yellow for larger particles. Similar to the case of depletion gilding (see above, Chapter 1.1.), this optical 'nano-effect' of gold, albeit unwittingly, was used by artisans many centuries ago. Roman artisans mixed gold chloride into molten glass, generating colloidal gold to impart a ruby stain to the glass. The famous Lycurgus Cup manufactured in about AD 300 by Roman artisans testifies to this early use of nanotechnology: depending on whether the light passes through the glass from the back or illuminates the front, it changes its color from ruby (back) to green (front). Further examples of glass staining using gold particles are the colorful glass windows of cathedrals used throughout the centuries in Europe. This effect originates from electrons in metals occupying delocalized bands. Coherent oscillations of conducting electrons along the surface determine the optical properties of metals. These oscillations are called surface plasmons. One type of surface plasmon resonance, localized surface plasmon resonance (LSPR), is pronounced in nanomaterials. When it is confined in a small space, as in the case of nanomaterials, a strong electromagnetic field is induced. The LSPR can be used among others for surface-enhanced Raman spectroscopy. The size, shape, and orientation of the nanostructures are key factors that strongly affect the LSPR. This optical property of gold nanostructures might even be used for novel optical-data storage and could exceed the limit of current storage media by several orders of magnitude.[39] By using gold nanorods with varying aspect ratios (the ratio of the length and diameter) and different orientations on a flat surface, the LSPR of single rods can be addressed, generating a bit density of 1.1 Tbit cm^{-3}, which corresponds to a data capacity of 1.6 Tbyte for a typical DVD.

More details regarding the advanced applications and the background of this exemplary gold nanomaterial will be focused on in the following chapters. The topics will range from the fundamentals of the formation of nano-patterns by

Au alloy corrosion on a general level (Chapter 2) as well as on an experimental level (Chapter 3) all the way to its mechanical characterization (Chapter 4), and synthesis and preparation using microfabrication (Chapter 5), for example. The final five chapters will cover the advanced applications of this material, including optics (Chapter 6), actuation (Chapter 7), catalysis and surface science (Chapter 8), electrocatalysis and energy-related applications (Chapter 9), and sensors (Chapter 10). This book will combine multidisciplinary topics describing this gold-based nanomaterial in its various aspects. This approach attempts to provide a comprehensive and up-to-date platform to help understand the interplay of various disciplines such as material science, physics, and chemistry as a critical step for the generation and development of novel nanomaterials for technical applications.

Acknowledgments

We are very grateful to Michael Bagge-Hansen (Lawrence Livermore Ntl. Laboratory) for many helpful discussions, inspiration, and critically reading the manuscripts. A portion of this work (AW and JB) was performed under the auspices of US Department of Energy by LLNL under Contract DE-AC52-07NA27344.

References

1. H. W. Pickering and P. R. Swann, *Corr.*, 1963, **19**, 373t.
2. A. J. Forty and P. Durkin, *Philos. Mag. A*, 1980, **42**, 295–318.
3. A. J. Forty, *Nature*, 1979, **282**, 597–598.
4. ISIWeb of Knowledge as of June/2011, Keyword 'nanoporous gold'.
5. J. Biener, A. M. Hodge, J. R. Hayes, C. A. Volkert, L. A. Zepeda-Ruiz, A. V. Hamza and F. F. Abraham, *Nano Lett.*, 2006, **6**, 2379–2382.
6. D. Lee, X. Wei, X. Chen, M. Zhao, S. C. Jun, J. Hone, E. G. Herbert, W. C. Oliver and J. W. Kysar, *Scr. Mater.*, 2007, **56**, 437–440.
7. J. Biener, G. W. Nyce, A. M. Hodge, M. M. Biener, A. V. Hamza and S. A. Maier, *Adv. Mater.*, 2008, **20**, 1211–1217.
8. A. M. Hodge, J. Biener, J. R. Hayes, P. M. Bythrow, C. A. Volkert and A. V. Hamza, *Acta Mater.*, 2007, **55**, 1343–1349.
9. G. W. Nyce, J. R. Hayes, A. V. Hamza and J. H. Satcher, *Chem. Mat.*, 2007, **19**, 344–346.
10. Y. Ding and M. W. Chen, *MRS Bull.*, 2009, **34**, 569–576.
11. R. Zeis, A. Mathur, G. Fritz, J. Lee and J. Erlebacher, *J. Power Sources*, 2007, **165**, 65–72.
12. M. M. Biener, J. Biener, A. Wichmann, A. Wittstock, T. F. Baumann, M. Bäumer and A. V. Hamza, *Nano Lett.*, 2011, **11**, 3085–3090.
13. A. Wittstock, A. Wichmann, J. Biener and M. Baumer, *Faraday Disc*, 2011, **152**, 87–98.
14. C. C. Jia, H. M. Yin, H. Y. Ma, R. Y. Wang, X. B. Ge, A. Q. Zhou, X. H. Xu and Y. Ding, *J. Phys. Chem. C*, 2009, **113**, 16138–16143.

15. A. N. A. Hesiod, *Hesiod; Theogony; Works and Days; Shield*, JHU Press, Baltimore, MD, 2004.
16. H. Renner, G. Schlamp, D. Hollmann, H. M. Lüschow, P. Tews, J. Rothaut, K. Dermann, A. Knödler, C. Hecht, M. Schlott, R. Drieselmann, C. Peter and R. Schiele, in *Ullmann's Encyclopedia of Industrial Chemistry*, Wiley-VCH, Weinheim, 2000.
17. M. Willbold, T. Elliott and S. Moorbath, *Nature*, 2011, **477**, 195–198.
18. H. E. Frimmel, *Science*, 2002, **297**, 1815–1817.
19. R. R. Loucks and J. A. Mavrogenes, *Science*, 1999, **284**, 2159–2163.
20. *Data from Mineral Commodity Report of US Department of the Interior and the US Geological Survey* (http://minerals.usgs.gov/minerals/pubs/mcs/).
21. K. Salazar and M. K. McNutt, ed. *US Dot. Interior and USG Survey*, USGS, Washington, DC, 2011, pp. 66–67.
22. T. W. G. Council, About gold, http://www.gold.org/about_gold/
23. J. Biener, A. Wittstock, T. Baumann, J. Weissmüller, M. Bäumer and A. Hamza, *Materials*, 2009, **2**, 2404–2428.
24. J. Erlebacher and R. Seshadri, *MRS Bull.*, 2009, **34**, 561–568.
25. J. Biener, A. Wittstock, L. A. Zepeda-Ruiz, M. M. Biener, V. Zielasek, D. Kramer, R. N. Viswanath, J. Weissmuller, M. Baumer and A. V. Hamza, *Nat. Mater.*, 2009, **8**, 47–51.
26. J. Weissmuller, R. N. Viswanath, D. Kramer, P. Zimmer, R. Wurschum and H. Gleiter, *Science*, 2003, **300**, 312–315.
27. W. Haiss, *Rep. Progress Phys.*, 2001, **64**, 591–648.
28. H. Ibach, *Surf. Sci. Rep.*, 1997, **29**, 195–263.
29. C. E. Bach, M. Giesen, H. Ibach and T. L. Einstein, *Phys. Rev. Lett.*, 1997, **78**, 4225–4228.
30. J. Biener, M. M. Biener, T. Nowitzki, A. V. Hamza, C. M. Friend, V. Zielasek and M. Baumer, *Chem. Phys. Chem.*, 2006, **7**, 1906–1908.
31. B. K. Min, A. R. Alemozafar, M. M. Biener, J. Biener and C. M. Friend, *Top. Catal.*, 2005, **36**, 77–90.
32. H. J. Jin, L. Kurmanaeva, J. Schmauch, H. Rosner, Y. Ivanisenko and J. Weissmuller, *Acta Mater.*, 2009, **57**, 2665–2672.
33. D. Kramer, R. N. Viswanath and J. Weissmuller, *Nano Lett.*, 2004, **4**, 793–796.
34. W. L. Yim, T. Nowitzki, M. Necke, H. Schnars, P. Nickut, J. Biener, M. M. Biener, V. Zielasek, K. Al-Shamery, T. Kluner and M. Baumer, *J. Phys. Chem. C*, 2007, **111**, 445–451.
35. J. Greeley, J. K. Nørskov and M. Mavrikakis, *Annu. Rev. Phys. Chem.*, 2002, **53**, 319–348.
36. J. K. Nørskov, T. Bligaard, B. Hvolbaek, F. Abild-Pedersen, I. Chorkendorff and C. H. Christensen, *Chem. Soc. Rev.*, 2008, **37**, 2163–2171.
37. J. K. Nørskov, T. Bligaard, A. Logadottir, S. Bahn, L. B. Hansen, M. Bollinger, H. Bengaard, B. Hammer, Z. Sljivancanin, M. Mavrikakis, Y. Xu, S. Dahl and C. J. H. Jacobsen, *J. Catal.*, 2002, **209**, 275–278.
38. H. Falsig, B. Hvolbaek, I. S. Kristensen, T. Jiang, T. Bligaard, C. H. Christensen and J. K. Nørskov, *Angew. Chem., Int. Ed.*, 2008, **47**, 4835–4839.
39. P. Zijlstra, J. W. M. Chon and M. Gu, *Nature*, 2009, **459**, 410–413.

Fundamental Physics and Chemistry of Nanoporosity Evolution During Dealloying

J. ERLEBACHER,[*a] R. C. NEWMAN[b] AND
K. SIERADZKI[c]

[a] Johns Hopkins University, Department of Materials Science and
Engineering, 3400 N. Charles St., Baltimore, MD 21218, USA; [b] University
of Toronto, Department of Chemical Engineering and Applied Chemistry,
200 College Street, Toronto, ON, Canada M5S 3E5; [c] Arizona State
University, Department of Materials Science and Engineering,
Tempe, AZ 85287-6106, USA
*Email: jonah.erlebacher@jhu.edu

2.1 Introduction

2.1.1 Context—Pattern-Forming Instabilities and Nanostructure Fabrication

Nanoporous gold (NPG) forms during electrochemical dealloying because
there is a kinetic competition between dissolution of silver from the base parent
alloy and surface diffusion of the remaining gold along the metal/electrolyte
interface. In this chapter, we will explore this competition from structural,
thermodynamic, and kinetic standpoints. We focus on the silver–gold system as

RSC Nanoscience & Nanotechnology No. 22
Nanoporous Gold: From an Ancient Technology to a High-Tech Material
Edited by Arne Wittstock, Jürgen Biener, Jonah Erlebacher and Marcus Bäumer

Published by the Royal Society of Chemistry, www.rsc.org

a model, but the same concepts can be applied to any dealloyable system. The morphology and composition profile of NPG varies greatly depending on how the kinetics are played against thermodynamics, making this material, and all porous materials made by dealloying, highly tunable and adaptable to the many applications discussed later in this book.

Often when there is atom-scale competition between a physical phenomenon that leads to roughening of a surface and a second physical phenomenon that leads to smoothening, a 'pattern-forming' morphological instability arises. Especially when the length scales over which these physical processes operate are different, a balance can appear that leads to a most-unstable mode that is the one which physically appears. Figure 2.1 shows the basic idea in the form of a dispersion relation—if we plot the growth rate of a feature with wavenumber q ($= 2\pi/\lambda$, where λ is the physical wavelength) *vs.* q, then roughening effects lead to positive growth rate, and smoothening effects lead to negative growth rates. The sum of these two effects is negative for large q, usually a manifestation of capillary-induced smoothening that quickly flattens very rough surface patches, and positive above some threshold. The physically manifested length scale is that with the fastest growth rate.

Pattern-forming instabilities have been studied for many years, in many contexts. An early thorough study concerns the so-called Mullins–Sekerka instability that leads to dendrite formation during solidification.[1] Here, a coupled thermal and mass diffusion problem is expressed as capillarity action operating as a smoothing phenomenon in competition with fast-growing morphological perturbations that penetrate into a constitutionally supercooled liquid so as to increase their growth velocity.

A general lesson of pattern-forming instabilities is that rate-limiting kinetics are never found on the side of the interface that contains the growing phase.[2] With regards to solidification, the growing phase is the solid, and the

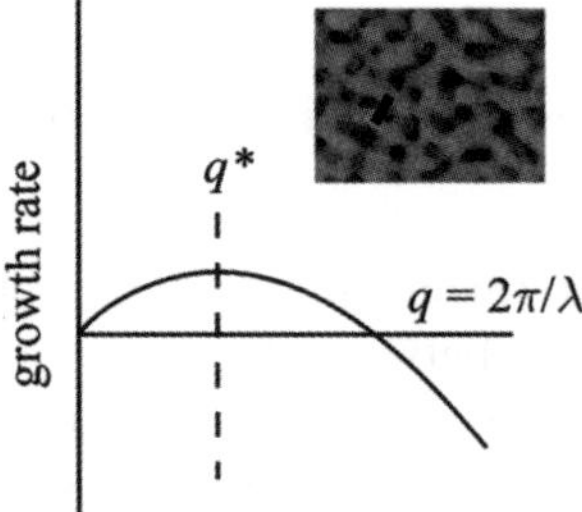

Figure 2.1   Pattern-forming instabilities and nanoporous gold. In any pattern-forming instability, there is a feature size that grows fastest ($\lambda^* = 2\pi/q^*$). Short-wavelength features generally get smoothened by capillary forces, and long-wavelength features grow slowly. In nanoporous gold, the fastest-growing wavelength might correspond to any of the following, which are all correlated: the ligament diameter, the void channel diameter, or the ligament/ligament spacing (black line in the inset SEM micrograph of NPG).

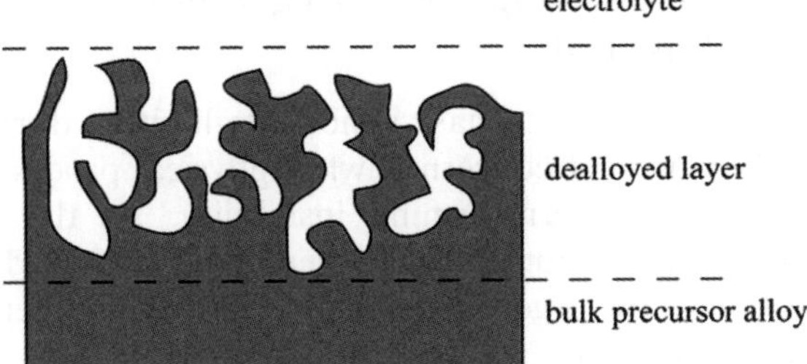

Figure 2.2   Multiscale complexity in dealloying. During selective dissolution into an electrolyte, a nanoporous dealloyed layer grows with some velocity in the bulk precursor alloy (Ag/Au for NPG). The thickness of the dealloyed layer may be hundreds or thousands of micrometers, but the width of the channels is only of the order of 10 nm. Because the growing phase is 'electrolyte dendrites', the rate-limiting step during porosity evolution cannot be transport through this phase.

rate-limiting behavior is primarily heat conduction through the liquid phase. If heat flow were rate-limiting on the solid side of the interface, the interface temperature would remain very high, so the temperature gradient in the melt would be low, and there would be no constitutional supercooling. Capillary forces would thus dominate, and the interface would solidify as a smooth, flat surface.

Figure 2.2 shows the multi-scale complexity of all the kinetics involved in the pattern-forming instability of nanoporous gold, including: (a) structural complexity of the alloy crystallography, with a length scale of ~ 2–$3\,\text{nm}$ when considering individual ligaments and voids, and length scales orders of magnitude larger when considering the depth of porosity may reach several millimeters; (b) kinetics of dissolution and surface diffusion at the atomic scale of the alloy/electrolyte interface, and kinetics of diffusion of silver through much-longer electrolyte filled channels out into the bulk electrolyte. This picture does not include the potential mechanical differences between the dealloyed layer and the substrate that can lead to cracking and other stress–corrosion-related phenomena.

However, in NPG, at least one complexity can be approximated away—as a pattern-forming instability, we identify the electrolyte side of the metal/electrolyte interface as the growing phase. As such, mass transport (liquid-phase diffusion) through the growing phase cannot be rate-limiting. For this reason, we can concentrate our attention at the metal/electrolyte interface directly, and not worry how silver gets through the channels. This conclusion is quite reasonable from our knowledge of diffusion coefficients in electrolyte, as well. For NPG grown via free corrosion in concentrated nitric acid, a typical growth rate is of the order of $1\,\text{nm}\,\text{s}^{-1}$.[3] Diffusion coefficients of ions in aqueous electrolyte tend to be between 10^{-6} and $10^{-4}\,\text{cm}^2\,\text{s}^{-1}$. Over 1 s, then, using the back-of-the-envelope $x = \sqrt{Dt}$ calculation, one would expect diffusion lengths of 10^{-3} to $10^{-2}\,\text{cm}$, many orders of magnitude larger than the 1 nm distance the interface has grown in the same period of time.

2.1.2   Basic Phenomenology—Parting Limit and Critical Potential

If mass-transport in the liquid phase is not rate-limiting during nanoporosity evolution in NPG, our next concern is what physical processes do take place that can lead to the pattern-forming instability. At the metal/electrolyte interface, two physical phenomena are operating—surface diffusion and dissolution. Both of these are largely dependent on the electrochemical potential at the metal/electrolyte interface, so we begin there.

Figure 2.3 shows a prototypical potential/dissolution-rate measurement that is used to characterize dealloying; measurements on real systems generally follow this kind of trend closely. This kind of description was first made by Pickering's pioneering studies in the 1960s and 1970s.[4,5] In a typical experiment, the potential is scanned at a fixed rate increasingly positive (anodic) relative to the standard reversible potential E^0_{Ag} of the least noble alloy component, silver in our case. For potentials a few hundred millivolts above E^0_{Ag}, the current remains small but measureable, and then quickly rises at the 'critical potential' E_c. A post-mortem analysis of the sample shows that in the region below the critical potential E_c, the surface is essentially smooth and uncorroded, but has become enriched in gold. Above the critical potential, the surface has become porous.

The critical potential itself is highly dependent on a number of different factors, primarily the composition of the base alloy. If the base alloy is rich in silver, then E_c lies closer to E^0_{Ag}, and increases with increasing alloy composition until about 55 at.% Au.[6] At this point (Figure 2.4), there is a step-function in the potential required to corrode the material, if at all possible. Unless the electrolyte being used can also dissolve gold, a passivating oxide will form. This composition threshold of ~55 at.% is quite sharp, and is known as a *parting limit*. The origins of the parting limit are discussed in detail in Section 2.2, but in short are associated with geometric features of the underlying crystal lattice

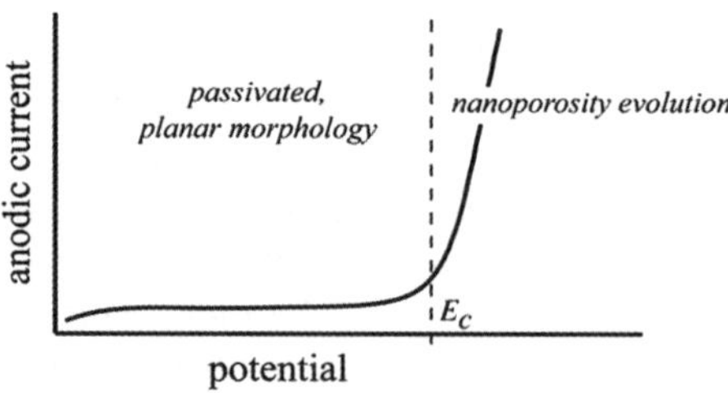

Figure 2.3   Typical current/potential behavior of a porosity-forming alloy during a potentiodynamic sweep. The origin here is the reversible potential E^0_{Ag} of the dissolvable alloy component. Upon sweeping the current positive of E^0_{Ag}, there is a small dissolution current associated with surface passivation that remains low until the potential reaches a critical value E_c, at which point the current rises quickly.

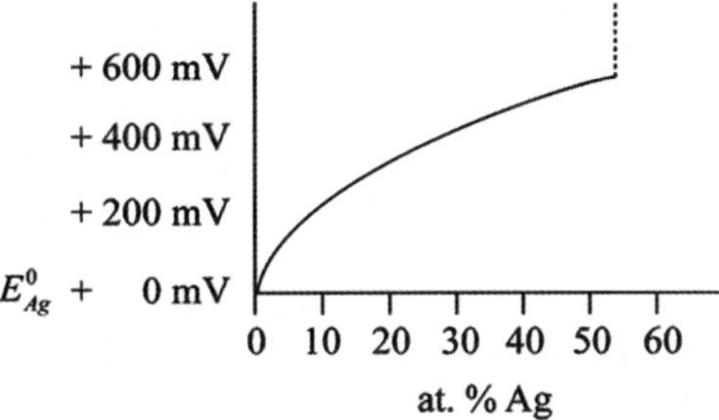

Figure 2.4 Critical potential versus composition in Ag/Au alloys, showing the parting limit at ~ 55 at.% Ag. This curve is meant to be approximate. The details of the critical potential change depending on the electrolyte, and any silver concentration pre-dissolved in the electrolyte—that is, the critical potential varies with any factor that can change the relative energy of silver in the alloy with silver in electrolyte, and any factor that can change the interfacial energy between the alloy and the electrolyte.

and the idea that one must have interconnected channels of silver to allow dissolution to proceed through the thickness of the material. Similar thresholds associated with a crystal geometry property are seen in other areas of corrosion science, including the very sharp minimum of ~ 13 at.% Cr required to form a passivating oxide on stainless steel.[7,8]

With regards to the critical potential, the historical problem here has been that early measurements of the critical potential did not produce an E_c that approached E^0_{Ag} as the composition approached pure silver; usually a ~ 200 mV gap remained. On one hand, there are irreversibilities associated with any electrochemical measurement, so one might expect that some sort of over-potential was required, but pure silver still dissolved at its reversible potential in the same electrolytes. As will be discussed in detail in Section 2.3, the problem was intrinsically associated with the measurement technique—scanning the potential during a critical potential measurement leads to a serious over-estimate for E_c, even using slow scan rates. For the formation of porous metals, this observation might be relatively innocuous, but it is quite important for the broader problem of identifying regimes of potential that lead to corrosion over long periods of time. As computationally analyzed by Erlebacher,[9] experimentally realized by Corcoran *et al.*,[10] and placed on firm theoretical footing by Sieradzki *et al.*,[11] the correct measurement to determine the propensity of an alloy to corrode or not should be first to equilibrate the material at a potential near or below E^0_{Ag}, and then step to a fixed higher potential and watch the current decay. For a fixed higher potential below E_c, the current will decay indefinitely, whereas for a fixed higher potential above E_c, the current will reach a steady-state dissolution current.

The origins of the critical potential are intrinsically thermodynamic in origin. For a fixed alloy composition, the potential E_c marks that potential above which there is a thermodynamic driving force for dissolution, and below which there is none. Note the careful distinction between 'driving force for dissolution' and porosity evolution. We now recognize that these are different concepts, and whether a pore forms or not is a subset of the 'alloy is susceptible to corrosion' problem. In Section 4 of this chapter, we discuss what is known about the kinetics of porosity evolution. This is a rich and complex area, and phenomena are still being discovered, such as the observation of 'bubbles' (voids not connected to the overall void space of the bicontinuous NPG structure) within ligaments. However, to date, all the basic geometric phenomena have been explained by a kinetic competition between dissolution of silver and surface diffusion of the remaining gold.

2.1.3 Short History of Theoretical Approaches for NPG Morphology Evolution

The first theoretical studies of NPG in the context of corrosion were made by Pickering's group and Forty's group in the 1970s.[12,13] Pickering and Wagner, in particular, developed a model that was widely adopted in the 1970s/1980s, and occasionally still appears to explain the formation of the corroded layer.[5] The idea of this model was to note (Figure 2.5) that the porous layer is poor in less-noble metal (silver in our case), whereas the undealloyed bulk is not. Thus, there is a concentration gradient in silver from the exterior surface to the interior, and thus a driving force for diffusion.

Pickering and Wagner recognized that diffusion of bulk silver to the electrolyte must require bulk diffusion mechanisms. Unfortunately, bulk diffusion mechanisms such as vacancy-mediated diffusion are too slow at room temperature. For this reason, Pickering and Wagner invoked the hypothesis of

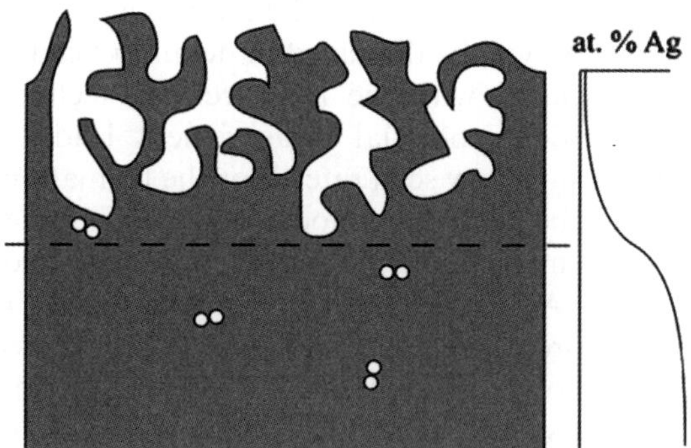

Figure 2.5 Pickering–Wagner model for dealloying. Pickering and Wagner noted that because the dealloyed layer is poor in silver, there is a driving force for bulk diffusion of silver to this region. As the vacancy diffusion is too slow, this invoked an exotic (and never-discovered) point defect (the 'di-vacancy', created at dealloyed layer/bulk layer interface) to mediate mass transport. In practice, there is virtually no composition gradient in either the dealloyed layer or the bulk, just a discontinuity at the etch front (dotted line).

fast-moving undercoordinated bulk defects ('divacancies'), formed at the rough metal/electrolyte interface, that could perform the job. Many recognized at the time that these too would move too slowly, and there is no evidence such defects exist.

The surface diffusion model of Erlebacher and Sieradzki explains porosity evolution, and most phenomena associated with porosity evolution, as explained in Section 4 of this chapter, but we do not want to dismiss Pickering and Wagner completely at this point. When the divacancy model was developed, there were two critical experimental observations that were not well understood. First, the nature of the dealloyed layer, *i.e.* a bicontinuous solid/void with pore sizes of order 10 nm was not clear. Although Forty looked at crystalline dealloyed films under TEM and saw 'porosity',[12] literature from this period generally talks of 'spongy' dealloyed layers, with speculation that these layers are amorphous. Only later did Sieradzki show that the dealloyed layers are very porous, when SEM resolution became high enough in the 1980s.[14]

Second, although bulk diffusion is clearly too slow, it was not at all clear that surface diffusion was fast enough. Forty suggested surface diffusion must play a role, but his model accounted for neither the correct crystallographic effects nor the required diffusivities. Consider that if the growth rate of a pore is ~ 1 nm s^{-1}, then gold atoms must retreat out of the pore within that same time, so as not to passivate it. Diffusion distances of 1 nm in 1 s lead to required surface diffusion coefficients greater than 10^{-14} cm^2 s^{-1}. Ultra-high vacuum measurements of surface diffusion coefficients on gold and similar metals had yielded values typically less than 10^{-20} cm^2 s^{-1}.[15]

It is only in the last 20 years or so that we have developed an understanding that surface diffusion in electrolytes can increase the diffusivity to the required values. Giesen, for instance, has shown most tangibly using *in situ* scanning tunneling microscopy that surface self-diffusion of copper in electrolyte can be faster even than 10^{-12} cm^2 s^{-1}.[16] Similar enhancements are seen for gold, too, and are quite dramatic in chloride containing electrolyte.[17] In electrolyte, it seems that the diffusing species is not the lone adatom or even kink-site atoms on step edges, but rather atoms somewhat solvated by anions in solution—the energy of this interaction is not enough to dissolve the atom, but enough to weaken its binding to the surface and thus increase the migration rate. As a result, in electrolyte surface diffusion coefficients are of values sufficient for porosity evolution.

One last observation regarding surface diffusion: because the diffusivity is structurally related to anion adsorption effects, it should follow that the nature of porosity evolution should strongly depend on the potential, and also on the particular anion used. This is indeed observed, and there are differences (qualitative differences primarily; we do not know of any detailed studies) in porosity evolution in different acids. The surface diffusion rate also depends on the applied potential. Usually the trend is that diffusivity rises with increasing potential, unless a surface oxide forms. However, this change in diffusivity is small relative to the exponential increase in dissolution rate and, for that reason, is usually justifiably neglected.

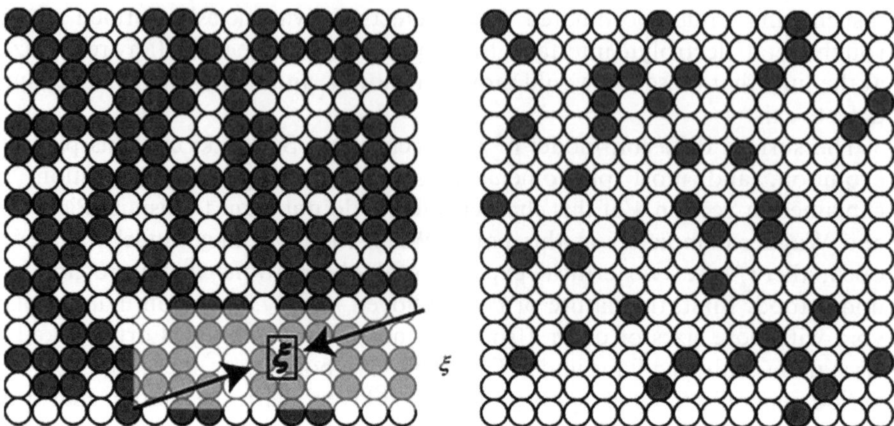

Figure 2.6   Basic notions of percolation theory. In the lattice on the left, the silver atoms (dark) form a percolating network because a chain of these can be formed from one side to the other. A length scale, ξ, can be associated with the 'width' of a percolating chain, as shown. In the lattice on the right, the silver atoms do not percolate.

2.2   Structural Considerations

2.2.1   Percolation Basics

The primary structural feature that accounts for the parting limit—the composition of gold over which selective dissolution cannot occur—is percolation. Percolation is a well-defined mathematical concept, and is described in the following way (Figure 2.6)[18]: consider a lattice occupied by A and B atoms (*e.g.*, A = Au, B = Ag). We say that B atoms percolate through the lattice if, for any system size, the probability of there being a chain of B atoms from one side of the system to the other is unity.

The threshold composition of B below which B atoms do not percolate is called the percolation threshold p_c. The value of p_c depends on the particular lattice under consideration. For face-centered cubic (fcc), $p_c = 0.198$; for body-centered cubic, $p_c = 0.256$; and for diamond cubic, $p_c = 0.43$.[19,20] For all of these lattices, the value of p_c remains below 50%, *i.e.* at a much lower concentration than parting limits in real materials. We will address this issue below.

A second geometric aspect of percolation that is useful in understanding porosity evolution is the percolation length, ξ. This length represents the average cluster diameter of a percolating channel, as also shown in Figure 2.6. It is related to the probability that any one particular silver atom has another silver atom next to it. As such, ξ is a strong function of composition. This dependence can be easily derived for 1-D percolation by considering the statistics of building a percolating network on a lattice with nearest-neighbor distance *a*. We cite the one-dimensional result—if, given an alloy with composition X_B, $\xi(X_B)$ is given by

$$\xi(X_B) = \frac{(1 + X_B)a}{1 - X_B}. \tag{2.1}$$

In using this for ξ, we assume that the three-dimensional lattice structure is in effect isotropic in lattice spacing. Clearly, this is an approximation,, but there are no closed form solutions available for $\xi(X_B)$ in three dimensions. This approximation works probably better than it should, not only for the dealloying problem discussed here but also for the alloy underpotential deposition problem. One way of addressing this issue in the future would be to perform simulations measuring the average cluster size of the 'percolation backbone' as a function of composition. In many ways, this type of measurement is related to high-density percolation as described below.

2.2.2 Percolation Applied to Nanoporous Gold: The Parting Limit

Gold–silver forms an fcc lattice, so, according to the discussion above, silver should start to percolate through a sample of any size at a composition of ~20 at.% Ag. Therein sits the first difficulty, as the parting limit in Au–Ag alloys is close to 60 at.% Ag. Why such a discrepancy? One idea is associated with kinetics. Consider the low-energy (111) surface of Ag–Au being attacked by an electrolyte. If a percolating chain one atom wide is attacked, this will form a terrace vacancy. If there are mobile gold atoms nearby, these may diffuse across the terrace and fill this terrace vacancy. To attack the percolating network further, the acid must penetrate more quickly into the bulk, dissolving subsurface atoms, than the time required to be filled by a gold atom. The overall effect of this should be to suppress the parting limit, as seen at the beginning of this paragraph.

However, one implication of this idea is that the parting limit should be a strong function of temperature, as the concentration of mobile gold atoms on the surface should change exponentially with temperature like any thermally activated species in equilibrium, but the parting limit does not seem to be strongly temperature-dependent. A related explanation that is more aligned with experimental observation has been discussed by Sieradzki, Newman, and co-workers, who used the concept of 'high-density percolation' (HDP) thresholds.[21,22] The physical idea is that for a network of silver atoms to be dissolvable in an electrolyte, it must have sufficient width so that anions in electrolyte can solvate it. A silver atom with 10 or 11 near-neighbors, for instance, should not be solvatable, whereas a silver atom sitting at a kink site on a step edge (six near neighbors) should be.

Kinetic Monte Carlo (KMC) simulations have been employed to test this model.[22] In KMC, kinetic rules for dissolution and surface diffusion are applied to a system of atoms that evolves in time. HDP can be tested by simply allowing a silver atom to dissolve or not, depending on its local coordination. It was found that when 10- and 11-coordinated silver atoms were not allowed to dissolve, the effective percolation threshold dropped to the observed 55 at.% Au, close to the real parting limit for these alloys.

2.3 Thermodynamic Origins of the Critical Potential

The critical potential, E_c , introduced above, is the potential at which silver is thermodynamically favored to dissolve into electrolyte rather than remain in the alloy. The ingredients of a model for E_c as it varies with composition are shown in Figure 2.7, and the description of the model begins with the following thought experiment: consider a surface of Ag–Au below the parting limit. Regardless of composition, silver atoms will dissolve copiously at first, as the alloy composition is rich in Ag. However, the gold atoms will quickly accumulate on the surface, passivating all step edges (the driving forces for such mass transport are further discussed below), so it is reasonable to consider simply a (111) terrace of material, and not worry about its boundaries. For dissolution to penetrate further into the bulk, a 9-coordinated silver atom must be dissolved from the terrace, leaving behind a terrace vacancy. As discussed below, this is the rate-limiting step in dissolution, but for now we need only recognize that the next atoms to be attacked by electrolyte are not below the terrace vacancy but rather are lateral to it. That is, those silver atoms that coordinated the first atom to dissolve will be attacked next; in fact, they will be attacked relatively quickly, since their coordination is less than 9. Thus, dissolution occurs somewhat stepwise—regions of surface silver will get attacked all at once.

The size of these silver patches will depend on the composition, and their diameters on average correspond to the percolation length, ξ. With this in mind, the critical potential naturally arises via a thermodynamic energy balance. During dissolution, the system lowers free energy by dissolving away silver from the terrace, but there is an energy penalty associated with creating the new surface that is the boundary of the silver patch. The critical potential is that potential for which the free energy reduction during dissolution exactly balances that required for the creation of a new surface.

Quantitatively, these ideas are expressed in the following model (Figure 2.7).[11] As dealloying occurs stepwise, we may consider the free energy change

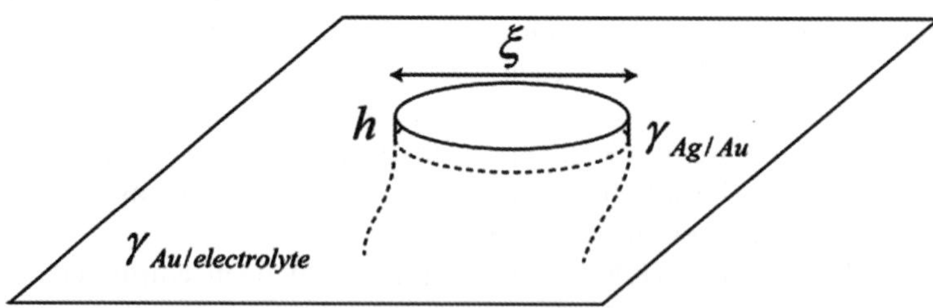

Figure 2.7 Ingredients to the theory for the critical potential for dealloying. The meaning of symbols are discussed in the text.

associated with dissolution of a region of surface area, A, and height, h; more specifically, A will be the diameter of a region of silver atoms whose width equals the percolation length, ξ. Dissolution of this patch of material will create a new surface area along the perimeter of the pit, with an area equal to $h\pi\xi$. On the surface side, energetically, this corresponds to destruction of the interface with energy $\gamma_{Ag/Au}$ (an energy penalty) and the formation of a new interface with energy $\gamma_{Au/elec}$. On the electrolyte side, the free energy change upon dissolution of this region will be $NkT \ln (a_{Ag^+}/a_{Ag})$, where $N = \pi h\,(\xi/2)^2/\Omega$ (*i.e.* the number of Ag atoms dissolved), where Ω is the atomic volume, and a_{Ag^+}, a_{Ag} are the activities of silver in the electrolyte and in the bulk alloy, respectively. Equilibrium is established when the free energy difference $\Delta F = 0$, where

$$\Delta F = NkT \ln\left(a_{Ag^+}/a_{Ag}\right) - h\pi\xi\gamma_{Ag/Au} + h\pi\xi\gamma_{Au/elec}. \tag{2.2}$$

The free energy difference is translated into an overpotential by noting that the activity of silver in the electrolyte corresponds to an electrochemical potential measured relative to the standard reversible Ag/Ag^+ potential, $\Delta E = -(kT/nq)\ln\left(a_{Ag^+}\right)$, where q is the elementary charge, and $n = 1$ is the charge transferred per cation formation. Within factors of integer multiples of π, these observations lead to the following expression for the overpotential as a function of percolation length, ξ:

$$\Delta E(\xi) = E_c(\xi) - E_{Ag^+/Ag} = \frac{2\left(\gamma_{Au/elec} - \gamma_{Ag/Au}\right)\Omega}{nq\xi} - kT \ln a_{Ag}. \tag{2.3}$$

There is excellent quantitative agreement between the predictions of this model with experimental dissolution studies in bulk three-dimensional alloys, model two-dimensional multilayered structures, and KMC simulations.

There are a few observations we must make with regard to the above analysis. First, experimental agreement with the model for E_c cannot be made using the potential sweep experiments such as that discussed by Pickering. There, the history of the sample, and the passivation associated with sweeping the potential through values above E_{Ag}^0 but below E_c, affects the measurement, and the timescales are too short. A potentiostatic measurement where the potential is pulsed from below E_{Ag}^0 to one higher, E, and then the current decay is tracked, is the correct method. If $E < E_c$, then the current will decay forever, whereas if $E > E_c$, dissolution will occur. For this reason, E_c as discussed here is an *intrinsic* critical potential. In the case of Ag–Au alloys E_c in fact sits a few hundred millivolts below any critical potential measured using a potential sweep.

Second, the critical potential as derived does not separate regimes of potential that lead to porosity evolution or not. In fact, while dissolution occurs at a potential slightly positive of E_c, porosity formation does not occur as usually observed when making NPG.

2.4 Kinetic of Porosity Evolution

Dealloying at potentials sufficiently above the critical potential (~ 100 mV; this is not a well-established value) leads to porosity evolution. The nature of porosity *vs.* adjustable parameters such as electrolyte, potential, and temperature confirms that porosity evolution is a richly variable phenomenon: at low temperatures (one can dealloy below 0 °C in concentrated nitric acid), pores are very small,[23] and pores are also very small at a very high potential: the amount of residual silver in the pores increases with decreasing pore size, and can be varied from almost zero to ~ 25 at.% Ag.[24]

2.4.1 KMC Simulations

The understanding of the evolution of porosity has been strongly influenced by examination of KMC simulation models.[9,25] In KMC, the time evolution of an ensemble of atoms is tracked as atoms diffuse and dissolve according to simple rate laws meant to mimic realistic surface diffusion and/or electrochemical dissolution events. These rate laws are simple Arrhenius expressions, a product of an attempt frequency v and a Boltzmann factor $\exp\,(-E^*/k_\mathrm{B}T)$, but the values for v and E^* are variable.

For diffusion, one particularly simple choice is to set E^* according to a 'bond-breaking model,' where the $E^* = nE_b$ associated with a particular event is the product of number of bonds n broken, each with energy E_b, in the event. For diffusion in NPG, a value of $E_\mathrm{b} = 0.15$ eV with $v = 10^{13}\,\mathrm{s}^{-1}$ leads to realistic coarsening of simulated structures, realistic in the sense that the time evolution of simulated porous structures of a particular scale roughly matches the experimental timescale of coarsening. Details such as anion-absorption influences on diffusion coefficients have been included in KMC models, but these are secondary effects as discussed above. The primary inputs to KMC models of porosity evolution have tried to capture the basic effects in which bulk vacancies diffuse more slowly than atoms on step edges, which in turn diffuse more slowly than adatoms, all without disallowing any one particular diffusing species.

For dissolution, expression of the Butler–Volmer equation in terms of a bond-breaking model yields an activation barrier in the form $E^* = nE_\mathrm{b} - \phi$, where ϕ is the applied potential. This expression applied to potentials implicitly in the Tafel regime, *i.e.* well above the reversible potential, and also well above the intrinsic critical potential. As such, deposition need not be included in the model (ignoring solution side-effects). For dissolution, $v = 10^4\,\mathrm{s}^{-1}$ has worked very well and corresponds to an exchange current density converted to a per-atom basis.

2.4.2 Working Model for Porosity Evolution

KMC simulations of simultaneous dissolution and diffusion in NPG according to the above rules have been spectacularly successful in modeling most effects associated with dealloying, including:

- the formation of passivation and porosity forming regimes in simulated potential sweeps;

- a parting limit of $\sim$55 at.% Au;
- the variation of pore size with increasing potential;
- the identification of intrinsic and extrinsic critical potentials;
- the formation of bubble voids inside of solid ligaments.

These successes have led to a qualitative working model for porosity evolution, shown in Figure 2.8.[26,27] In the initial stages of dissolution, silver is dissolved in a quasi-layer-by-layer manner. The first step is to form a terrace vacancy. This vacancy then grows laterally as the neighboring, percolating, and low-coordination silver atoms are dissolved. Gold atoms could be left behind as laterally uncoordinated adatoms, but a simple regular solution model for the interface shows that their equilibrium concentration of this kind of surface species should be of the order of 10^{-8}, *i.e.* much higher than the bulk. Therefore, gold atoms tend to diffuse and accumulate along the receding step edges, covering mounds and passivating steps rather than being left as non-equilibrated adatoms. This is a kind of uphill diffusion that can be modeled by the Cahn–Hilliard equation, discussed below.

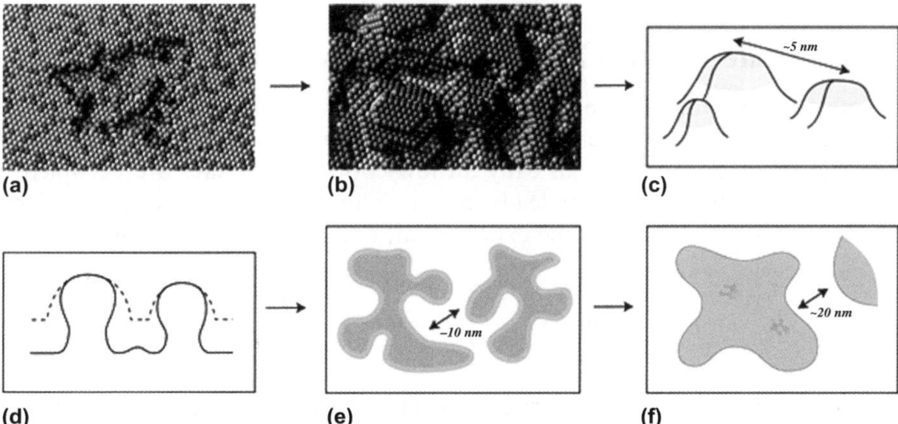

Figure 2.8   Working model for porosity evolution during dealloying. (a) The rate-limiting step is the formation of a terrace vacancy in the low-energy (111)-oriented surface that grows laterally into a vacancy cluster, the step edges of which are passivated with gold. (b) Dissolution proceeds layer by layer, in which surface regions between islands are dissolved, and all low-coordination sites are passivated with gold. (c) As the mounds grow in height, their bases increase in diameter, requiring more gold to passivate them; however, less gold is available as the width of exposed alloy between mounds decreases. (d) Mounds become gold-poor at their base and are undercut, leading to a surface-area increase and the formation of ligaments. New gold clusters can now nucleate between existing ligaments, bifurcating the porosity. (e) As porosity extends into the bulk, the intrinsic structure of porous gold is a silver-gold alloy with a surface passivated with gold. (f) Secondary coarsening leads to an increase in length scale and dissolution of further amounts of silver.

The mechanism of surface-area increase is that gold-covered mounds become undercut. Gold atoms that are liberated upon the surface in the middle of these undercut regions will sometimes nucleate new mounds rather than diffuse to existing ligaments. Thus, as porosity extends into the bulk, an intrinsic porous network is formed that has a core-shell microstructure comprising the base alloy core, and a gold skin.

In the acids in which fast surface diffusion of gold atoms occurs, a secondary coarsening process will be in effect that will grow the ligament/pore size behind the dissolution front. This will lead to further etching of silver and a reduction in the overall silver content. This idea is consistent with two steps associated with strategies to reduce the mobility of gold. The first was to dealloy in neutral pH solution, creating a kind of pitting corrosion/dealloying that acidified only the etch front. A short distance behind the etch front, the surface passivated with an oxide. This had the effect of reducing the ligament size to ~ 5 nm, and also of 'trapping' residual silver, up to 25 at.%.[24] The second step was to add a small amount of lower-mobility Pt to the parent alloy (5 at.%).[28] During dealloying of this material, the ligament size was < 5 nm in diameter, and stable to both thermal and electrochemical annealing, even with a residual silver fraction greater than 20 at.%.

2.4.3 Rate-Limiting Behavior

Note that the rate-limiting step in the porosity-evolution process is the dissolution of a terrace silver atom to form a terrace vacancy, because after the first layer of dissolution, basically all low-coordination sites are occupied by gold atoms. This observation leads to a prediction that the rise in current J with potential ϕ ($\phi > E_c$) should be exponential, and have the following form:

$$\left(\ln(J) - \frac{\Phi}{k_B T} \right) = \ln(v'_E) - \frac{E_T^*}{k_B T} \qquad (2.4)$$

Here, E_T^* is the activation barrier associated with dissolution of silver from a terrace site, forming a terrace vacancy. A point of confusion has informally been raised here, namely, E_T^* is not a formation energy, *i.e.* the difference in energy of the system with and without the terrace vacancy, but the activation barrier.

Recently, this expression was probed via a temperature-dependent measurement of the slope of the dissolution current *vs.* potential curve.[29] It was shown that indeed the dissolution current can be easily fitted to eqn (2.4), and that the effective activation energy from dissolution from a terrace site was reasonably found to be within a range of values, the limits of which are set experimentally by vacancy diffusion (on the high end) and step-edge diffusion (on the low end). Although this measurement is essentially indirect, it represents to our knowledge the only method to measure the activation barrier associated with dissolution from a terrace site.

2.4.4 Analytical Models for Porosity Evolution

While the KMC model for dealloying has been quite successful, it is still a description that focuses on the microscopic physics of diffusion and dissolution, and not larger-scale behavior such as the capillary forces that should tend to smoothen the surface. For this reason, there have been a few attempts to write analytical equations that describe surface evolution during dissolution.[26] The earliest of these was a one-dimensional model that expressed the coupled interface velocity, v_n (normal to the surface), surface concentration of gold, C, and surface diffusion current, J_S (see Figure 2.9):

$$\frac{\partial C}{\partial t} = v_n C_0 - v_n \kappa C - \nabla \cdot J_S \tag{2.5}$$

$$J_S = -M(C)\left(\frac{\partial^2 f}{\partial c^2}\right)\nabla C + 2M(C)w\nabla^3 C \tag{2.6}$$

$$v_n(C) = V(C)\left[1 - \left(\frac{\gamma\Omega}{k_B T}\right)\kappa\right]; \quad V(C) = V_0(\phi)\exp\left(\frac{-C}{C^*}\right). \tag{2.7}$$

The first equation here controls the rate of change in gold concentration on the surface. Here, C_0 is the bulk alloy composition, so the term $v_n C_0$ is the rate at which gold is coming out of the bulk and accumulating. This accumulation is opposed by two effects. The first is purely geometric—a convex surface with curvature κ growing normal to itself will increase its surface area, diluting surface gold concentration and vice versa. This effect is reflected in the second term.

The divergence of the surface current also leads to changes in gold on the interface, but here the challenge is to capture this feature in the working model that gold tends to diffuse along with step edges rather than be left behind as adatoms on the surface. This is a kind of 'uphill diffusion' because gold adatoms are diffusing so as to accumulate, increasing their local concentration, and not moving so as to distribute themselves uniformly over the entire surface. Simple Fickian diffusion is unable to capture this effect, but it is captured by the

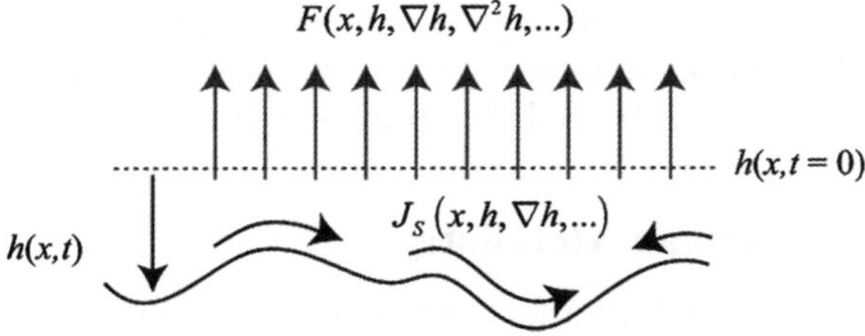

Figure 2.9 Kinetic processes at the metal/electrolyte interface during dealloying. The dissolution flux F may depend on lateral position x, pore depth h, and surface gradients and curvature of the pore depth. The surface diffusion flux J_S is non-uniform as well, also potentially dependent on position and pore depth.

Cahn–Hilliard equation, the second of the above equations. Here, the Cahn–Hilliard equation has been written in its completely general form including a concentration-dependent mobility, $M(C)$, gradient energy coefficient, w, and curvature of the local free energy, f, with respect to surface mole fraction, c.

The interface velocity itself is controlled by dissolution, and here a quasi-empirical expression has been used. KMC simulations suggested that the interface velocity of a planar surface with concentration C was given by the exponential relationship given above, where $V_0(\phi)$ and C^* were fitted to the measured values of $V(C)$. To correct this expression for a curved surface, a slowing-down force associated with the capillarity must be included. This Gibbs–Thomson effect must have the form shown in the third equation, above, where γ is the surface energy, Ω is the atomic volume, and $k_B T$ is the product of Boltzmann's constant and temperature.

Perturbation analysis of the above series of equations is rendered difficult because all planar surfaces start off with an excess of silver. Thus, all surfaces initially roughen, regardless of composition.[30] Such roughness will increase until enough gold has accumulated such that capillary forces lead to smoothening on timescales of the order of, or shorter than, the dissolution rate. Nonetheless, numerical solution of these equations has been studied using advanced analytical tools, and it has been confirmed that the expressions do indeed lead to porosity evolution.[31]

There are many challenges left in this area. To test these analytical models, comparison to experiment is of course critical, but in the short term it will be necessary to compare the evolution of a simulated (analytical) structure with an atomistic (KMC) model. This comparison is quite difficult, as it requires translation of a discrete set of atoms into a smooth mesh. Recently, several methods have been developed to do just this,[31] but analytical modeling of ligament pinch-off, a coarsening effect, remains a significant challenge.

2.5 Nanoporous Gold Throughout History

While dealloying is enjoying a resurgence as a nanofabrication tool, it has primarily been studied as a detrimental material corrosion problem, one to be avoided. In the course of studying this material, we have run across a number of interesting historical contexts where aspects of the physics and chemistry of nanoporous gold have appeared. We provide a brief summary here.

2.5.1 Pre-Columbian Metallurgy

It has long been known that a form of depletion gilding was practiced by pre-Columbian artesans, who made gold–copper alloys, packed them in mineral salts that dealloyed the near surface region of copper, and then burnished the surface to produce an object that appeared to be pure gold. An excellent review on this subject has been provided by Lechtmann.[32] Undoubtedly, the primary motivation for this kind of dealloying was economic. This probably upset the conquistadores, who brought such artifacts back to Spain and melted them down only to discover they were primarily copper, but the subject of who had more basis for feeling aggrieved is beyond the scope of our expertise.

2.5.2   Parting limits and Leonardo da Vinci

The Italian Renaissance saw the first mentions of NPG and NPG-related concepts in Western civilization. Cennino Cennini, in the first 'how-to' book on art, 'Il libro del'arte' (1430s), mentions that one should never use silver–gold leaf in outdoor applications, as it turns black.[33] This might be tarnish but could also be the darkening of alloyed gold upon dealloying. Also, in the Italian Renaissance, Leonardo da Vinci, in the Codex Atlanticus (*ca.* 1508), discusses parting silver from silver/gold alloys using potassium nitrate (saltpeter).[34]

2.5.3   Origins of Modern Dealloying Theory

In the late 19th and early 20th century, the thermodynamics of ordered *vs.* disordered alloys was a lively subject of interest that was mainly investigated using electrochemical methods, since X-rays had barely been discovered (1901), and Bragg's law (1913) had not yet been developed. The system Cu_3Au was particularly useful for studying these questions, because a disordered alloy can be be easily quenched with high temperatures, whereas an ordered alloy appeared with sufficiently slow cooling. Measurements of the electrochemical potential required to dealloy copper from such alloys revealed that potentials almost 200 mV more positive were required to dissolve Cu from ordered alloys compared to disordered ones. Such measurements introduced concepts such as the critical potential and parting limit.[35,36]

2.6   Summary

The theoretical foundations of the study of nanoporous gold have made significant progress in the past 20 years. This is a particularly rich area of materials science, touching on the branches of electrochemistry, surface science, thermodynamics, and applied mathematics. Successes include clarification of the nature of percolation in describing the parting limit, the nature of thermodynamics in predicting the critical potential, and a working model for porosity evolution that includes only realistic surface diffusion kinetics. An important aspect to note is that the rise in interest in nanoporous gold, and nanoporous metals made by dealloying generally, could not have been made without these advances in the fundamental areas.

Acknowledgment

We are grateful to the NSF for financial support under program DMR-1003901 (J.E.).

References

1. W. W. Mullins and R. F. Sekerka, *J. Appl. Phys.*, 1964, **35**, 444.
2. J. S. Langer, *Science*, 1989, **243**, 1150.
3. Y. Ding, Y. J. Kim and J. Erlebacher, *Adv. Mater.*, 2004, **16**, 1897.
4. H. W. Pickering, *Corr. Sci.*, 1983, **23**, 1107.
5. H. W. Pickering and C. Wagner, *J. Electrochem. Soc.*, 1967, **114**, 698.
6. K. Sieradzki, N. Dimitrov, D. Movrin and J. Erlebacher, *J. Electrochem. Soc.*, 2002, **149**, B370.
7. K. Sieradzki and R. C. Newman, *J. Electrochem. Soc.*, 1986, **133**, 1979.
8. Y. M. Kolotyrkin, in *Stress-Corrosion Cracking and Hydrogen Embrittlement of Iron Base Alloys*, ed. R. W. Staehle, J. Hochmann, R. D. McCright and J. E. Slater, National Association of Corrosion Engineers (NACE), Houston, TX, 1977, p. 946.
9. J. Erlebacher, *J. Electrochem. Soc.*, 2004, **141**, C614.
10. A. Dursun, D. V. Pugh and S. G. Corcoran, *Electrochem. Sol. State Lett.*, 2003, **6**, B32.
11. J. Rugolo, J. Erlebacher and K. Sieradzki, *Nat. Mater.*, 2006, **5**, 946.
12. A. J. Forty, *Nature*, 1979, **282**, 597.
13. A. J. Forty and G. Rowlands, *Philos. Mag. A*, 1981, **43**, 171.
14. R. Li and K. Sieradzki, *Phys. Rev. Lett.*, 1992, **68**, 1168.
15. E. G. Seebauer and C. E. Allen, *Prog. Surf. Sci.*, 1995, **49**, 265.
16. S. Baier, S. Dieluwei and M. Giesen, *Surf. Sci.*, 2002, **502**, 463.
17. J. M. Dona and J. Gonzalez Velasco, *J. Phys. Chem.*, 1993, **97**, 4714.
18. D. Stauffer, *Introduction to Percolation Theory*, Taylor & Francis, London, 1986.
19. C. D. Lorenz and R. M. Ziff, *J. Phys. A*, 1998, **31**, 1998.
20. S. C. van der Marck, *Int. J. Mod. Phys. C*, 1998, **9**, 529.
21. K. Sieradzki, R. R. Corderman, K. Shukla and R. C. Newman, *Philos. Mag. A*, 1989, **59**, 713.
22. D. M. Artymowicz, J. Erlebacher and R. C. Newman, *Philos. Mag. A*, 2009, **89**, 1663.
23. L. H. Qian and M. W. Chen, *Appl. Phys. Lett.*, 2007, **91**, 083105.
24. J. Snyder, K. Livi and J. Erlebacher, *J. Electrochem. Soc.*, 2008, **155**, C464.
25. S. A. Policastro, J. C. Carnahan, G. Zangari and R. C. Kelly, *J. Electrochem. Soc.*, 2010, **157**, C328.
26. J. Erlebacher, M. J. Aziz, A. Karma, N. Dimitrov and K. Sieradzki, *Nature*, 2001, **410**, 450.
27. J. Erlebacher and R. Seshadri, *MRS Bull.*, 2009, **34**, 561.
28. J. Snyder, A. Piyapong, A. B. Dalton and J. Erlebacher, *Adv. Mater.*, 2008, **20**, 4883.
29. J. Snyder and J. Erlebacher, *J. Electrochem. Soc.*, 2010, **157**, S21.
30. J. Erlebacher and K. Sieradzki, *Scripta Mater.*, 2003, **49**, 991.
31. C. Eilks and C. M. Elliott, *J. Comp. Phys.*, 2008, **227**, 9727.
32. H. Lechtman, *Sci. Amer.*, 1984, **250**, 56.

33. C. Cennini, *The Craftsman's Handbook: The Italian 'Il libro dell' arte.'* Translated by Daniel V. Thompson., ed. D. V. Thompson, Dover Publications, New York, 1960.
34. L. Reti, *Isis*, 1965, **56**, 307.
35. G. Tammann, *Z. Anorg. U. Allg. Chem.*, 1919, **107**, 144.
36. G. Masing, *Z. Anorg. Allg. Chem.*, 1921, **118**, 293.

CHAPTER 3

Mechanistic Studies of Initial Dealloying

FRANK UWE RENNER

Max-Planck-Institute for Iron Research, Max-Planck-Straße 1, 40237
Düsseldorf, Germany
*Email: renner@mpie.de

3.1 Introduction

Understanding mechanistic relations during processes usually aims for finally
directing the respective processes towards a desired result. Unraveling corrosion mechanisms, for example, will thus help to protect materials from
deterioration, *e.g.* by developing inhibition schemes. Dealloying or selective
dissolution is a corrosion process but is also employed for producing
nanoporous metals, which can be used in a number of different applications.[1–7] A more sustainable society needs more reliable and enduring
materials as well as smarter and cheaper technical processes; and materials
are essentially made up from atomic building blocks, as well as chemical and
electrochemical reactions taking place on a molecular level. For this reason,
mechanistic studies need to be performed on small-nanoscopic-length scales
and ultimately on the atomic level, which is also true for accompanying
(atomistic) simulations.

Regarding the direct atomistic observation, surface science and surface
reactions probably have an advantage of accessibility for their observation.
Historically, the first individual atomic structures have been observed from
surfaces of sharp tips by field ion emission,[8,9] and later, the invention of the

RSC Nanoscience & Nanotechnology No. 22
Nanoporous Gold: From an Ancient Technology to a High-Tech Material
Edited by Arne Wittstock, Jürgen Biener, Jonah Erlebacher and Marcus Bäumer
© Royal Society of Chemistry 2012
Published by the Royal Society of Chemistry, www.rsc.org

scanning tunneling microscope (STM)[10–12] and subsequent scanning probe techniques[13,14] have boosted mechanistic studies on surfaces, from observing surface diffusion to film deposition and growth studies to catalytic reactions. The award of the 2007 Noble Prize for Chemistry to Gerhard Ertl for his contributions[15] mainly for ultrahigh vacuum (UHV) and low-pressure experiments on model systems underlined the success of mechanistic surface science studies. Using and detecting electrons that are inherently surface-sensitive, UHV techniques offer a variety of experimental probes.[16,17] The success and achievements of surface science relied in the past mainly on such UHV-based work. Yet, next to high-pressure catalytic surface reactions, solid–liquid and solid–electrolyte interfaces (*i.e.* solid surfaces immersed in liquid) are the next necessary challenge,[18] although electron-based techniques cannot generally be employed. Material formation processes such as film deposition, ion transfer at battery electrodes, catalytic fuel-cell reactions, or dealloying are at the heart of electrochemical surface science. Many different aspects and applications of nanoporous gold produced by dealloying are reflected in the different chapters of this book. For a fundamental understanding of dealloying, as well as other electrochemical processes, a serious interplay of theory, atomistic simulations of electrochemical processes, and, primarily, *in situ* atomic and nanometer-scale observations is required.

This chapter focuses on reviewing the initial steps of dealloying of Cu_3Au (111). This surface revealed a number of subsequent initial structural stages[19–21] during dealloying that were consequently used to trace the effects of additives to the electrolyte[22–24] and of surface modifications.[21,25,26] Cu–Au alloys possess constituent elements of largely different equilibrium potentials (E_{eq}) and a large difference (13%) in respective lattice parameters. With the pronounced differences in lattice constants, the initial processes of dealloying in this system become naturally addressable by X-ray diffraction.[27] We have employed surface-sensitive X-ray diffraction using highly brilliant synchrotron radiation as an *in situ* tool. This allows high-resolution reciprocal space studies of the surface-structure evolution of such binary alloys during the initial steps of dealloying. The insight into the mechanisms thus obtained enables new ways of controlling the surface and film morphology, in bottom-up as well as top-down scenarios. These detailed mechanistic studies and accompanying atomistic simulations will be continued in the future. Here, a summary of the main results obtained to date is given.

3.2 Sample Preparation, Experimental Techniques, and Simulation

3.2.1 Preparation of Cu_3Au (111) Surfaces

In our studies, we mainly use well-defined single-crystal surfaces (supplied by MaTecK, Germany, and SPL, Netherlands) which were prepared and characterized in ultra-high vacuum (UHV) by ion sputtering and annealing cycles (sputter-cleaning by argon ions at 500 eV for about 1 h and subsequent annealing at about 600 °C). Cu_3Au crystallizes at room temperature in an $L1_2$ structure, where the four available sub-lattices are occupied by Cu and Au, respectively. The lattice constant a_0 is 0.375 nm. Nevertheless, to obtain a well-ordered crystal superstructure, the sample has to be tempered for several hours at 20–30 K below the order–disorder transition temperature of Cu_3Au at 390 °C to allow for a noticeable ripening of the ordered domains with a high driving force and rate. Although Cu_3Au thus possesses an ordered structure at room temperature, we note that our experiments are not sensitive to the influence of the ordered or disordered state of the Cu_3Au substrate, as we did not attempt to obtain a well-ordered superstructure with large domains that would require very accurate control of annealing temperature over many hours. The samples typically have been cooled down over several hours. The surface quality was verified by low-energy electron diffraction (LEED), Auger electron spectroscopy (AES), and atomic force microscopy (AFM) in air. For repeated experiments, the single crystal sample surfaces were re-prepared using sputtering and annealing cycles in UHV. We therefore refrained from approaching the critical potentials for extended time periods in order to avoid a strong corrosion attack and respective irreversible change in surface morphology of the single crystal surfaces. After more severe corrosion attacks of the crystals, the surfaces were re-polished.

The Teflon, polychlorotrifluroethylene, and glass parts of the electrochemical cells were cleaned in a highly oxidizing mixture of H_2SO_4 and H_2O_2 (Caro's acid), and the reference electrodes were rinsed by ultrapure water. The 0.1 M H_2SO_4 and 0.1 M $H_2SO_4 + 0.1$ mM HCl (1 mM KBr, 1 mM and 10 mM KI)

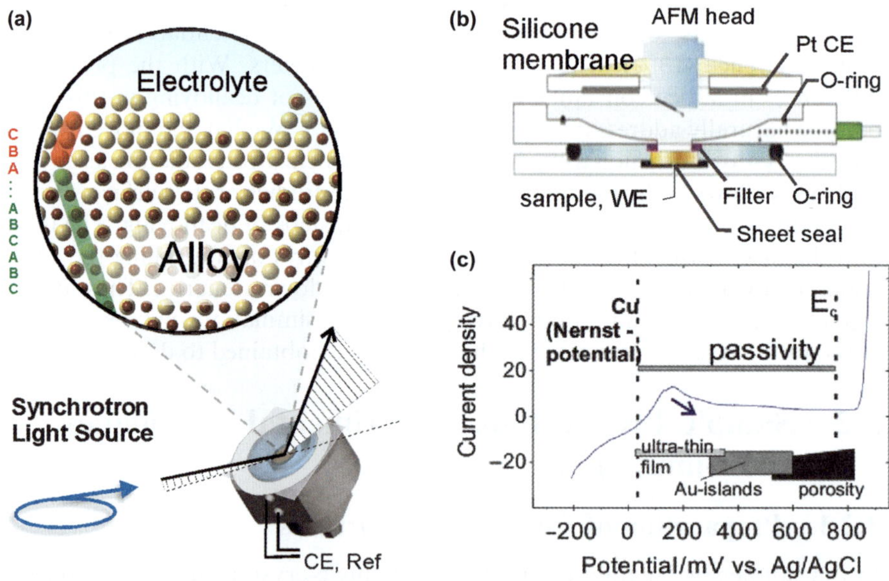

Figure 3.1 Experimental methods for *in situ* studies. (a) Surface-sensitive *in situ* X-ray diffraction employing synchrotron light addressing crystallographic changes during dealloying. (b) Sketch of a home-build *in situ* AFM electrochemical cell.[30] (c) Schematic current–potential curve for Cu_3Au (111).

electrolyte solutions used were mixed from suprapure H_2SO_4 and HCl, KBr, and KI (Merck) with ultrapure water (18.2 M cm, TOC < 3 ppb) and deaerated using argon gas. The current–potential curves were recorded in a common electrochemical cell.

3.2.2 Main Experimental Techniques

One of the greatest obstacles in continuing the success story of surface science from UHV conditions towards solid–liquid and electrochemical interfaces is simply that electron-based techniques cannot be (easily) applied. Nevertheless, studies on electrochemical interfaces do benefit also from a tremendous progress in UHV-based techniques such as scanning Auger electron microscopy (SAEM) if they are applied in an *ex situ* mode. On the other hand, in the last two decades a limited number of powerful *in situ* methods were developed. Optical and spectroscopic methods such as infrared spectroscopy are one such example.[28] Here, we mainly employed *in situ* scanning probe techniques, namely AFM and STM, and *in situ* surface-sensitive X-ray diffraction employing synchrotron light. These two complementary methods have been quickly developed since their first appearance in the late 1980s and are now among the most important *in situ* techniques available for addressing electrochemical interfaces.

The STM included a commercial electrochemical cell made from polychlorotrifluroethylene and a bi-potentiostat (Molecular Imaging, Pico STM). Reference electrodes were self-made microelectrodes. We used Hg/Hg_2SO_4 electrodes in pure sulfuric acid. For better clarity, all potentials are given with respect to Ag/AgCl. The STM tips[29] were prepared from Pt/Ir wire and covered with an organic coating (BASF, Münster, Germany). The STM images were taken at different, subsequently higher potentials and thus correspond to a mix of potential and time. A typical STM image was recorded within 120 s. As the single crystals had to be prepared before the experiments in UHV and transferred through air to be mounted in the electrochemical *in situ* STM, an initial surface oxidation could not be avoided. The STM study thus focuses on morphological changes during dissolution within the passive regime rather than the very initial dissolution steps. For the atomic force microscope (AFM) studies presented we used a commercial JPK NanoWizard I AFM and a home-made *in situ* electrochemical AFM cell[30] made from Kel-F (Valtiner cell, Figure 3.1(b)).

X-ray diffraction is one of the most fruitful methods used in (bulk) materials science. With the inauguration of dedicated, so-called third-generation, synchrotron light sources with their highly brilliant beams, grazing-incidence surface-sensitive X-ray diffraction[31–33] has become a powerful technique for studying electrochemical interfaces *in situ*. Grazing-incidence diffraction maximizes the signal-to-noise ratio of surface structures, and even the diffraction signal from sub-monolayer structures on surfaces can be detected. Due to a break in symmetry at surfaces, X-ray intensity is also observed away from Bragg-points in reciprocal space. This extra intensity occurs along the direction of the surface normal (also called asymptotic Bragg scattering), and for low-index surfaces, the intensity tails connect to a continuous rod of intensity (crystal truncation rod (CTR)). The exact (weak) intensity profile can be measured given the high incident intensity at synchrotron light sources. The CTR is very sensitive to the detailed surface structure, which can be subsequently fitted to the X-ray data.

The described surface-sensitive X-ray diffraction methods allow for *in situ* studies of electrochemical interfaces. In detail, we aim here to follow the structural evolution during initial dealloying (Figure 3.1(c)). For our *in situ* X-ray diffraction experiments, we used a thin-film electrochemical cell[34] design with the common, three-electrode setup as shown in Figure 3.1(a). The cell contains connections for the working electrode (sample, WE), the counter (CE) and reference electrodes (REF), and for the supply and outlet of the electrolyte. Cells of this kind have been widely used and described in the literature.[35,36] During the X-ray experiments, the alloy surface was exposed to film of electrolyte, only a few micrometers thin, between the surface and a thin Mylar® foil sealing the electrochemical cell. To allow mass transport at well-defined potentials, especially when initiating the dissolution of Cu during the stepwise change in applied potential, the volume of electrolyte above the sample surface was increased to a thickness of several millimeters by applying a slight overpressure to the solution in the cell. In our *in situ* cell, the counter-electrode (CE, a Pt wire) and the Ag/AgCl reference electrode (REF) are arranged below the sample surface plane. The sample represents the working electrode (WE) and is held in place inside the electrochemical cell by under-pressure with the help of an aspiration hole in the sample holder underneath the WE. This way of mounting assures clean electrochemical conditions and allows an easy exchange of the disk-like WE. We note that for used X-ray energies of 20 keV and higher, a larger volume of liquid can be penetrated, and droplet cells[37,38] can be used that exhibit a much better geometry with respect to mass transport reactions.

To describe the reciprocal space of the $L1_2$-Cu_3Au crystal, and the scattering vector $\boldsymbol{q}$, we adopted a commonly used surface unit cell[20] with two unit vectors $\mathbf{a_1}$, $\mathbf{a_2}$ lying in the surface plane and the third unit vector $\mathbf{a_3}$ pointing along the surface normal ($a_1 = a_2 = \sqrt{2} \times a_0 = 531$ pm, $a_3 = \sqrt{3} \times a_0 = 650$ pm, $\alpha = \beta = 90°$, $\gamma = 120°$). This leads to a hexagonal reciprocal lattice unit cell of size $a_1^* = a_2^* = 13.670$ nm^{-1}, $a_3^* = 9.666$ nm^{-1} and $\alpha^* = \beta^* = 90°$, $\gamma^* = 60°$. The reciprocal L-coordinate was thus oriented along the surface normal. A sketch of the reciprocal space is shown in the inset of Figure 3.2(a). The reciprocal positions are thus indicated in relative reciprocal units (rlu) with respect to Cu_3Au substrate coordinates. The (2, 0, L) and (–2, 0, L) substrate CTR as well as the (1.9, 0, L) and (–1.9, 0, L) film rods (and their equivalents) are sensitive to the stacking sequence of the close-packed {111} atomic layers along the surface normal. The Cu_3Au substrate shows in addition to the superstructure reflections fundamental Bragg peaks at (–2,0,2) and (2,0,1), and we define the associated orientation as 'ABC' stacking, *i.e.* the sequence of the substrate. We can thus distinguish fcc 'ABC' unambiguously from 'CBA' stacking sequences, as well as a hexagonal 'ABAB' sequence from the respective positions of the corresponding Bragg reflections. The 'CBA' stacking is equivalent to a rotation around the surface normal by 180° or 60° relative to the 'ABC' substrate orientation.

Auger electron maps were recorded using a JAMP-9500F scanning Auger microprobe (JEOL, Japan) with an electrostatic hemispherical analyzer. Depending on the signal-to-noise ratio, an electron beam of 25 kV and 10 nA or 15 nA at an angle of incidence of 30° was applied. These acquisition conditions led to a high spatial resolution with probe diameters of about 10 nm.

In order to gather a deeper and on-atomic-scale understanding of the strain relaxation mechanisms underlying the Cu_3Au/Au system, we performed large-scale calculations[39,40] using embedded atom method (EAM) potentials. For the interatomic interactions of the Cu–Au alloy system, the parametrization proposed by Barrera *et al.* was used. We modeled the $Cu_3Au/Au(111)$ surface using a slab geometry consisting of 6 monolayers (MLs) of Cu_3Au in the $L1_2$ crystal structure and 3 MLs of an Au adlayer. The Au adlayer was biaxially compressed by 5%, while the Cu_3Au substrate was strain-free. The atomic positions of the top 6 MLs were relaxed by a simulated annealing procedure, while the bottom 3 MLs of Cu_3Au were kept fixed. In addition, *ab initio* calculations within the density functional theory, using PBE exchange correlation and VASP package were currently performed[40] to address the structure and orientation of Au surface films on Cu_3Au (111). We briefly discuss here the first results of these simulations.

3.3 Earlier Mechanistic Studies on Initial Dealloying

Dealloying is one of the earliest model processes studied in corrosion science.[41] In recent decades, this has been frequently studied, and new available experimental methods have always been applied for advanced characterization. Electrical currents can be measured with great accuracy, making current measurements in electrochemistry surface-sensitive, *e.g.* in cyclic voltammetry or current transients. Addressing mainly AgAu, CuAu, and CuPd model systems, voltammetry has also been used to address passive-like noble-metal enriched surface layers at subcritical potentials. Vvedenskii has used different sequences of applied potential pulses to infer mechanisms of surface restructuring during initial dealloying of AgAu alloys. Nevertheless, direct studies of the related surface structures were not available to confirm any discussed mechanistic models. Forty addressed the initial evolution of the surfaces during dealloying by TEM and introduced an island growth model based on surface diffusion.[42] With the emergence of powerful surface science tools in the 1970s and 1980s, the detailed surface structure could finally be addressed in more detail in post-process *ex situ* measurements,[43] although the results were often still not sufficiently detailed for a clear test of multi-parameter mechanistic models of dealloying.

A most promising surface-sensitive method appeared with the invention of the STM and its application for *in situ* studies of surface immersed in liquids.[12,44–47] The initial dissolution from surfaces could be followed, but mostly alloys with a high noble metal content (below the so-called 'parting limit') were studied to avoid severe roughening due to bulk dealloying. Despite the inspiring atomistic view on the surfaces during dealloying, STM lacks in most cases a chemical sensitivity. Moffat,[44] Oppenheim,[45] and Chen[46] addressed alloy dissolution by STM. Despite these early pioneering success stories, alloy dissolution has still been rarely studied. Alloy electrochemistry and alloy surface science remain challenging.

The accuracy and speed of atomistic calculations are ever increasing, and calculations and simulations are required to accompany and demand for experimental results. So far, simulations performed on dealloying of Ag–Au alloys by Erlebacher *et al.*[48,49] are still outstanding. Their special approach did not take into account any vertical diffusion. Solely surface diffusion was allowed, and major characteristics of dealloying such as a pseudo-passive regime and a clearly defined critical potential were reproduced. These simulations so far do not allow for development of strain during dealloying.

3.4 Initial Dealloying of Cu₃Au (111)

During our recent detailed characterization[19–21] of initial dealloying, Cu_3Au (111) emerged as a prolific model system because of the unique structural features developing during the initial dealloying steps. Cu–Au alloys possess constituent elements of largely different equilibrium potentials (E_{eq}) and a large difference (13%) in respective lattice parameters. As a consequence of the different lattice parameters, the developing interface of the substrate and the expected Au-rich phase during dealloying might be influenced by strain effects. With the pronounced differences in lattice constants, the initial processes of dealloying in this system can also be addressed by X-ray diffraction. We have employed surface-sensitive X-ray diffraction using highly brilliant synchrotron radiation as an *in situ* tool. This allows high-resolution reciprocal space studies on the surface-structure evolution of binary alloys during the initial steps of corrosion. In addition, a number of other surface-sensitive methods such as *in situ* STM, (*in situ*) AFM, Auger electron spectroscopy, and electron microscopy have been employed.

The equilibrium potential for Cu in pure 0.1 M H_2SO_4 electrolyte is approximately 100 mV *vs.* Ag/AgCl, and the critical potential approximately 800 mV. The potential regime directly above the equilibrium potential of Cu up to the critical potential is characterized by a (passive-like) low current. As will be described later, an epitaxial ultra-thin Au-rich layer forms in this region with an inverted CBA … stacking sequence compared to the $L1_2$ Cu₃Au (111) substrate (ABCAB …). At lower overpotentials ('low overpotential regime') the surface is covered by an ultra-thin Au-rich film, up to 3–4 atomic ML thick. At medium overpotentials of about 350 mV to 600 mV ('medium overpotential regime') this ultra-thin Au-rich film transforms into almost pure Au islands, thus with a thickness of about 10–15 ML and inheriting the inverted stacking sequence. Finally, above 600 mV (close to the critical potential), a new substrate-oriented Au structure appears, and the initial stacking-inverted layer is vanishing. In what follows, we review our recent studies of experimental characterization and first results on *ab initio* and large-scale molecular dynamics simulation of initial stages of dealloying of Cu_3Au (111). Furthermore, the influence of halide additives to the sulfuric acid solution and the effect of a thiol-modified starting surface are described.

3.4.1   Clean Cu_3Au (111) Starting Surface

The selective dissolution of Cu from Cu–Au alloys in electrolyte occurs from the top-most atomic layer. Many earlier studies used as-polished surfaces with a relatively unknown state of preparation on the atomic level, but residual defects, surface roughness, or partial amorphization after mechanical polishing, to name a few possible consequences, can have a major influence on the initial dissolution behavior. In addition, surfaces are often subjected to surface reconstructions as well as segregation phenomena. Exact knowledge of the initial surface structure and composition is therefore a prerequisite for atomic-scale understanding of the initial steps of dealloying. For the Au-rich system Au_3Cu, as an example, Eckstein and coworkers explained the higher initial dissolution current of the (001) surface compared to the (111) surface that was observed in current–potential sweep experiments by a higher content of top-surface Cu after the initial UHV preparation. A combination was used of X-ray diffraction (CTR measurements) employing synchrotron light and STM to characterize the clean surface before electrochemical experiments. The $Au_3Cu(001)$ surface has been found to contain a small amount of small Cu clusters ($\sim$5%) in the two topmost surface layers.[50] With the high Au content bulk, dealloying is not observed for Au_3Cu.

For the Cu-rich Cu_3Au system, bulk dealloying is well established above the critical potential.[27,41,51,52] We characterized in detail the initial surface structure by surface-sensitive X-ray diffraction[53] in addition to the more standard LEED and AES in UHV, and *ex situ* AFM in air. The fitting of a structural model to the obtained X-ray CTR data revealed that Cu_3Au (111) does not show surface segregation or reconstruction. The top and second surface layer showed a small relaxation of –2.3% of their interlayer distance. The average composition and superstructure order were present up to the topmost layer.

3.4.2   Low and Medium Overpotential Regime

3.4.2.1   In Situ *Surface-Sensitive XRD*

We first focus on the voltage regime, starting from the Cu Nernst potential at 100 mV up to 350 mV *vs.* Ag/AgCl ('low overpotential regime'). This regime is far below the critical potential E_c of about 800 mV. Directly after bringing the sample into contact with the deaerated 0.1 M H_2SO_4 electrolyte (at a potential of −100 mV *vs.* Ag/AgCl), we observed by *in situ* X-ray diffraction only Cu$_3$Au (111) substrate reflections (inset, Figure 3.2(a)). Above + 100 mV, an additional in-plane Bragg peak can be observed close to the fundamental substrate peak (2, 2, 0) and its symmetry equivalent peaks. The inset of Figure 3.2(a) shows a sequence of radial in-plane scans for increasing potentials. A new in-plane Bragg reflection appears at (1.9, 1.9, 0), which corresponds to an in-plane lattice constant between the values for Cu$_3$Au and Au, and can be explained by a gold-rich alloy layer.[19,20] The corresponding diffraction peaks are centered along the <1,1,0> radial directions of the substrate. From the width of the diffraction peak in the radial (*HK*) scans, a lateral domain size l of about 12 nm (l = 2π/Δq) can be deduced. As shown in Figure 3.2(a), the peak intensities grow with increasing voltage (they also grow slowly with time), while the peak position shifts slightly towards a smaller *HK*, *i.e.* larger in-plane lattice parameters. Figure 3.2(a) shows *HL*-maps close to (2, 0, 1) where the stacking of the grown layers at different potentials is visible. Due to the different in-plane lattice constant of the growing film, the respective surface rod of the film is well separated from the crystal truncation rod (CTR) of the substrate. The distinct diffraction features indicate the structural development of the surface during dealloying. A first streak (not shown here) is the clear feature of the formation of monolayer patches of a Au-rich film on the surface at lower overpotentials.

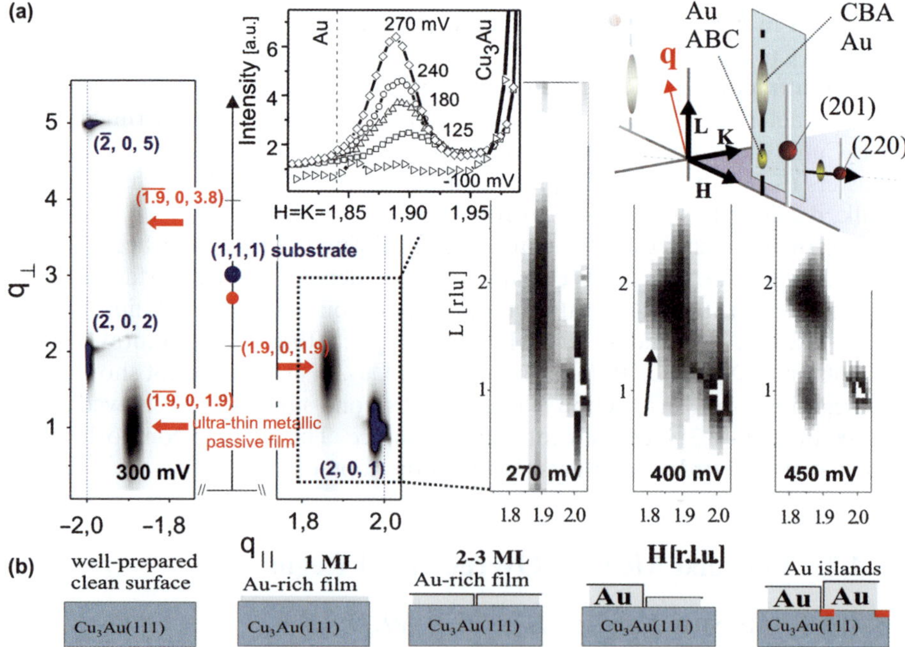

Figure 3.2   *In situ* X-ray diffraction on Cu$_3$Au (111) during initial dealloying. (a) Out-of-plane diffraction scans.[19,20] The insets show the in-plane diffraction scans and a sketch of the reciprocal space. (b) Sketch of the developing surface morphology.

This film grows and shows with the forming third layer of the fcc-like Au-rich film a clearly reversed stacking with respect to the substrate.[20] The stacking sequence is immediately visible from the reversed positions of the Bragg reflections in the reciprocal space depicted in Figure 3.2(a). The elongated diffraction features in reciprocal space are a clear sign of an ultrathin layer. In the L-direction, perpendicular to the surface, the respective reflection spreads along an extended broad peak ($\Delta L \sim 1$ rlu), and the width of these intensity steaks gives an estimation of the layer thickness of about 3 ML, which was confirmed by more detailed CTR or 'surface-rod' data from the ultrathin surface film.[19] Up to about 350 mV, this ultrathin film is stable on the timescale of tens of hours of a typical diffraction experiment. Above 350 mV ('medium overpotential range'), the reversed ultrathin film is also transformed to stacking-reversed nearly-pure Au islands, and these Au islands thus inherit the orientation from the initial ultrathin film. The respective Bragg reflection moves to the expected position for Au (see arrow in Figure 3.2(a) at 400 mV), and the reflection is much narrower, indicating an island thickness of about 10–15 ML. After the complete initial film has vanished and transformed into the thicker Au islands at 450 mV, a new intensity becomes visible at a position (1.84, 0, 0.92) close to the substrate peak. The sequence of subsequent surface structures and morphology is sketched in Figure 3.2(b).

3.4.2.2 Local Surface-Sensitive Probes

While XRD is averaging over a relatively large area (so called 'footprint' formed by the incoming X-ray beam and field of view for the detector), STM and SAEM give us important local information.[21] Figure 3.3 shows (*ex situ*) an SAEM analysis after the application of two selected potentials for 1 h, respectively. The high resolution of the SAEM of about 10 nm enables the forming islands at medium overpotentials and the boundary region in between them to be probed. The surface subjected to 200 mV for 1 h shows a homogeneous enrichment of Au, which is in line with a single ML of Au. While the SAEM images (imaging mode) have been obtained by quickly scanning a limited number of energy values over a certain Auger signal (and thus show a higher statistical fluctuation), additional high-resolution scans do not reveal any differences in composition at different points. The SEM image (Figure 3.3(a)) depicts the area of Auger signal mapping shown in Figure 3.3(b) for Au and Figure 3.3(c) for Cu. The spectra shown in the insets of Figure 3.3(b) and (c) have been recorded at the positions indicated by arrows and show the enrichment of Au (and depletion of Cu) on the surface together with the corresponding Auger peaks for a clean Cu_3Au (111) surface. The surface subjected to a higher potential of nominal 300 mV for 1 h has developed islands on the surface, and the Au and Cu signals of the Auger measurements follow the respective topography. Even the signal in between two adjacent islands shows an increased Au signal (and lower Cu signal), and the islands show only a very weak residual Cu signal. These measurements can be interpreted by assuming

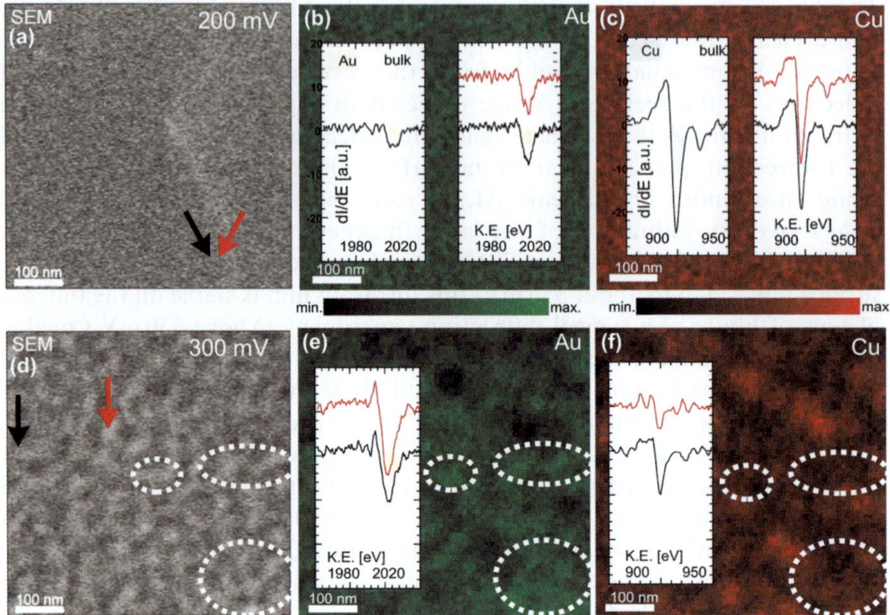

Figure 3.3 *Ex situ* SEM and SAEM after initial dealloying of selectively dissolved Cu_3Au (111) surfaces. Color code is from minimum to maximum signal at each map and potential. The specimens were polarized 1 h at lower potentials. SEM (a), Cu signal (b), and Au signal (c) at 200 mV. SEM (d), Cu signal (e), and Au signal (f) at 300 mV. The SEM images and respective elemental maps of the same area of Au and Cu along with corresponding spectra (insets) are shown. The colored arrows in the SEM images indicate the positions for the respective inset Auger spectra. Potentials are referenced to an Ag/AgCl electrode. Reprinted from *J. Am. Chem. Soc.*, 2011, **133**, 18264. Copyright 2011 American Chemical Society.

islands approximately 3 nm thick Au and a film approximately 1 nm thick in between the islands with the respective Cu signals originating from the substrate below. This interpretation is supported by the described X-ray diffraction results, which showed that no Cu-rich Cu–Au alloys are formed during the initial dissolution. The SAEM data illustrate well that already at an early stage of the initial selective dissolution of Cu from Cu_3Au (111), an Au-rich film covers the entire surface. The step of Au island formation during initial dealloying is thus terminated at medium overpotentials with the completion of a continuous film of stacking-reversed Au islands.

In situ STM measurements[21] were performed in the same lower-potential regime to address the changes in morphology of the Cu_3Au (111) surface during Cu dissolution. Figure 3.4 shows a selection of potential-dependent STM images of the dissolving alloy surface. After preparation in UHV, the alloy surface consists of flat mono-atomic terraces. Probably due to dissolution of

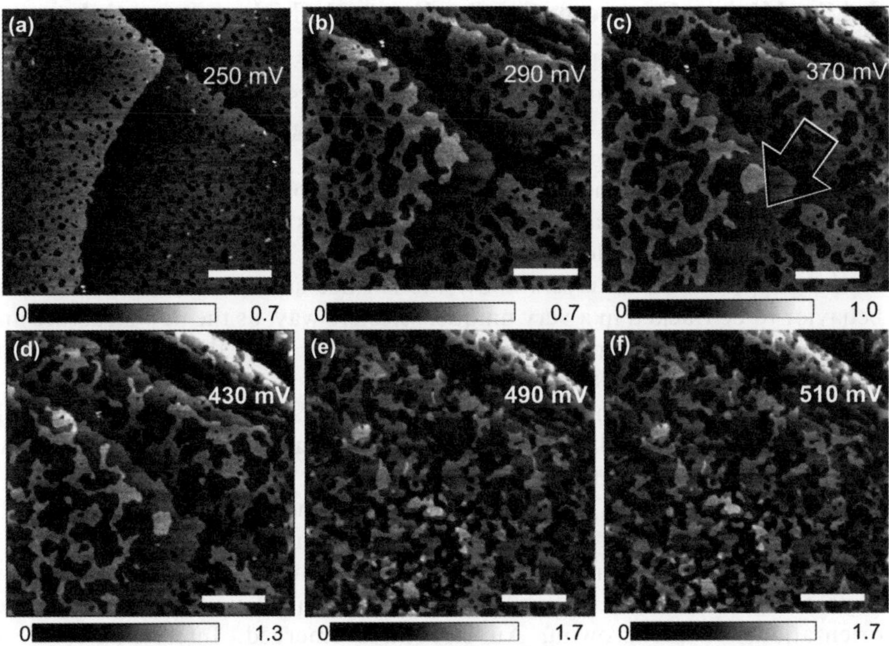

Figure 3.4   Selection of *in situ* STM images (500 nm × 500 nm) during dealloying. (a) 250 mV, $t = 161$ min, (b) 290 V, 181 min. (c) 370 mV, 197 min, (d) 430 mV, 209 min, (e) 490 mV, 261 min, (f) 510 mV, 267 min. The images reveal a stable second layer (arrow) and pronounced roughening at about 450 mV. Potentials are referred to Ag/AgCl. Height scale in nanometers. Reprinted from *J. Am. Chem. Soc.*, 2011, **133**, 18264. Copyright 2011 American Chemical Society.

surface oxides formed during the short transfer in air, the terraces and step edges show small pits of monolayer depth after immersing the sample in electrolyte. The terraces become smoother again, reflecting the high Au surface diffusion in electrolyte with the increase in potential. Noteworthy is the relative compactness of areas of the second atomic layer after partial dissolution of the topmost surface layer (Figure 3.4(a)–(d), compare arrow). The presumed enrichment in Au atoms and the associated relative stability of the lower terrace points directly to an exchange of Cu and Au atoms within the top and second layer during dissolution of the top layer (interlayer exchange). At slightly elevated potentials (Figure 3.4(e) and (f)), the surface showed substantial roughening with an evident increase in depth of the nanometer-scale pits. This is also the exact medium potential regime where the formation of Au islands occurs. These islands inherited the stacking-reversed structure of the ultrathin film. The detailed Auger microprobe data complement the morphological information from the *in situ* STM and the earlier (laterally averaging) X-ray data in an unprecedented way: even on the nanoscopic scale, a laterally continuous almost pure Au layer is already formed on the surface at low overpotentials.

3.4.3 Higher Overpotential Regime and Critical Potential

Techniques such as STM or SAEM become more difficult to use at elevated potentials because pronounced dealloying irreversibly alters, roughens, and eventually destroys the elaborately prepared single crystalline surface. The initial film of Au islands at that stage covers the entire surface, and critical processes might take place at the buried film–substrate interface. X-ray diffraction here offers clear advantages, as it addresses the topmost surface as well as buried structures below.[21,23] For the special case of Cu_3Au (111), the stacking-reversed initial layers allow the surface structures and their growth behavior to be tracked in a very unique, selective way, as the new structure and its signature diffraction peak are clearly separated from the (larger) substrate peak and background intensity.

At medium overpotentials, the entire surface is covered by a structurally reversed but epitaxial film (in the sense of a defined orientation relation Au (111) ∥ Cu_3Au (111) and the azimuthal orientation $<-2, 2, 0> \parallel <2, -2, 0>$). Ongoing Cu dissolution and further growth of the surface Au film could be expected to lead to Au structures that follow the initial Au film in orientation, *i.e.* a template effect for the further growth. Considering the low stacking fault energy in bulk Au, this could also be easily understood if further orientations of the growing Au structures emerged. Interestingly, X-ray results at higher overpotentials close to the critical potential showed that the initial reversed structures (and their Bragg peaks) vanish, and only substrate-oriented Au ligaments form during continuing dealloying. This surprising result might be understood by a new mechanism during the dissolution process producing substrate-oriented Au. As the surface diffusion in the Au/electrolyte system is high, a ripening process can explain the vanishing reversed surface layer.

Large-scale simulations and detailed calculations offer to shed light on the observed mechanisms. Our own efforts[39,40] target the stability of the initial heterophase interface and the energetic origin of the observed initial stacking reversal. The in-plane lattice parameters of the film (H = K = 1.9 rlu) and the substrate (H = K = 2 rlu) are different by 5%. Here, we select smaller sub-regions of the resulting larger heterophase film/substrate superstructure and treat these in a supercell with a pseudomorphic approximation for different possible stacking sequences. Preliminary results indicate that actually phonon interactions are necessary to clearly select the reversal of the stacking sequence across the forming interface. Also, large-scale simulations using EAM potentials address larger atom ensembles but are less accurate. The ensembles include the entire superstructure necessary to describe the Moirée-like (heterophase) film/substrate interface. First results indicate here that the expected misfit dislocations at the interface are mainly located within the buried top-most substrate atomic layer. The resulting disturbed interface structure should also facilitate a vertical transport within the affected zone of Cu towards the buried film/substrate interface (and finally to the solid–electrolyte interface). Residual Au atoms will not easily move away from their respective sites, and the created vacancies might be filled up with Au atoms. Nucleation of substrate-oriented Au clusters within the substrate might finally lead to the disappearance of the initial reversed Au layer in a ripening process.

3.4.4 Influence of Halide Additives

The sequence of subsequent surface structures was unaltered in experiments performed with additives of halides.[23–25] Chloride[22] or bromide[24] ions to the 0.1 M sulfuric acid caused, in addition to a cathodic shift in the critical potential, a cathodic shift in the potential of the structural transformation within the passive-like region from a (reversed) ultrathin film towards thicker (reversed) islands. Halides are known to increase the surface mobility of Au adatoms,[54,55] and the results thus support the large influence of surface diffusion on dealloying.

Also, after contact of the Cu_3Au (111) surface with 0.1 M $H_2SO_4 + 1$ mM KI electrolyte, the described ultrathin stacking-reversed Au-rich film (marked 'CBA' in Figure 3.5(a)) forms[23] and grows. The rapid growth of this initially ultrathin film reveals, in contrast to pure sulfuric acid, that the layer now transforms directly at lower potentials into thicker islands (with Bragg peaks at $L \sim 1.84$ rlu of a much narrower profile in L-direction). The initial ultrathin Au-rich film can be easily distinguished from the corresponding narrow in-plane peak at about $H = K \sim 1.9$ rlu. At almost the same stage also substrate-oriented

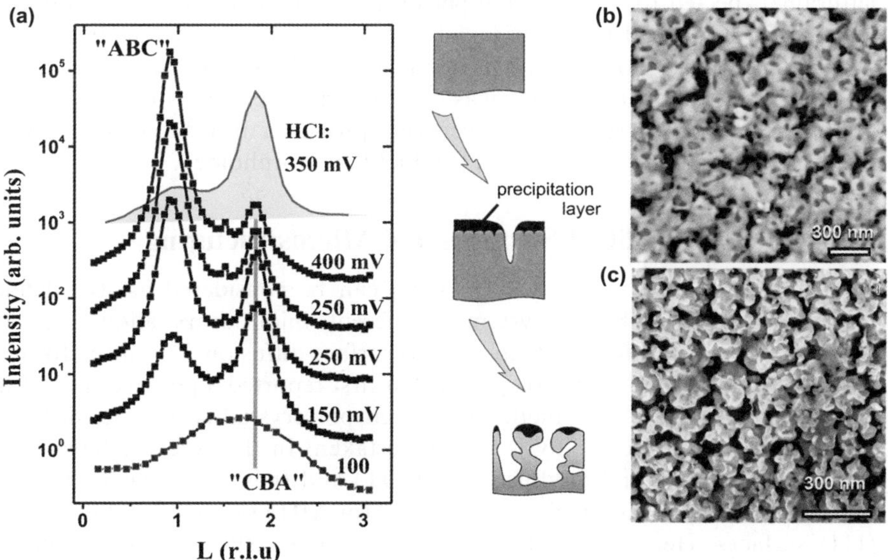

Figure 3.5 SEM images after dealloying under influence of a CuI precipitation layer. (a) X-ray diffraction data. Reprinted from *Surf. Sci.*, 2012, doi: 10.1016/ j.susc.2012.01.013. Copyright Elsevier 2012. (b)–(c) SEM images and sketch of the reaction path.

ligaments ('ABC'-stacking), with their out-of-plane Bragg peak at $L \sim 0.92$ rlu in Figure 3.5(a), start to grow and replace the (vanishing) initial film with its associated Bragg peak at $L \sim 1.84$ rlu. The initial ultrathin film is not stable, even at lower overpotentials. The L-scans show a clear narrowing and transformation from CBA to ABC structures, reflecting the expected increase in thickness of the initial films and later ligaments. Compared to our earlier experiments in pure H_2SO_4 and in the presence of HCl, the dealloying process does not slow down but continuously proceeds. It is noteworthy that the process therefore does not show any passive-like behavior but proceeds continuously. In parallel, additional copper iodide Bragg peaks indicate in the case of a higher 10 mM KI concentration the presence of two epitaxial CuI domains on the substrate. This precipitation process was further confirmed by *ex situ* AFM experiments, where at a higher potential of 200 mV, the surface morphology is clearly changed. In general, adsorbates or precipitation layers have a strong impact on surface reactivity such as metal dissolution processes.[56,57]

A high dissolution rate of Cu ions results in an ion concentration above the related solubility limit of CuI in the electrolyte and leads to precipitation of a film that appears to cover the whole surface. The CuI precipitation can be initiated by either a high concentration of iodide in the solution, or by quickly offering a large quantity of Cu ions. Both cases lead to precipitation of a layer of CuI on the surface. In turn, the precipitate influences the dealloying processes and mechanisms, markedly changing the obtained morphology of the produced nanoporous gold surfaces. The detailed pore morphology and connectivity are naturally expected to influence the performance of the respective nanoporous material. Figures 3.5(b) and (c) depict two surfaces after different treatments. One surface was dealloyed in a one-step process at higher potentials for several minutes, and the second surface was cleaned once from precipitated layers in acetone in between. Precipitation layers thus provide a means to further control the dealloying process and the final product film morphology.

3.4.5   Thiol-Modified Surfaces and Microstructuring

Thiol self-assembled monolayers (SAMs) can be considered as well-defined model corrosion-inhibition systems. Different thiol layers allow here for directly probing the effect of (suppressed) surface diffusion[28,41] on the forming surface structures and morphology. Moffat first reported a positive shift of the critical potential of a pentanethiol-modified polycrystalline Cu_3Au surface,[44] but no detailed surface analysis was undertaken. In this line, we have so far studied hexadecanethiol-modified[21,25] (HDT) and (next to a plasma polymer film) mixed aminobenezenethiol-modified[26] (m-ABT) single crystalline Cu_3Au (111) surfaces. Here we shortly review the recent results on HDT-modified starting surfaces.

Voltammograms for hexadecanethiol-modified surfaces show in the first cycle a 'critical' potential with a value of about 1 V, which is shifted anodically by about 100–200 mV compared to the behavior in pure sulfuric acid where the critical potential was observed at about 800–900 mV. In a second cycle, a lower critical potential is obvious, and the current rises steeply after reaching 900 mV (close to E_c for a clean surface in pure sulfuric acid). Atomic-scale information on such thiol-modified Cu₃Au (111) starting surfaces could be obtained during dealloying for the entire passive potential range by *in situ* SXRD. Structural information is revealed by the out-of-plane *in situ* SXRD L-scans in Figure 3.6(a). An initial scan after immersion at low potential already indicates an initial ultra-thin CBA (stacking-reversed) Au-rich film that is likely induced by initial Cu dissolution and Au segregation due to a preference of the thiol–Au (S–Au) *vs.* the S–Cu bond. The signature of the ultrathin Au layer in real space is the broad peak ($\Delta L \approx 1$, 3 ML) in reciprocal space. With an initial increase in the applied potential, further growth of the signal from the initial CBA layer could be observed together with a gentle narrowing of the peak width. The first signal from ABC (substrate-oriented) Au ligaments emerges only at a well-elevated potential of 750 mV. The peak width is immediately much sharper for the Au ligaments. The transformation potential on a hexadecanethiol-modified surface is about 300 mV higher than that of the corresponding bare Cu₃Au (111) surface. Furthermore, the initial almost-pure Au layer is still stable at this high potential and consists of several atomic monolayers. Thicker Au islands are not formed. The analysis of the diffraction peaks reveals a thickness of 5–7 ML of an almost-pure Au layer, in contrast to the thiol-free surfaces with 3 ML (ultrathin Au-rich layer) or the subsequent Au

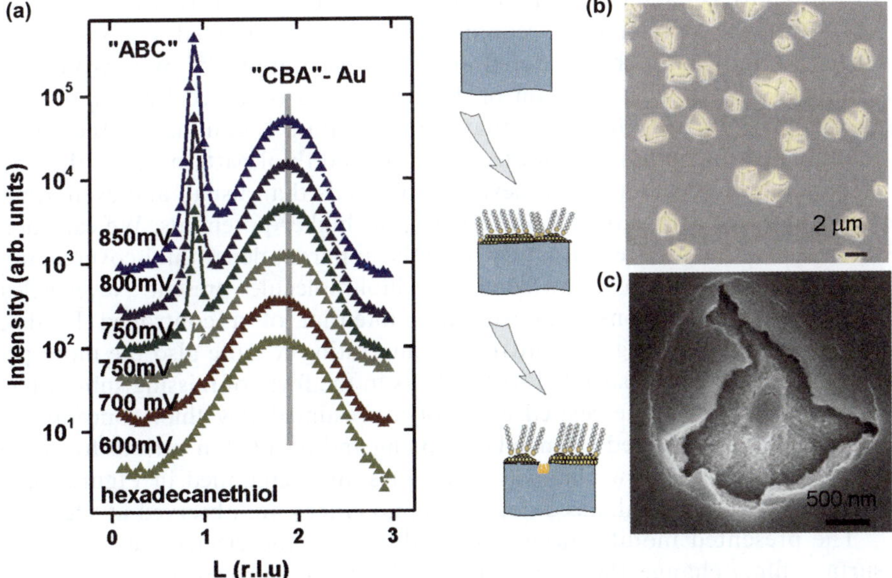

Figure 3.6 Dealloying of a thiol-modified surface. (a) *In situ* XRD L-scans. (b) STM after application of a dodecanethiol film before dealloying. (c)–(e) SEM after dealloying showing micro-cracks. Reprinted from *J. Am. Chem. Soc.*, 2011, **133**, 18264. Copyright 2011 American Chemical Society.

islands of about 12–15 ML thickness. In addition, the intensity from the initial CBA Au film (inverted stacking) does not decrease with the formation of porosity at potentials as high as 900 mV. The initial surface film structure is stabilized such that the initial inverted film is present even above the apparent 'critical' potential E_c. Figures 3.6(b) and (c) show SEM images of the surface after applying potential above the 'critical' potential. The surface is covered by defects or micro-cracks with a nanoporous interior, instead of the observed homogeneous porosity on clean surfaces. Clearly, the SAMs considerably stabilize the surface, as no island formation, nanometer-scale pit formation, or porosity is observed on the thiol-covered, defect-free surface, even after applying these high potentials. The dissolution is concentrated to a few crack sites, which are formed every few micrometers. The 'critical' potential here is a 'pitting' potential. The respective current–potential cycles for the thiol-modified surfaces and the apparent 'pitting' potential thus reflect an entirely new mechanism and also a unique surface morphology compared to the usual creation of homogeneous nanoporosity in this system. Thiols are known to suppress the surface diffusion,[28,43] and a ripening process is thus strongly decreased. The nature of the initial defects at the sites of crack formation here is not completely clear, but their average distance is far greater than the typical domain size of the SAMs and thus not associated with typical structural defects like the domain boundaries of the SAMs. Compared with the otherwise homogeneous formation of nano-porosity, the dealloying mechanism is clearly altered for the thiol-modified surfaces. As a consequence, such thiol inhibition layers might be applied to structure surfaces and create nanoporous regions only at the desired locations.

To test the possibility to confine dealloying to limited regions on the surface, a pattern of thiol-covered areas has been applied[30] by micro-contact printing (µ-cp[58]). The initiation of dealloying was observed by *in situ* AFM. In Figures 3.7(a)–(d), it is clear that the dissolution is confined to the regions that are at a maximum distance from the applied thiol patterns. Still, the onset of dissolution occurs at more elevated potentials than usual, and even apparently thiol-free areas are still protected up to higher potentials. We can therefore assume that also regions away from the apparent dense thiol-covered areas are still affected, *e.g.* by a low density of thiol molecules present. Nevertheless, the experiment demonstrates well the potential of a controlled localized dealloying process (Figure 3.7(d)). It is interesting to note also the more pronounced dealloying associated with defects in the form of missing sub-patterns of the stamp used: the related areas of a presumed low thiol concentration (contamination) showed extended dealloying and resulted in various forms and features such as hollow channels or cavities, as exemplified in Figures 3.7(e) and (f). Despite the enhanced dealloying, no cracks are observed at such sites.

The presented modifications with additives to the electrolytes or applied surface films change the homogeneous dealloying processes towards inhomogeneous or localized dealloying. The changes in mechanisms represent top-down and bottom-up approaches to control the morphology of the nanoporous Au films produced. The different schemes for this morphology control are depicted in Figure 3.8.

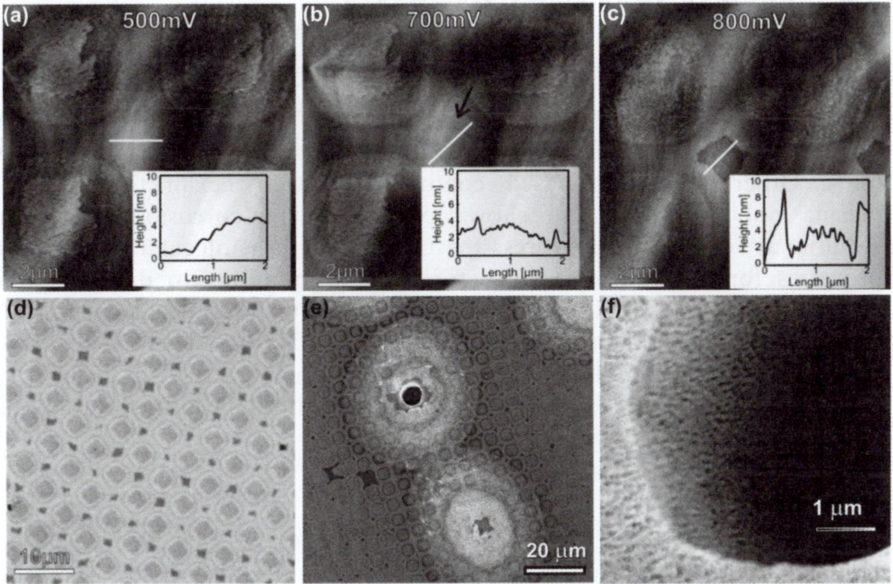

Figure 3.7 Nanoporous microstructures obtained by dealloying of µCP-modified surfaces. (a)–(c) *In situ* AFM at different subsequent potentials. (e)–(f) SEM after this AFM study. Reprinted from *Rev. Sci. Instr.*, 2011, **82**, 023703. Copyright 2011 American Institute of Physics.

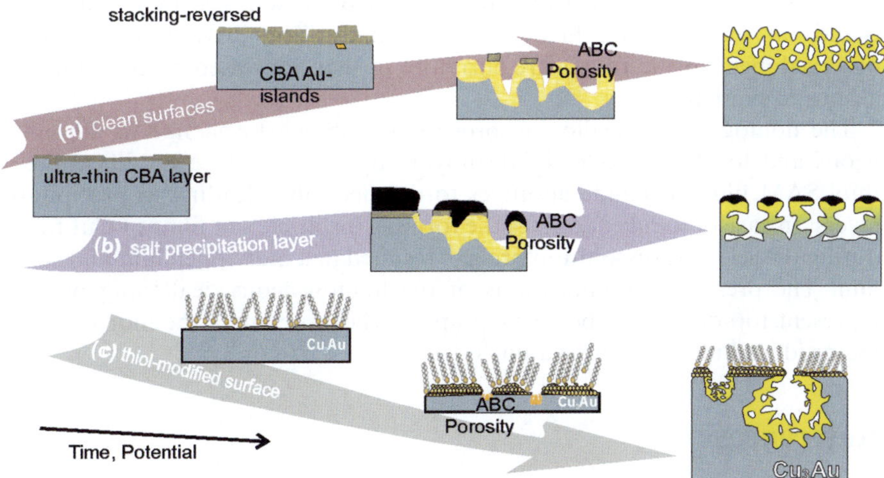

Figure 3.8 Sketch of the reaction pathways on clean and thiol-modified Cu_3Au (111) surfaces as well as with iodide additives to the electrolyte.

3.5   Further Work and Perspectives

The current work on Cu–Au alloys will be continued in the direction of new effects of different surface modifications. Recent experiments show, for example, that the stability of the passive-like surfaces can be enhanced even further by application of cross-linking mixed aminobenzenethiol-based SAM films. Other than thiol-based coatings, ultrathin hexamethyldisiloxane plasma polymer films also show good dealloying-corrosion protection. Additional activities will address the transfer of the molecular systems and processes for the control of structure and morphology to different alloy systems. Especially noble-metal alloys such as Pt alloys[59,60] or Pd alloys[61] are interesting candidates here, due to their widespread use in catalytic reactions. Further simulation work is expected here to shed more light on the driving forces for the observed peculiar structural surface evolution and related mechanisms.

3.6   Summary

By combining state-of-the-art surface analytical tools with *in situ* X-ray diffraction we could in great detail trace the structural evolution of Cu_3Au (111) surfaces during initial stages of dealloying. X-ray diffraction offers at high overpotentials clear advantages by addressing the topmost surface as well as buried structures below eventual surface layers. For Cu_3Au (111), the stacking-reversed initial layers offer to track the surface structures and their growth behavior in a very unique way and enabled unprecedented experiments on the effect of additives to the electrolyte and modified starting surfaces.

We showed that the Cu_3Au (111) surface is covered by a continuous layer of Au islands at an early stage of dealloying far below E_c and that substrate-oriented Au ligaments nucleate and grow below this closed film of initial Au islands at medium and higher overpotentials. The driving forces for the observed mechanistic routes are currently further addressed by computational atomistic simulations.

The homogeneous dealloying process was shown to change to inhomogeneous and localized modes of dealloying by either surface modifications by thiol SAM films or halide additives to the electrolyte leading to precipitation layers. In case of iodide additives the forming insoluble salt films of CuI formed during the selective dissolution process resulted in a more open bimodal surface film. The presented modifications of the homogeneous dealloying processes represent top-down and bottom-up approaches to control the morphology of the produced nanoporous Au films.

References

1. E. Schofield, *Trans. Inst. Met. Finish.*, 2005, **83**, 35.
2. J. Erlebacher and R. Seshadri, *MRS Bull.*, 2009, **34**, 561.
3. Y. Ding and M. Chen, *MRS Bull.*, 2009, **34**, 569.

4. J. Weissmuller, R. C. Newman, H. J. Jin, A. M. Hodge and J. W. Kysar, *MRS Bull.*, 2009, **34**, 577.
5. Y. Ding, M. W. Chen and J. Erlebacher, *J. Am. Chem. Soc.*, 2004, **126**, 6876.
6. J. Weissmuller, R. N. Viswanath, D. Kramer, P. Zimmer, R. Würschum and H. Gleiter, *Science*, 2003, **300**, 312.
7. A. Wittstock, V. Zielasek, J. Biener, C. M. Friend and M. Baeumer, *Science*, 2010, **327**, 319.
8. E. W. Müller, *Phys. Rev.*, 1956, **102**, 618.
9. T. T. Tsong, *Surf. Sci.*, 1994, **299/300**, 153.
10. G. Binnig and H. Rohrer, *Rev. Mod. Phys.*, 1987, **59**, 615.
11. H. Rohrer, *Surf. Sci.*, 1994, **299/300**, 956.
12. R. Sonnenfeld and P. K. Hansma, *Science*, 1987, **223**, 211.
13. G. Binnig, C. F. Quate and Ch. Gerber, *Phys. Rev. Lett.*, 1986, **56**, 930.
14. S. Manne, P. K. Hansma, J. Massie, V. B. Elings and A. A. Gewirth, *Science*, 1991, **251**, 133.
15. *J. Phys. Chem. B*, 2004, **108**, 14183.
16. C. B. Duke, *Surf. Sci.*, 1994, **299/300**, 24.
17. C. B. Duke, *J. Vac. Soc. Technol. B*, 1993, **11**, 1336.
18. S. Trasatti and K. Wandelt, *Surf. Sci.*, 1995, **335**, I; and contributions to this volume.
19. F. U. Renner, A. Stierle, H. Dosch, D. M. Kolb, T. L. Lee and J. Zegenhagen, *Nature*, 2006, **439**, 707.
20. F. U. Renner, A. Stierle, H. Dosch, D. M. Kolb, T. L. Lee and J. Zegenhagen, *Phys. Rev. B*, 2008, **77**, 235433–10.
21. A. Pareek, S. Borodin, A. Bashir, G. N. Ankah, P. Keil, G. A. Eckstein, M. Rohwerder, M. Stratmann, Y. Gründer and F. U. Renner, *J. Am. Chem. Soc.*, 2011, doi: 10.1021/ja2054644.
22. F. U. Renner, A. Stierle, H. Dosch, D. M. Kolb and J. Zegenhagen, *J. Electrochem. Comm.*, 2007, **9**, 1639.
23. F. U. Renner, G. N. Ankah and A. Pareek, *Surf. Sci. Lett.*, submitted.
24. G. N. Ankah, A. Pareek, M. Rohwerder and F. U. Renner, submitted.
25. A. Bashir, G. N. Ankah, A. Pareek and F. U. Renner, submitted.
26. A. Pareek, G. N. Ankah, P. Ebbinghaus, A. Erbe and F. U. Renner, *Phys. Rev. B.*, submitted.
27. L. Graf, *Z. Metallkd.*, 1955, **46**, 378.
28. T. Iwashita and F. C. Nart, *Prog. Surf. Sci.*, 1997, **55**, 271.
29. C. E. Bach, R. J. Nichols, W. Beckmann and H. Meyer, *J. Electrochem. Soc.*, 1993, **140**, 1281.
30. M. Valtiner, G. N. Ankah, A. Bashir and F. U. Renner, *Rev. Sci. Instr.*, 2011, **82**, 023703.
31. H. Dosch, B. W. Batterman and D. C. Wack, *Phys. Rev. Lett.*, 1986, **56**, 1144.
32. H. Dosch, *Phys. Rev. B*, 1987, **35**, 2137.
33. I. K. Robinson and D. J. Tweet, *Rep. Prog. Phys.*, 1992, **55**, 599.
34. F. U. Renner, Y. Gründer, P. F. Lyman and J. Zegenhagen, *Thin Solid Films*, 2007, **515**, 5574.

35. M. Fleischman and B. W. Mao, *J. Electroanal. Chem.*, 1987, **229**, 125.
36. O. R. Melroy, M. F. Toney, G. L. Borges, M. G. Samant, J. B. Kortright, P. N. Ross and L. Blum, *Phys. Rev. B*, 1988, **38**, 10962.
37. A. H. Ayyad, J. Stettner and O. M. Magnussen, *Phys. Rev. Lett.*, 2005, **94**, 066106.
38. F. U. Renner, Y. Gründer and J. Zegenhagen, *Rev. Sci. Instrum.*, 2007, **78**, 033903.
39. F. U. Renner, G. A. Eckstein, L. Lymperakis, A. Dakkouri-Baldauf, M. Rohwerder, J. Neugebauer and M. Stratmann, *Electrochim. Acta.*, 2011, **56**, 1694.
40. L. Lymperakis, J. Neugebauer and F. U. Renner, in preparation.
41. G. Tammann, *Z. Anorg. Allg. Chem.*, 1919, **107**, 1.
42. A. J. Forty and P. Durkin, *Philos. Mag. A*, 1980, **42**, 295.
43. J. Laurent and D. Landolt, *Electrochim. Acta.*, 1990, **36**, 49.
44. T. P. Moffat, F. R. F. Fan and A. J. Bard, *J. Electrochem. Soc.*, 1991, **138**, 3224.
45. I. C. Oppenheim, D. J. Trevor, Ch. E. D. Chidsey, P. L. Trevor and K. Sieradzki, *Science*, 1991, **254**, 687.
46. J. S. Chen, F. Sanz, D. F. Ogletree, V. M. Hallmark, T. M. Devine and M. Salmeron, *Surf. Sci.*, 1993, **292**, 289.
47. M. Stratmann and M. Rohwerder, *Nature*, 2001, **410**, 420.
48. J. Erlebacher, M. J. Aziz, A. Karma, N. Dimitrov and K. Sieradzki, *Nature*, 2001, **410**, 450.
49. J. Erlebacher, *J. Electrochem. Soc.*, 2004, **151**, C614.
50. G. A. Eckstein, S. Maupai, A. S. Dakkouri, M. Stratmann, M. Nielsen, M. M. Nielsen, R. Feidenhans'l, J. H. Zeysing, O. Bunk and R. L. Johnson, *Phys. Rev. B*, 1999, **60**, 8436.
51. H. Gerischer, *Werkstoffe und Korrosion*, 1961, **12**, 608–613.
52. H. W. Pickering and P. R. Swann, *Corrosion*, 1963, **19**, 373.
53. A. Stierle, A. Steinhauser, A. Rühm, F. U. Renner, R. Weigel, N. Kasper and H. Dosch, *Rev. Sci. Inst.*, 2004, **75**, 5302.
54. R. J. Nichols, O. M. Magnussen, J. Hotlos, T. Twomey, R. J. Behm and D. M. Kolb, *J. Electroanal. Chem.*, 1990, **290**, 21.
55. M. Giesen, *Prog. Surf. Sci.*, 2001, **68**, 1–153.
56. O. M. Magnussen and M. R. Vogt, *Phys. Rev. Lett.*, 2000, **85**, 357.
57. P. Broekmann, N. T. M. Hai and K. Wandelt, *Surf. Sci.*, 2006, **600**, 3971.
58. A. Kumar and G. M. Whitesides, *Appl. Phys. Lett.*, 1993, **63**, 2002.
59. R. Yang, P. Strasser and M. F. Toney, *J. Phys. Chem. C*, 2011, **115**, 9074.
60. M. Oezaslan and P. Strasser, *J. Power Sources*, 2011, **196**, 5240.
61. S. Meimandi and F. U. Renner, *ECS Trans.*, 2011, **33**, 31.

Mechanical Properties of Nanoporous Gold

ANDREA M. HODGE*[a] AND THOMAS JOHN BALK[b]

[a] University of Southern California, Department of Aerospace and Mechanical Engineering, Los Angeles, CA 90089-1453, USA; [b] University of Kentucky, Department of Chemical and Materials Engineering, Lexington, KY 40506-0046, USA
*Email: ahodge@usc.edu

4.1 Introduction

The mechanical behavior of nanoporous Au (np Au) has been a subject of great interest during the last decade since early results revealed ultra-high strengths and unique deformation mechanism.[1–8] Np Au is a very promising functional material, and therefore, it is important to understand its mechanical behavior and mechanical stability so that it can be incorporated into current and future applications as discussed throughout this book. Previous mechanical properties on porous solids with cell sizes above 1 μm (macroporous) have been extensively researched, and various scaling relations describing the mechanical properties have been developed.[9] In general, it has been found that the most important parameter in controlling the mechanical properties of large cell size foams is the relative density or porosity. Other factors include the mechanical properties of the solid material and the foam morphology such as ligament shape and connectivity.

RSC Nanoscience & Nanotechnology No. 22
Nanoporous Gold: From an Ancient Technology to a High-Tech Material
Edited by Arne Wittstock, Jürgen Biener, Jonah Erlebacher and Marcus Bäumer
© Royal Society of Chemistry 2012
Published by the Royal Society of Chemistry, www.rsc.org

However, at the nanoscale, many new mechanisms that deviate from conventional materials mechanics have been observed in nanostructured materials. Thus, similar changes are expected to arise in an np structure for which the internal length scale of the pores and ligaments have been shown to play a more predominant role in determining the mechanical properties. In this chapter, we will discuss the current available testing methods including results from experimental tests and modeling studies on the mechanical properties and behavior of open-cell np Au foams over a wide range of porosities (60–80% porosity) and ligament sizes (10–1000 nm).

4.2 Elastic-Plastic Deformation Behavior

4.2.1 Scaling Equations for Mechanical Properties

As mentioned in Section 4.1, several equations were previously developed for open-cell macroporous foams (with a pore size of micrometers to millimeters), which can serve as a basis for understanding porous material behavior in general. A complete discussion about open- and closed-cell macroporous foams can be found in the book *Cellular Solids* by Gibson and Ashby.[9] In the following sections, we will focus on the evaluation of two basic properties, hardness and/or yield strength (σ) and elastic modulus (E), and their relation to the foam's relative density (ρ^*/ρ_s) which is defined as the density of the foam (ρ^*) divided by the density of the material it is made of (ρ_s). From Gibson and Ashby, we have the following relationships:

$$\sigma^* = C_2 \sigma_s \left(\frac{\rho^*}{\rho_s}\right)^{3/2} \tag{4.1}$$

$$E^* = C_1 E_s \left(\frac{\rho^*}{\rho_s}\right)^2. \tag{4.2}$$

In these equations, the asterisk refers to foam properties, and s refers to the bulk properties; $C_1 = 1$ and $C_2 = 0.3$ are fitting constants. It is important to note that these scaling relations contain no explicit length-scale dependence and assume that the material properties of the ligaments such as the yield strength and the Young modulus are size-independent and equal to the bulk value. However, for np foams, eqns (4.1) and (4.2) do not fully capture the behavior, since, as stated by Weismüller *et al.*,[8] there are three distinct size effects at the nanoscale that are not fully accounted for: (1) the action of capillary forces, (2) the plastic behavior of the ligaments, which is different from that of the bulk material, and (3) the pore ligament sizes, which can be very different from the crystal size. It should also be mentioned that the current scaling equations were based on the assumption that the ligament size is much smaller than the cell size. For np foams, the ligament and the cell size are approximately the same size; therefore, some fundamental cell/pore deformation behaviors used for deriving eqns (4.1) and (4.2) need to be revised and will be discussed in Section 4.3. As a first attempt to identify a scaling equation for np Au, Hodge *et al.*[4] proposed a revised equation for the yield strength derived from empirical data, as follows:

$$\sigma^* = C_\text{s}\left[\sigma_0 + kL^{-1/2}\right]\left(\frac{\rho^*}{\rho_\text{s}}\right)^{3/2}, \tag{4.3}$$

where C_s is a fitting coefficient, σ_0 is the bulk material yield strength (σ_s), k is the Hall–Petch-type coefficient for the theoretical yield strength of Au in the regime (10 nm to 1 μm), and L is the ligament size. For macroporous foams ($L \gg 500$ nm), $kL^{-1/2}$ becomes negligible, and eqn (4.3) yields eqn (4.1). To date, a revised empirical equation for the elastic modulus has not been formulated, but experimental details will be discussed throughout this chapter as well as recent numerical modeling efforts.

With the goal of developing a fundamental understanding of the mechanical behavior of np Au, here we present several different testing techniques and results that highlight their mechanical properties and deformation behavior. A comparison to the predicted values from eqns (4.1) and (4.2) will also be evaluated as a function of the porosity. To calculate the predicted yield strength and Young modulus of np Au as a function of relative density, one needs the bulk properties of Au as an input. The yield strength of Au is very much dependent on the sample history, and values as low as 2 MPa for well-annealed crystals[10] and as high as $\sim$200 MPa for severely worked single crystals[11] have been reported. For this chapter, we will take the highest reported yield strength value $\sim$200 MPa,[12,13] and the average value of the elastic modulus, $\sim$80 GPa.[12]

4.2.2 Compression Tests

Given the typical thickness for an np Au sample (nanometers to micrometers range), compression tests allow for relatively straightforward testing, and as such most tests on np Au have been performed by either nanoindentation or pillar compression.[3,4,14–17] Other types of tests are also available and will be discussed in Sections 4.2.2.1 and 4.2.3. In the following sections, the mechanical properties of a wide range of porosities (60–80%) and ligaments sizes (40–900 nm) are evaluated under various compression techniques.

In general, it has been observed that under compression, np Au foams are up to 10–15 times stronger than predicted, and as the ligament size decreases, the yield value approaches the theoretical strength of Au.[4] Thus, we have two conflicting factors: high porosity and high strength. The exact reasons for such high-yield strength have been discussed in previous publications and range from dislocation-based models to defect-type models.[1,4,8,18] In addition, the comparison between hardness and yield strength is a source of discrepancy between high- and low-hardness values, and a recent article has presented low-strength np Au foams.[19] However, the most of the data clearly show a higher-than-expected strength, but this is somewhat dependent on the testing method and will be discussed further in Section 4.2.2.1.

4.2.2.1 Indentation Tests

Indentation is a frequently used technique to assess hardness and yield-strength data for cellular solids.[20–24] In the case of low-density foams ($\rho^*/\rho_s \leq 0.3$), it can be generally assumed that the indentation hardness (H) is approximately equal to the yield strength, $H \sim \sigma_y$, rather than following the $H \sim 3\sigma_y$ relationship typically observed for fully dense materials.[9] In fact, low-density nanoporous metal indentation has been shown to be mostly plastic (plastic Poisson ratio v_p near 0)[22,25–28], and thus mostly compressible; regions outside the indent remain undeformed, as shown in Figures 4.1(a) and (b). Yet, the relation between hardness and yield strength is not unanimously supported,[8,17] and this is still an unresolved topic within the community (see Section 4.2.3). Please note that the indentation data here will be presented as hardness (raw data) or yield strength (assuming $H \sim \sigma_y$) as discussed elsewhere.[1,4]

In Figure 4.2, the hardness values from nanoindentation tests comparing the role of relative density and ligament sizes are presented for (a) 30% and 35% relative density foams as a function of displacement and (b) for a wide range of ligament sizes (40–900 nm). Specifically, in Figure 4.2(a), several critical issues regarding the tests can be observed: first, there is a surface contact effect, and as such the test will lead to higher-than-expected values at the initial stages of the tests;[4] second, both relative density and ligament size appear to affect the average values. In order to identify clearly which critical parameter controls the strength of the material, Figure 4.2(b) includes the yield-strength values for a range of relative densities between 25 and 40% (60–80% porosity), as well as a wide range of ligament sizes. For simplicity, samples have been labeled as having ligaments either smaller or larger than 100 nm. The calculated values from eqn (4.1) are also shown, which predicted yield strength values between 7.5 and 16 MPa. Clearly, the experimental data do not follow the prediction made by the current scaling relation, and the results are generally higher than expected.

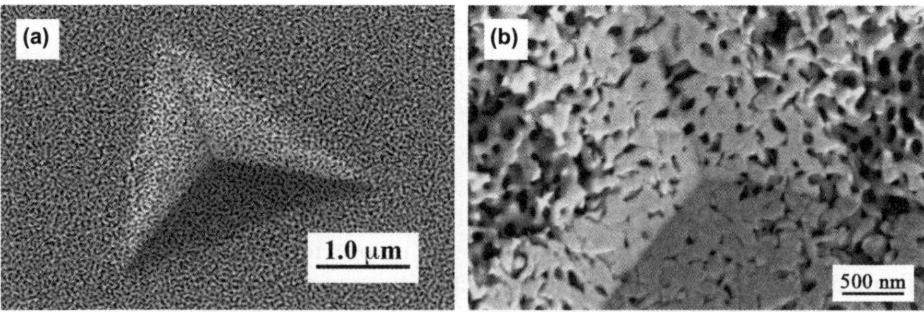

Figure 4.1 Scanning electron micrographs of nanoindentation on a fractured surface of nanoporous gold using a Berkovich tip with a curvature of ~200 nm. (a) Representative single indent and (b) zoom view of the area underneath the indenter. Note that the plastic deformation is confined to the area under the indenter, and adjacent areas are virtually undisturbed.

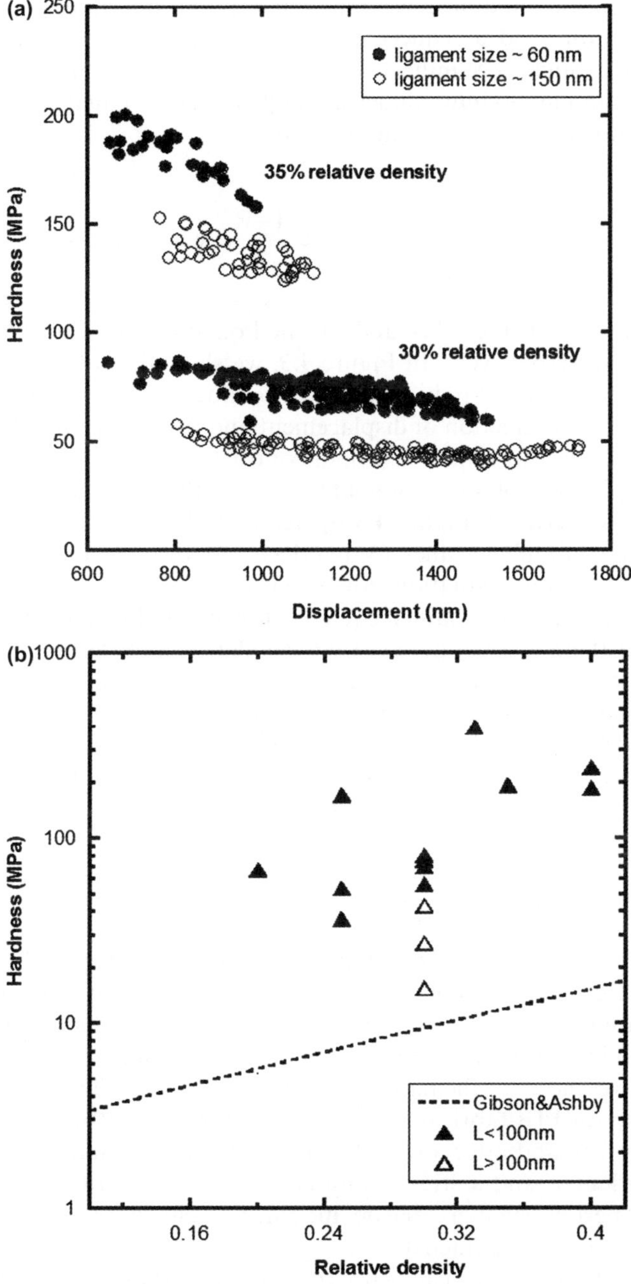

Figure 4.2 Effect of nanoindentation depth on measuring the reduced modulus of the nanoporous foam. The 30% and 35% relative density foams have ligament diameters of 50 nm (as-prepared) and 150 nm (after heat treatment). (b) Plot of foam relative density *versus* reduced elastic modulus measurement for 25–40% relative density nanoporous Au foams. The dashed line presents the Gibson and Ashby prediction.

For the elastic modulus, we must first point out that the results from nanoindentation yield a reduced modulus value (E_r) rather than the typical modulus (E). Since a significantly stiffer diamond indenter is used during testing, the sample modulus values are approximately equal to the reported E_r values, as shown in the following equation[29]:

$$\frac{1}{E_r} = \frac{1-v_i^2}{E_i} + \frac{1-v_s^2}{E_s} \approx \frac{1-v_s^2}{E_s} \text{ when } E_i \gg E_s, \tag{4.4}$$

where E is the elastic modulus, and v is the Poisson ratio of the indenter (i) and the sample (s), respectively. In Figure 4.3, we show the role of relative density and ligament size on the reduced elastic modulus for (a) 30% and 35% relative density foams as a function of displacement and (b) a range of relative densities and ligament sizes. In Figure 4.3(a), in contrast to Figure 4.2(a), it is clear that relative density still plays a dominant role, yet the effect due to ligament size seems to be less critical. Figure 4.3(b) clearly highlights the difference between the predicted modulus values (5–14 GPa) and the test data. Note that the conventional scaling equation for macroporous foams was used for the modulus, as there is no modified equation for the modulus of np materials. It is evident that there is a strong density dependence, and there is a much clearer trend than for the hardness values. However, no particular theory regarding the effect of ligament size has been developed. Though a recent publication by Mathur and Erlebacher[30] indicates an increase in modulus as a function of the ligament size, in general, changes in the elastic modulus are not usually observed or expected for nanoscale ligaments.[1,31]

4.2.2.2   Pillar Compression Tests

By using a method first developed by Dimiduk, Florando, and Nix,[32] submicrometer np Au columns can be fabricated by focused ion-beam micromachining. Several variations of this technique can also be called *in situ* nanomechanical testing or microcompression testing.[3,16,17,33,34] In Figures 4.4(a)–(c), we show a schematic of the process where Figure 4.4(a) depicts a typical setup for FIB-fabricated pillars, while Figures 4.4(b) and (c) show an np Au pillar before and after a test. The FIB-fabricated columns are compressed by a flat indenter, and a stress/strain curve can be generated. From the data, several properties such as elastic modulus, yield strength, and the elastic Poisson ratio can be extracted.

The size of the pillars can vary from hundreds of nanometeres to a few micrometers, and a size relationship can be extrapolated as a function of pillar diameter. Recent developments from tests performed on ultra-fine and micrometer-sized (180 nm to 1 μm) single-crystal gold and np Au pillars prepared by FIB have demonstrated a strong size dependence on the material strength.[17] For example, Volkert and Lilleodden[16] reported yield-strength

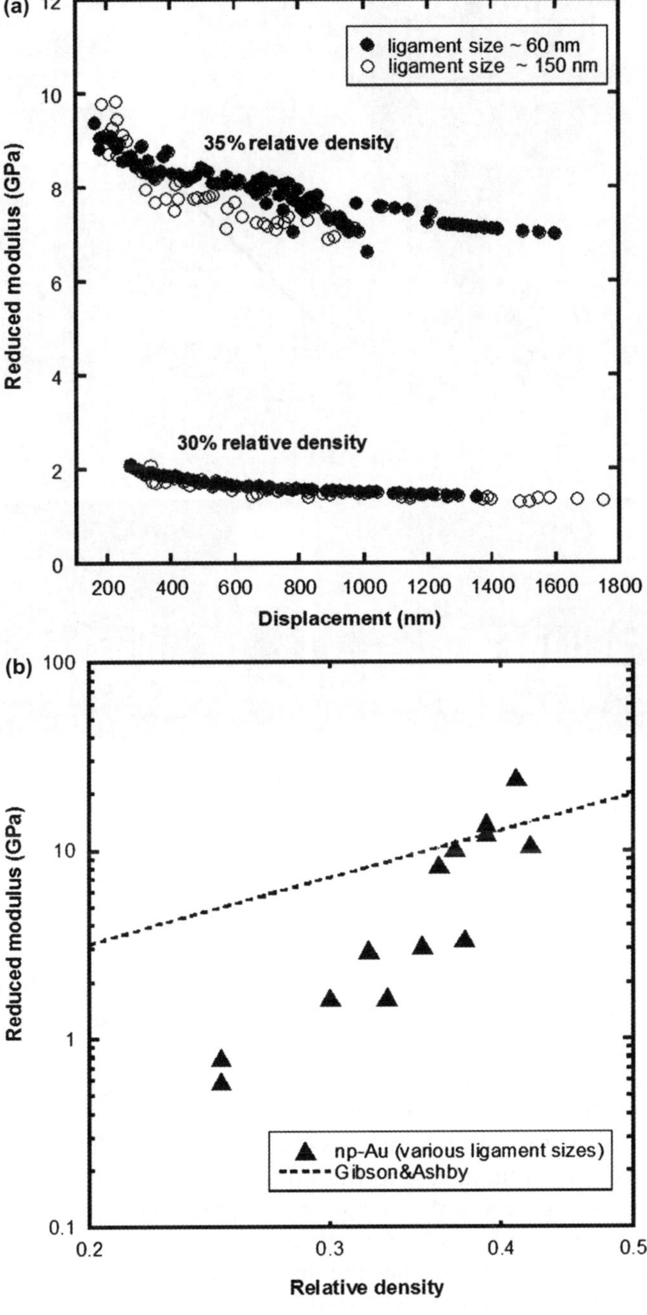

Figure 4.3 Nanoindentation data for 30% and 35% relative density sample foams plotted as a function of indentation depth given two distinct ligament sizes: 60 nm and 150 nm, and (b) experimental values for foam hardness values over a wide range of ligaments and relative densities (25–40%). The dashed line presents the Gibson and Ashby prediction.

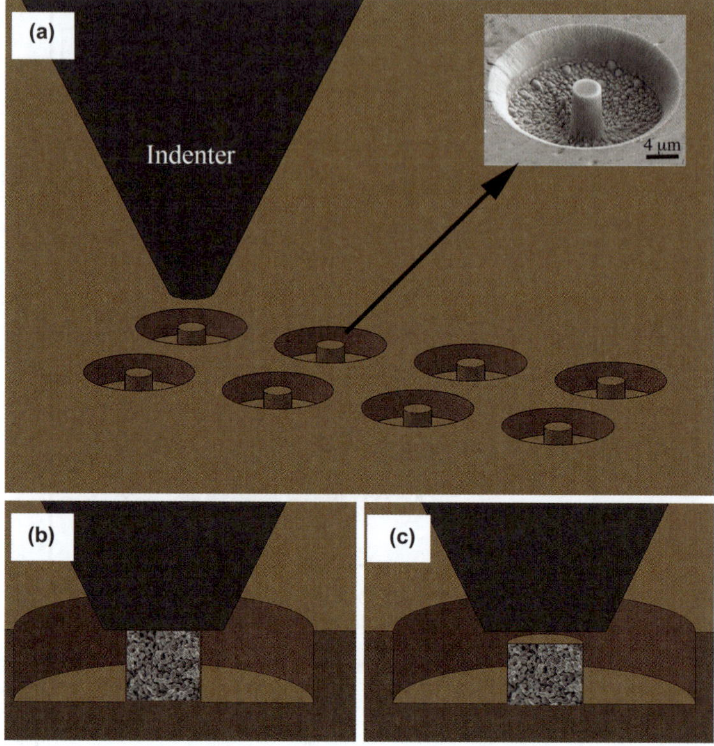

Figure 4.4 Schematic of *in situ* compression of FIB-fabricated nanoporous Au micropillars based on Gianola *et al.*[33] (not to scale). (a) Representative test setup showing columns and indenter. The inset shows a scanning electron micrograph of a FIB-fabricated np Au column. (b) *In situ* test configuration before and (c) after compression.

values of submicrometer Au columns following a yield-strength power law of d^{-n}, where d is the column diameter, and n is 0.6; this trend of size to strength is similar to results from nanoporous Au ligaments. From our discussion of Figures 4.2(a) and (b), it is observed that the ligament sizes have a strong influence on the properties, and thus, we can think of foam as a system of interconnected wires or columns. As such, we can directly compare the microcompression column test to the np foam as a function of the ligament/column size. In Figure 4.5, the normalized ligament/column yield stress *vs.* the ligament/column size shows that as the pillar size decreases, the yield strength of the pillar/ligaments approaches the theoretical yield strength of Au.[16,35] Due to FIB size fabrication limits, the tested columns are larger than a typical foam ligament size, thus preventing a complete size comparison between both methods. However, in the overlapping length-scale range (200 nm to 1 μm), we observe a good agreement between the calculated ligament yield strength and the single-crystal column micro-compression tests.[17]

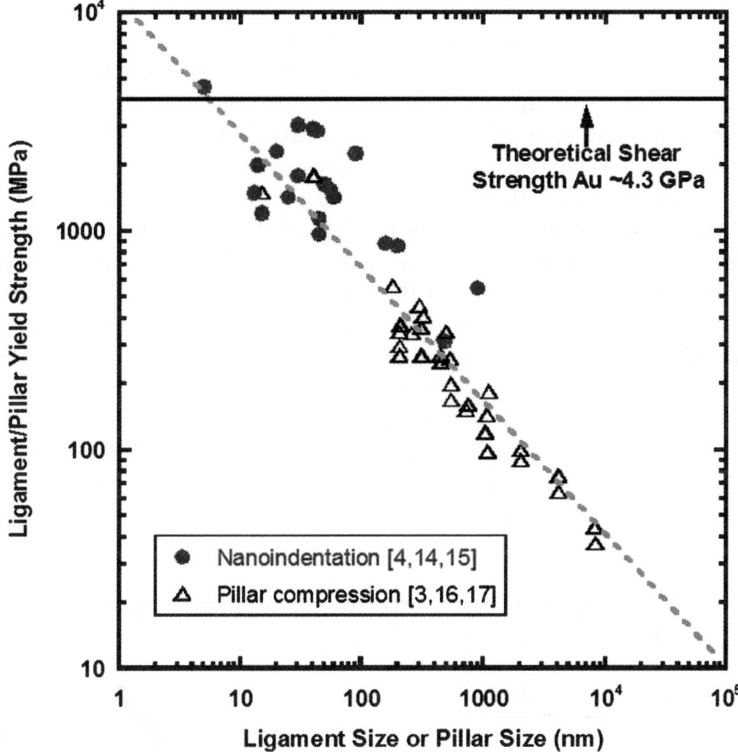

Figure 4.5   Relationship of ligament size to ligament yield stress for nanoporous gold foams obtained by nanoindentation and by column micro-compression testing. The foam yield stress was calculated assuming $H \sim \sigma_y$.

4.2.3   Tensile Testing

While most of the published literature on mechanical behavior of np Au is based on nanoindentation, some studies using small-scale variants of traditional mechanical test techniques have also been reported.[14,19,36,37] Microspecimen tests performed in tension and compression showed that the initial yield point of np Au is approximately 10–20 MPa, which is one order of magnitude lower than values determined by nanoindentation[2,4,17,38] and by the tensile test of thin np Au films.[14,37] However, this finding is somewhat expected, since in macroporous foams, there is wide discrepancy in strength values that can be attributed to changes in the synthesis process.[21,39–42] For example, strength values obtained under compression are typically higher than those under tension, due to various types of defects in the cell walls that become dominant under tension.[40,42,43]

A fundamental and as-yet unanswered question regarding np Au is how inherently brittle or ductile the material is. However, given the paucity of tensile data in the literature, ductility is often inferred from microstructural observations of np Au following indentation. For these samples, the surfaces of indents are typically smeared, with extensive deformation of np Au ligaments suggesting a high amount of ductility. Nonetheless, ductility can be measured in tension, and the available tensile data for bulk np Au[36] indicate only limited ductility, albeit accompanied by nanoscale plasticity within individual ligaments.[25] Tensile tests of np Au show elastic behavior followed by fracture microspecimens, as can observed in the stress–strain curves in Figure 4.6.[44] No plastic deformation is apparent from the test data. Samples fail via rapid crack propagation through the gage section, with cracks typically initiating at a grain boundary on the sample edge and generally following a path along grain boundaries to the opposite side of the sample. An example of a ruptured tensile sample is shown in Figure 4.7(a).[44] The final stage of crack propagation corresponds to a small amount of overall sample strain, with gage elongation on the order of 100 nm. This distance is equivalent to 1–2 cells of the nanoporous structure (where a cell represents a single pore cage and therefore corresponds to the sum of pore diameter and ligament width). The morphology of np Au in the vicinity of the crack face reveals that ligaments experience extensive plastic deformation before they rupture. This can be seen in

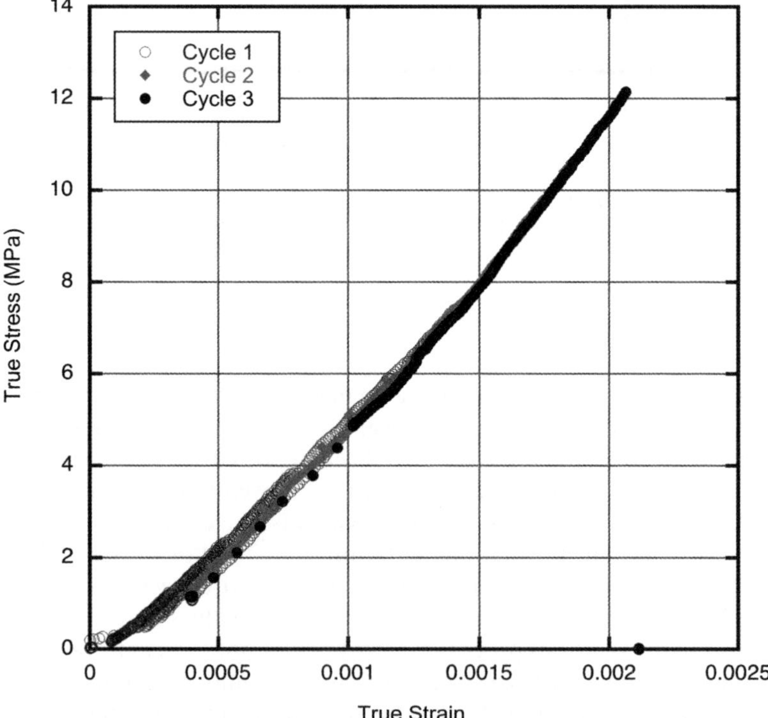

Figure 4.6   Stress–strain curve for a microspecimen tested in tension. The sample deformed elastically, during repeated load–unload cycles, and eventually fractured at a stress level of 12 MPa with no indication of plastic deformation.

Figure 4.7(b), which shows the fracture surface of a tensile microspecimen near the sample surface. Numerous ligaments neck down, but this necking occurs only within a narrow region close to the crack face. The underlying nanoporous structure does not appear to have undergone plastic deformation. This general observation agrees with the small amount of overall deformation that occurs during crack propagation through the sample. Thus, crack propagation appears to involve significant localized plastic deformation and rupture of ligaments that span the crack plane. The Au ligaments in np Au are still ductile, but they are unable to support permanent deformation throughout a sample, resulting in apparent macroscale brittleness.

Additionally, tensile testing allows the determination of the Poisson ratio, which is an important elastic property for the application of scaling laws to other measured mechanical properties. In contrast, during indentation, the assumption is that $H \sim \sigma_y$, which characterizes a fully compressible material.[17,43] This is because during indentation, we are specifically referring to the 'lateral expansion coefficient' or 'plastic Poisson ratio (v_p)' which typically approaches zero for low-density foams.[45,46] For example, the elastic Poisson ratio of low-density open-cell aluminum (Duocel) foams has been shown to be ~ 0.25,[43] while the plastic Poisson ratio is reported to be ~ 0.052.[45] In the case of np Au, Jin *et al.* reported a Poisson value of 0.08 in compression, for samples that exhibited extensive plastic deformation.[19] Microspecimen testing indicated an elastic Poisson ratio of 0.16 under both tensile and compressive loading.[36] Further studies are needed to establish reliable values for both the elastic and plastic Poisson ratio of np Au, but it is important to note the consequences of a non-zero plastic Poisson value, namely in the hardness-to-strength relationship. If the plastic Poisson ratio increases from zero, the ratio of hardness to yield strength rapidly increases from 1 to ~ 3.[9] For $v = 0.16$, this ratio can be as high as ~ 2.5. Therefore, there is a need for a comprehensive study to measure both the elastic and plastic Poisson ratio similar to previous work performed on macroporous foams.[43,45]

4.3 Fracture Behavior

The fracture toughness of np Au can also be calculated from microspecimen testing and was found to be low.[44] Taking the stress before fracture of a tensile specimen and the initial crack length, as observed with optical microscopy during the tensile test[36], one can estimate the K_{Ic} value for bulk np Au. The relevant results from that study include: fracture strength = 11 MPa; crack length = 80 μm. Calculating fracture toughness as $K_{Ic} = \sigma \sqrt{\pi a}$ yields a value of 0.17 MPa.m$^{1/2}$. This is an extremely low value, given the inherent ductility of Au, but it agrees with the macroscopic brittleness of np Au. However, preliminary studies performed by Volkert *et al.*[47] using FIB-fabricated cantilever beams yield values around $K_{Ic} = 0.4$ MPa.m$^{1/2}$, which are significantly higher than those measured by Balk *et al.*

The macroscopic brittleness of np Au is well established and has been systematically evaluated by Li and Sieradzki[48] and others,[1,3] who found a correlation between ligament size and fracture behavior. Although np Au appears to exhibit brittle fracture, observation of the ligament structure reveals evidence for nanoscale plasticity leading to rupture of ligaments that are less than 100 nm in diameter.[25,36,49] This is evident in SEM micrographs of pencil-like ligaments that have undergone extensive necking, presumably due to dislocation motion (see Figure 4.7(b)). However, this plasticity is limited to within approximately one cell width of the crack plane, which results in macroscopically brittle behavior.

Images of fractured np Au by Hodge and Doucette[50] illustrate general aspects of fracture in np Au as well as effects of residual Ag that is not removed

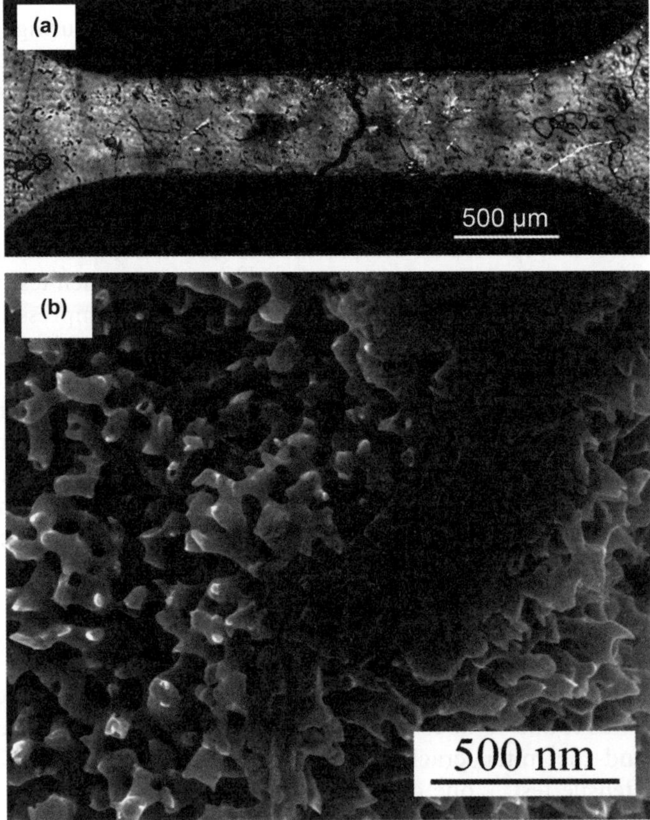

Figure 4.7 (a) Tensile microspecimen after testing to failure, with fracture in the gage section. Sample cracking initiated at a grain boundary on the top edge of the gage, and the crack path followed grain boundaries as it progressed through the sample. (b) SEM micrograph of a microspecimen fracture surface following tension testing. Ligaments located along the crack faces underwent extensive necking before rupture, as shown in the bottom right of this image. However, ligament deformation was confined to the crack plane, as indicated by the intact and seemingly undeformed np-Au structure below the ruptured ligaments (shown most clearly on the left of this image).

during the dealloying process. Figure 4.8(a) is an SEM micrograph showing the fracture surface of a fully dealloyed sample that originally contained 35 at.% Au (65 at.% Ag) and after dealloying contained less than 1 at.% Ag. This composition is representative of fully dealloyed bulk np Au, which typically exhibits a residual Ag content between 0 and 1 at.% as determined by energy-dispersive X-ray spectroscopy,[50] *i.e.* the Ag content is at the lower limit of detection by this method. For this fully dealloyed specimen, the fracture surface appears somewhat smooth, at least at the micrometer length scale, indicating that crack propagation was primarily intragranular. In contrast, a

Figure 4.8   (a) SEM micrograph showing the fracture surface of fully dealloyed bulk np-Au. The fracture surface appears smooth and indicates that crack propagation proceeded straight through the sample, with only slight deviations along grain boundaries lying close to the crack plane. (b) SEM micrograph of fracture surface in bulk np-Au with high residual Ag content (15 at.%) after dealloying. The fracture morphology is primarily intergranular, with some grains having been cleaved.

bulk np Au sample dealloyed from the same precursor material, yet containing a higher residual Ag content, exhibits noticeably different fracture behavior. The SEM micrograph in Figure 4.8(b) shows the surface of a fractured np Au sample containing 15 at.% Ag, where the crack path followed the grain boundaries through most of the sample thickness.

The exact reasons for this change in fracture behavior are not known, but the correlation of higher residual Ag content with a shift to intergranular fracture suggests one possible cause. In the np Au with a high Ag content, the grain boundaries are relatively weak (compared to the nearly pure np Au in Figure 4.8(a)). This could be due to the grain boundary regions losing Ag more quickly during dealloying, with the grain interior regions not being fully dealloyed and thus exhibiting a lower porosity and higher strength. However, the higher Ag content does not cause a noticeable change in pore/ligament morphology, so an argument based solely on the uniformity of ligament shape/dimension cannot explain the fracture behavior. The remnant Ag is likely trapped within the ligaments of np Au, since the ligament surfaces can be expected to lose Ag very rapidly. The ligaments therefore should have an Au–Ag alloy core, with a higher strength than pure Au ligaments of the same dimension. Increasing the ligament strength while holding the grain boundary strength constant (or at least not allowing it to increase as quickly as the ligament strength) could explain the change in fracture behavior. Given that grain boundaries appear to be natural weak points in np Au, it may be more difficult to increase their strength than to strengthen the ligaments. Additional tensile testing of np Au, especially with a focus on evaluating the strength of grain boundaries, should help explain the fracture behavior of this material.

4.4 Modeling and Simulation Studies

Several important deformation behaviors of np Au foams can be explored further by the use of simulations and models. To date, studies using molecular dynamics (MD) simulations and numerical models have aimed at understanding the stability and deformation behavior as a function of ligament, and pore shapes and sizes.[51–54] For example, Wang and Xia[52] and Feng et al.[55] implemented new numerical deformation models on np Au with ligaments smaller than 100 nm in order to modify the Gibson and Ashby equations for elastic modulus. Both studies present the addition of surface effects such as elasticity and tension in order to modify the typical unit cell by including modifications from surface elasticity models. Overall, it is found that the ligament size must be below 10 nm in order to have a significant influence on the modulus, similar to previous results on Au nano wires.[31]

In regard to the yield strength, MD simulations have been employed for a variety of studies including ligament stability and length-scale effects in plasticity. Several of the studies focus on nanoscale Au columns and films, which can be used to understand the behavior of individual ligaments in a foam.[31,35,56–58] Preliminary results show that it is indeed difficult to incorporate stable dislocation sources (such as a Frank–Read source) in nanometer-sized ligaments.[53] These simulations also reveal that plastic deformation of nanometer-sized Au columns is still dislocation-mediated, and that these dislocations nucleate at free surfaces at stress levels close to the theoretical strength of the material. Additional modifications to eqns (4.1) and (4.3) have been performed by a numerical model that incorporates the effect of cell corners to the overall ligament size, leading to the conclusion that the maximum relative density that can yield high strengths is restricted by the elastic buckling collapse of the ligaments.[54] A different model by Dou and Derby[18] presented a similar approach to the original cell/pore deformation mechanism-based models from Gibson and Ashby. Their model includes the influence of strain gradients and geometrically necessary dislocations, which yielded a final equation for strength in relation to the ligament size rather than cell size. Overall, simulations and models are still lagging behind, compared to experimental data, and more work designed to understand the mechanical properties of nanoporous foams is still needed.

4.5 Summary

In this chapter, we have reviewed the most current results for the mechanical deformation and properties of nanoporous Au, as determined by a variety of testing techniques, as well as micromechanical models and simulations. Several different material properties such as yield strength, elastic modulus, Poisson ratio, and fracture toughness have been presented and evaluated. The macroscopic brittleness in np Au is a major factor that affects the accuracy of the current mechanical measurements; the degree of the brittleness differs slightly with composition and dealloying technique, and therefore a systematic study including both compression and tension tests is still needed. By incorporating new deformation models as well as simulations, it is possible to develop a new theoretical approach to predict the mechanical properties of np Au. At this point, the view of ligaments as individual columns or wires seems to neglect the complexity of interconnection between ligaments, which can be a major contributor to the measured or calculated values. Np Au is indeed a very complex material that exhibits both micro-ductility and macro-brittleness, and two levels of nanoscale components: ligaments and pores. Further studies are still needed that will allow the prediction of more general trends such as those found for macroporous foams.

Acknowledgments

Parts of this work were performed under the auspices of AFOSR grant number FA9550-10-1-0478. The authors would like to acknowledge Prof. C. Volkert at University of Göttingen, Germany for providing original results. This material is also based upon work supported by the National Science Foundation under Grant No. DMR-0847693 (T.J.B.).

References

1. J. Biener, A. M. Hodge and A. V. Hamza, in *Micro and Nano Mechanical Testing of Materials and Devices*, ed. F. Yang and J. C. M. Li, Springer, New York, 2008.
2. J. Biener, A. M. Hodge, A. V. Hamza, L. M. Hsiung and J. H. Satcher, *J. Appl. Phys.*, 2005, **97**, 024301.
3. J. Biener, A. M. Hodge, J. R. Hayes, C. A. Volkert, L. A. Zepeda-Ruiz, A. V. Hamza and F. F. Abraham, *Nano Lett.*, 2006, **6**, 2379–2382.
4. A. M. Hodge, J. Biener, J. R. Hayes, P. M. Bythrow, C. A. Volkert and A. V. Hamza, *Acta Mater.*, 2007, **55**, 1343–1349.
5. K. Sieradzki and R. C. Cammarata, *Phys. Rev. Lett.*, 1994, **73**, 1049.
6. K. Sieradzki and R. Li, *Phys. Rev. Lett.*, 1986, **56**, 2509–2512.
7. K. Sieradzki and R. C. Newman, *Philos. Mag. A Phys. Condens. Matter Struct. Defects Mech. Prop.*, 1985, **51**, 95–132.
8. J. Weissmüller, R. C. Newman, H.-J. Jin, A. M. Hodge and J. W. Kysar, *MRS Bull.*, 2009, **34**, 577–586.
9. L. J. Gibson and M. F. Ashby, *Cellular Solids: Structure and Properties*, Cambridge University Press, Cambridge, 1997.
10. A. S. Argon, *Mechanical Behavior of Materials*, Addision-Wesley, Reading, MA, 1966.
11. S. Lapman (Ed.), *ASM Handbook, Volume 2*, ASM International, 1990.
12. J. R. Davis, ed., *Metals Handbook*, ASM International, Materials Park, OH, 1998.
13. E. M. Savitskii, ed., *Handbook of Precious Metals*, Hemisphere Publishing Corporation, New York, 1989.
14. D. Lee, X. Wei, X. Chen, M. Zhao, S. C. Jun, J. Hone, E. G. Herbert, W. C. Oliver and J. W. Kysar, *Scr. Mater.*, 2007, **56**, 437–440.
15. Y. Sun, J. Ye, Z. Shan, A. M. Minor and T. J. Balk, *J. Miner. Metals Mater. Soc.*, 2007, **59**, 54–58.
16. C. A. Volkert and E. T. Lilleodden, *Philos. Mag.*, 2006, **86**, 5567–5579.
17. C. A. Volkert, E. T. Lilleodden, D. Kramer and J. Weissmuller, *Appl. Phys. Lett.*, 2006, **89**.
18. R. Dou and B. Derby, *J. Mater. Res.*, 2009, **25**, 746–753.
19. H.-J. Jin, L. Kurmanaeva, J. r. Schmauch, H. Rösner, Y. Ivanisenko and J. R. Weissmüller, *Acta Mater.*, 2009, **57**, 2665–2672.
20. E. W. Andrews, G. Gioux, P. Onck and L. J. Gibson, *Int. J. Mech. Sci.*, 2001, **43**, 701–713.
21. Z. Liu, C. S. L. Chuah and M. G. Scanlon, *Acta Mater.*, 2003, **51**, 365–371.
22. M. Wilsea, K. L. Johnson and M. F. Ashby, *Int. J. Mech. Sci.*, 1975, **17**, 457–460.
23. Y. Toivola, A. Stein and R. F. Cook, *J. Mater. Res.*, 2004, **19**, 260–271.
24. U. Ramamurty and M. C. Kumaran, *Acta Mater.*, 2004, **52**, 181–189.
25. J. Biener, A. M. Hodge and A. V. Hamza, *Appl. Phys. Lett.*, 2005, **87**, 121908.

26. E. W. Andrews, G. Gioux, P. Onck and L. G. Gibson, *Int. J. Mech. Sci.*, 2001, **43**, 701–713.
27. P. S. Kumar, S. Ramachandra and U. Ramamurty, *Mater. Sci. Eng.*, 2003, **347A**, 330–337.
28. A. D. Brydon, S. G. Bardenhagen, E. A. Miller and G. T. Seidler, *J. Mech. Phys. Solids*, 2005, **53**, 2638–2660.
29. W. C. Oliver and G. M. Pharr, *J. Mater. Res.*, 2004, **19**, 3–20.
30. A. Mathur and J. Erlebacher, *Appl. Phys. Lett.*, 2007, **90**, 061910.
31. B. Wu, A. Heidelberg and J. J. Boland, *Nat. Mater.*, 2005, **4**, 525–529.
32. M. D. Uchic, D. M. Dimiduk, J. N. Florando and W. D. Nix, *Science*, 2004, **305**, 986–989.
33. D. S. Gianola, A. Sedlmayr, R. Mönig, C. A. Volkert, R. C. Major, E. Cyrankowski, S. A. S. Asif, O. L. Warren and O. Kraft, *Review of Scientific Instruments*, 2011, **82**, 063901.
34. Y. Sun, J. Ye, A. M. Minor and T. J. Balk, *Microsc. Res. Technique*, 2009, **72**, 232–241.
35. J. R. Greer and W. D. Nix, *Appl. Phys. A Mater. Sci. Process.*, 2005, **80**, 1625–1629.
36. T. Balk, C. Eberl, Y. Sun, K. Hemker and D. Gianola, *J. Miner. Metals Mater. Soc.*, 2009, **61**, 26–31.
37. Y. Sun, K. P. Kucera, S. A. Burger and T. J. Balk, *Scr. Mater.*, 2008, **58**, 1018–1021.
38. M. Hakamada and M. Mabuchi, *Scr. Mater.*, 2007, **56**, 1003–1006.
39. M. E. Cox and D. C. Dunand, *Mater. Sci. Eng. A*, 2011, **528**, 2401–2406.
40. P. S. Liu and G. F. Chen, *Mater. Sci. Eng. A Struct. Mater. Prop. Microstruct. Process.*, 2009, **507**, 190–193.
41. P. H. Thornton and C. L. Magee, *Metallurg. Trans. A Phys. Metallurg. Mater. Sci.*, 1975, **6**, 1253–1263.
42. C. M. Ford and L. J. Gibson, *Int. J. Mech. Sci.*, 1998, **40**, 521–531.
43. E. Andrews, W. Sanders and L. J. Gibson, *Mater. Sci. Eng. A Struct. Mater. Prop. Microstruct. Process.*, 1999, **270**, 113–124.
44. T. J. Balk, N. Briot, D. S. Gianola, T. Kennerknecht and C. Eberl, unpublished, 2010.
45. G. Gioux, T. M. McCormack and L. G. Gibson, *Int. J. Mech. Sci.*, 2000, **42**, 1097–1117.
46. R. E. Miller, *Int. J. Mech. Sci.*, 2000, **42**, 729–754.
47. C. A. Volkert, E. T. Lilleodden, D. Kramer and J. Weißmüller, *Appl. Phys. Lett.*, 2006, **89**, 061920.
48. R. Li and K. Sieradzki, *Phys. Rev. Lett.*, 1992, **68**, 1168–1171.
49. A. M. Hodge, J. R. Hayes, J. A. Caro, J. Biener and A. V. Hamza, *Adv. Eng. Mater.*, 2006, **8**, 853–857.
50. A. M. Hodge, R. T. Doucette, M. M. Biener, J. Biener, O. Cervantes and A. V. Hamza, *J. Mater. Res.*, 2009, **24**, 1600–1606.
51. D. A. Crowson, D. Farkas and S. G. Corcoran, *Scr. Mater.*, 2009, **61**, 497–499.
52. X. S. Wang and R. Xia, *Europhys. Lett.*, 2010, **92**, 16004.

53. L. A. Zepeda-Ruiz, B. Sadigh, J. Biener, A. M. Hodge and A. V. Hamza, *Appl. Phys. Lett.*, 2007, **91**.
54. H. L. Fan and D. N. Fang, *Mater. Des.*, 2009, **30**, 1441–1444.
55. X.-Q. Feng, R. Xia, X. Li and B. Li, *Appl. Phys. Lett.*, 2009, **94**, 011916.
56. Y. Gan and J. Chen, *Appl. Phys. A Mater. Sci. Process.*, 2009, **95**, 357–362.
57. F. d. r. Lançon, J. Ye, D. Caliste, T. Radetic, A. M. Minor and U. Dahmen, *Nano Lett.*, 2010, **10**, 695–700.
58. S.-Y. Lin, T. H. Wang, T.-H. Fang and W.-L. Chen, *Curr. Appl. Phys.*, 2010, **10**, 5–9.

CHAPTER 5

Microfabrication of Nanoporous Gold

OYA OKMAN* AND JEFFREY W. KYSAR

Department of Mechanical Engineering, Columbia University, New York, USA
*Email: oo2128@columbia.edu

5.1   Introduction

The idea behind the fabrication of nanoporous gold (NPG) relies on a simple principle: the less noble component of a gold alloy is removed selectively, while the more noble material in the alloy, *i.e.* Au, remains behind, forming a bicontinuous structure. This process is called dealloying. Some suitable precursor alloys are AuAg,[1,2] AuCu,[3,4] and AuNi,[5] as well as a ternary alloy AuAg(Pt).[6] The most commonly preferred precursor is the AuAg alloy. The AuAg alloy is particularly useful in NPG fabrication, since both alloying elements have face-centered cubic structures with almost identical lattice constants.[7] The alloying elements form a homogeneous solid solution over the entire range of compositions, leading to a uniform porous structure.

The mechanism of dealloying is based on selective dissolution of the less noble component of the precursor alloy along with surface diffusion of the remaining Au.[8,9] Upon removal of the secondary element in the first atomic layer of the alloy, the remaining low-coordinated Au atoms diffuse and form clusters on the surface.[10] As the process continues, Ag is removed from the underlying atomic layers to form a three-dimensional nanoporous structure (Figure 5.1). For

RSC Nanoscience & Nanotechnology No. 22
Nanoporous Gold: From an Ancient Technology to a High-Tech Material
Edited by Arne Wittstock, Jürgen Biener, Jonah Erlebacher and Marcus Bäumer
© Royal Society of Chemistry 2012
Published by the Royal Society of Chemistry, www.rsc.org

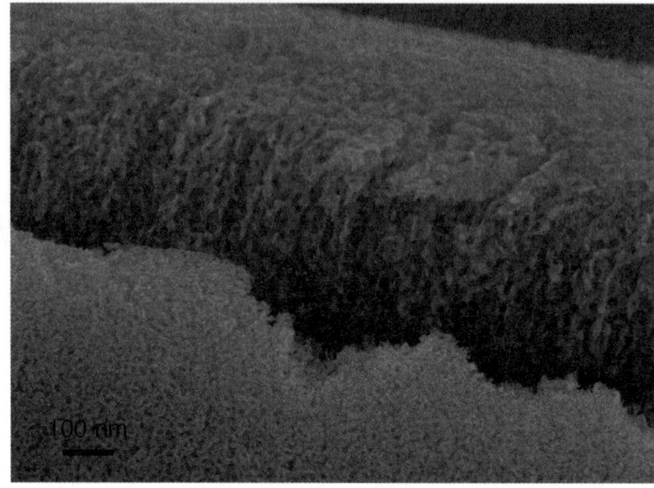

Figure 5.1 Nanoporous gold film fabricated from an $Au_{0.32}Ag_{0.68}$ alloy. The cross-section was exposed when the film ruptured in two.

dealloying to take place, the less noble component should exist at a compositional lower limit that is higher than the percolation threshold. This is called the parting limit. If the secondary alloying element is below this composition, there may still be selective dissolution from the first atomic layers, but the remaining Au passivates the alloy surface.[9,11] Experimental studies by Newman *et al.*[12] on the parting limit in Au_xAg_{1-x} alloys suggest an approximate threshold value of 0.40 for x. Atomistic simulations using the kinetic Monte Carlo method by Artymowicz *et al.*[13] show that this threshold can be as high as 0.45.

During dealloying, the less noble component of the precursor alloy is removed either by free corrosion or by electrochemical means. Free corrosion is a simple process, in which the precursor alloy is exposed to an acidic solution, generally aqueous nitric acid.[7] In this process the molarity and temperature of the electrolyte are the main adjustable process parameters. The second common fabrication method is electrochemical dealloying,[11,14] where the alloy serves as the anode of a two- or three-electrode electrochemical cell. The potential value is selected to be between the oxidation potentials of the two alloying elements, allowing the less noble element to dissolve and Au to remain and diffuse on the surface. Sustained formation of a bicontinuous porous structure is possible above a critical potential, which depends on the type and composition of the alloy.[14,15] Electrochemical dealloying provides more flexibility in process control. Adjustable process parameters include the value and temporal history of the applied potential, as well as the type and temperature of the electrolyte. Nanoporous gold always contains residual amounts of the less noble alloying element of the original alloy. In the films made from Au–Ag alloys, the least reported concentration of residual Ag is less than 2 at.% Ag.[16,17] Residual Ag content may be treated as an effectiveness measure to assess the dealloying as well as a means to tailor the functionality of NPG in certain applications.

Varying amounts of residual Ag content determine the functionality of NPG as catalysts[18] or substrates for surface-enhanced Raman scattering.[19]

The general principles of NPG fabrication hold, regardless of the length scale and the geometry of the precursor alloy. However, given the propensity for volume reduction in the NPG during dealloying,[20] the quality of the final porous product depends strongly on the applied physical constraints among other parameters. Fabrication of crack-free blanket NPG films on silicon substrates and constrained NPG structures is particularly challenging, due to the evolution of film stress during the complex surface reconstruction in dealloying[17,21–24] as the secondary component is removed.

In this chapter, fabrication techniques for NPG fabrication are described in terms of the final product quality and applicability in microscale fabrication. In addition, the residual Ag content and propensity of cracking are discussed. NPG has received attention as a promising candidate for sensors,[25] actuators,[18,26,27] catalysts,[28,29] and biocompatible materials.[30] For potential applications as a functional material in microelectromechanical system (MEMS) devices, clean-room techniques should be adapted to enable incorporation of NPG on silicon devices using scalable methods. It is therefore imperative to establish a thorough understanding of the formation mechanisms and fabrication techniques to provide the porosity and quality of NPG necessary to meet the specific service requirements. Here, we look at NPG fabrication from an experimenter's point of view and present common challenges, provide practical suggestions, and relate the observations to the theory of the complex nature of surface reconstruction.

5.2 Challenges in Fabrication of NPG Thin Films

The morphology and the integrity of NPG thin films are closely related to the morphology and the composition of the precursor alloy. For alloy compositions between 28 and 35 at.% Au, uniform NPG films can be obtained with minimum surface damage (*cf.* Figure 5.2(a)).

Blanket thin films prepared by conventional film deposition techniques, *i.e.* sputtering, *etc.*, inherently have nanovoids within the film. During diffusion of Au in dealloying, those voids nucleate into crack-like features within the nanoporous structure.[21,22] Such damage may be observed in many thin films, depending on the deposition and dealloying parameters; however, the damage is more pronounced if the Au content of the initial alloy is below 25 at.% Au (*cf.* Figure 5.2(b)). Thermal treatment of AuAg alloy thin films and of suspended beams by rapid thermal processing (RTP) at 300 °C is an effective means to eliminate microvoids in NPG films.[21,31] At first glance, these observations seem contradictory with the elevated tensile residual stress level in the thin alloy films after annealing; however, heat treatment at sufficiently high temperatures causes grain growth and removes nanoscale voids that appear in sputter-deposited films. This method is most efficient for alloy films around the parting limit.

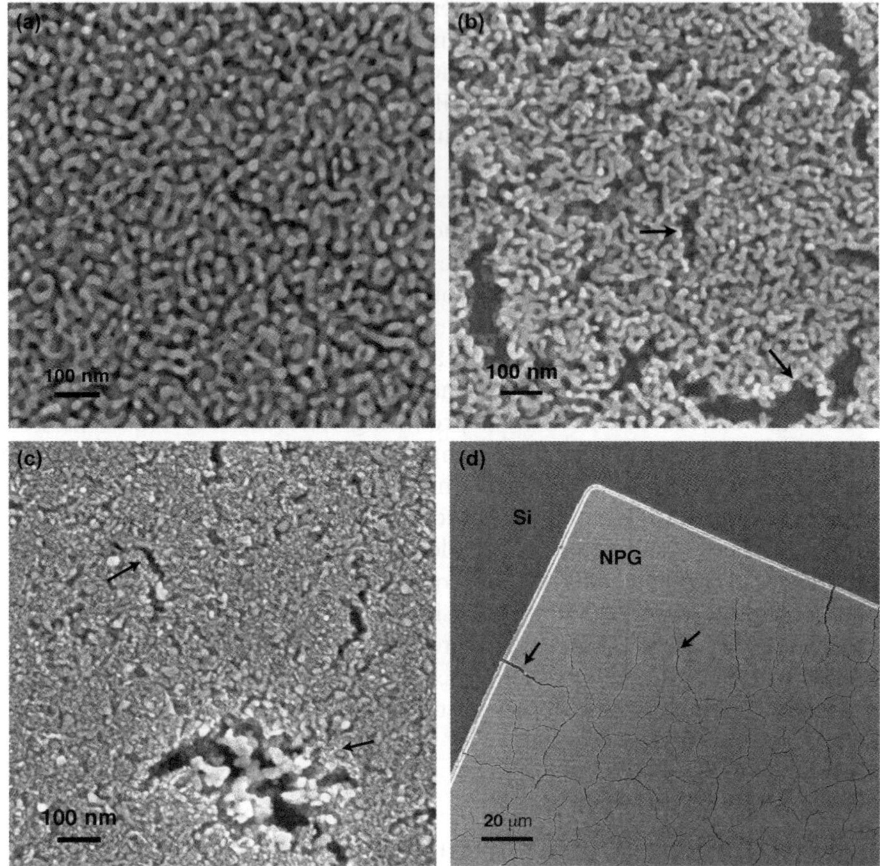

Figure 5.2 Nanoporous gold thin film: (a) fabricated from $Au_{0.35}Ag_{0.65}$, with no surface damage; (b) fabricated from $Au_{0.22}Ag_{0.78}$, with crack-like features on the surface; (c) fabricated from $Au_{0.30}Ag_{0.70}$, with extensive surface damage owing to rapid Ag removal; (d) fabricated from $Au_{0.30}Ag_{0.70}$, with through-thickness cracks owing to high film stress during dealloying.

It is important to distinguish between stress-driven cracks and the crack-like features appearing due to surface evolution when seeking means to improve the techniques for fabrication of blanket NPG films. Wide grooves between nanoporous islands on a substrate are commonly seen that may not have formed through loading of the film beyond its fracture toughness. These grooves are often formed due to the existence of microvoids in the precursor alloy, which grow into larger-scale defects during surface construction. They often appear in dealloying of precursor alloy films with less than 25 at.% Au (*cf.* Figure 5.2(b)).

The second common type of defects is surface damage, *i.e.* small-scale cracks or pits observed on the surface (*cf.* Figure 5.2(c)). Cracks of this class are generally in the order of a couple of micrometers in size, and they do not extend through the thickness of the film. They may appear during rapid dealloying schemes, such as application of constant potential.[22] These cracks initiate at the grain boundaries or impurities, where the local stress is high. Improving the film quality as well as the process itself can eliminate this type of cracking.

The third kind of defect is the through-thickness cracks (*cf.* Figure 5.2(d)). They are generally much larger than the film thickness. This type of defect is closely related to the overall stress state in the alloy. Understanding the relationship between critical film thickness for cracking and film stress is essential to define the strategies for effective dealloying techniques. Such cracks grow in a thin film when the film thickness exceeds a critical value given by[32]

$$h_{\mathrm{f}}^{\mathrm{cr}} = \frac{2\Gamma_{\mathrm{f}}E_{\mathrm{f}}}{\pi c_{\mathrm{f}}^2 \sigma_{\mathrm{m}}^2},\tag{5.1}$$

where c_{f} is a numerical coefficient related to the morphology of the crack, Γ_{f} is the energy of fracture of the thin film, E_{f} is the plane strain modulus of the thin film, and σ_{m} is the mean biaxial film stress induced due to strain mismatch between the film and the substrate. Eqn (5.1) is strictly valid only when the elastic properties of the substrate are the same as those of the film; however, the functional form of the expression shows that the critical thickness is inversely proportional to the square of the stress, σ_{m}. Most precursor films have a non-zero σ_{m} induced during the film deposition process. However, σ_{m} may also increase during dealloying due to the mechanisms that cause shrinkage of the precursor alloy.

The conventional dealloying techniques used for suspended thin leaves or bulk alloys may not be effective for constrained precursor alloys. The additional physical constraints induce a high film stress, leading to cracking. Thus, in microdevice fabrication incorporating NPG structures, the conventional dealloying techniques must be tailored, and the parameters should be selected carefully, to avoid surface defects and cracking.

5.3   Dealloying by Free Corrosion

The simplest method for selective material removal from an alloy is via free corrosion. The process consists simply of submersing the alloy in an electrolyte that facilitates selective dissolution. Aqueous nitric acid (HNO_3) is commonly used in dealloying AuAg alloys in ingot or thin film form. It provides a corrosive environment, in which Au is noble, and selective ionization of Ag is the dominant mode of dissolution.[7] Early studies on surface morphology during exposure to nitric acid show silver dissolution from the upper surface of the alloy and Au-rich cluster formation. The size of the clusters grows upon further exposure to the acid. The silver nitrate formed through the corrosion is soluble and does not interfere with the evolution of the alloy surface.[33]

The extent of dealloying below the alloy surface depends strongly on the alloy composition. In alloy films with 50 at.% Au, Au diffusion sufficiently passivates the surface to impede selective material removal below the surface. However, surface pits form due to incomplete surface coverage of the remaining gold islands, which deepen with prolonged exposure to the acid.[33] As the Au content in the alloy falls below the parting limit dealloying proceeds further below the alloy surface, and the resulting architecture becomes a bicontinuous porous structure.[7] For deposited thin films, which have a high grain boundary density, material removal takes place predominantly on the grain boundaries if the Au content is more than ~ 36 at.% Au.[34] A three-dimensional porous Au structure with low residual Ag content is possible with the choice of proper acid concentration and alloy composition (*cf.* Figure 5.3(a)).

The pore and ligament sizes are the smallest at the early stages of dealloying, and they increase during the dealloying process. Even after the Ag is depleted, Au diffusion continues, and the feature sizes continuously increase as long as the NPG structure is exposed to the electrolyte. This is referred to as *coarsening*. The dealloying time affects the porosity and the surface area, and thus is an important process parameter to tailor the porous structure. Through extended exposure, the ligaments eventually coarsen into wormlike fibrils. In thin films of NPG, further coarsening leads to a shift from homogeneous three-dimensional porosity to nonhomogeneous two-dimensional porosity.[34,35] The NPG film (115 nm thick) in Figure 5.3(b) demonstrates the morphological shift towards two-dimensional porosity. The pores extend though the thickness and are vertical to the surface after dealloying the film in 32.5% HNO_3 for 8 h. For thin films less than 80 nm in thickness, the pores tend to break up upon extended exposure to the electrolyte.[23]

Low residual Ag content can be achieved in free corrosion of AuAg alloys. Etching in concentrated aqueous nitric acid of thin blanket $Au_{0.35}Ag_{0.65}$ and $Au_{0.27}Ag_{0.73}$ films results in NPG films with a residual Ag content of ~ 5 at.% after 6 h of exposure to the etchant. The residual Ag content stabilizes with further exposure. This concentration is highest on the surface and drops down to less than ~ 1 at.% below the surface of the NPG film.[23]

The geometry and the mechanical constraints of the alloy film determine the effectiveness of the dealloying process. The precursor material can be ingots, thin alloy leaves, or thin films deposited on stiff substrates, such as silicon. Free corrosion is a particularly effective method to dealloy free-floating AuAg alloy leaves, which are typically used for decorative purposes and can be purchased from art supply stores. The suspended alloy leaves are as thin as 100–120 nm[20,35,36] and prove to be an excellent parent material in terms of producing uniform, crack-free NPG gold samples. Dealloying of 100 nm thick $Au_{0.35}Ag_{0.65}$ leaves in 0.7 M HNO_3 results in ligament sizes varying from 3 nm to 40 nm, with uniform three-dimensional porosity.[35] During dealloying, the leaves float on the electrolyte surface and spread uniformly due to the surface tension. The dealloyed leaf is then transferred onto a silicon substrate or elsewhere, depending on the application. However, the leaves are very fragile,

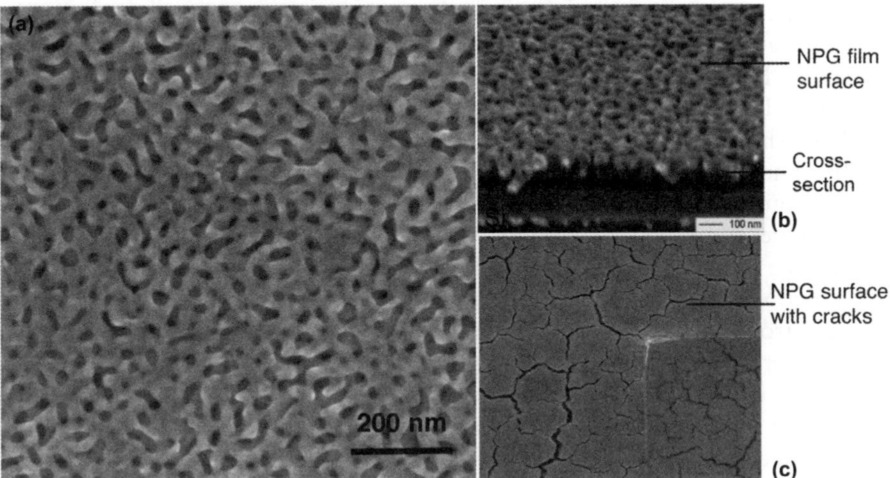

Figure 5.3   NPG thin film fabricated by free corrosion: (a) 120 nm thick $Au_{0.30}Ag_{0.70}$ alloy leaf is etched in 70% nitric acid solution for 10 min. to obtain an intact NPG structure; (b) from 115 nm thick $Au_{0.30}Ag_{0.70}$ blanket film by dealloying in 32.5% HNO_3 for 8 h (reprinted from Lu *et al.*,[34] Copyright 2007, with permission from Elsevier); (c) from 250 nm thick $Au_{0.30}Ag_{0.70}$ blanket film by dealloying in 70% HNO_3, which leads to extensive cracking.

difficult to handle, and require manual intervention to incorporate them into microscale devices.

Within the thickness range of 50–200 µm, intact nanoporous structures can be obtained by dealloying of alloy sheets in aqueous nitric acid solution.[37] Thicker alloy sheets are prepared by cold forging of alloy ingots. The alloy preparation involves annealing and cold rolling, which leads to a larger grain size, *i.e.* several micrometers, than in deposited thin films. In the design of NPG structures, the fabrication method should be tailored with regard to the typical grain size specific to the alloy fabrication method.

For the case of alloy ingots or thin films adhered to a stiff substrate, free corrosion leads to extensive cracking. The underlying structure constitutes a constraint such that the alloy at the surface cannot relieve the tensile stress by shrinkage (*cf.* Figure 5.3(c)). Examples of this problem are reported in many individual studies, such as for $Au_{0.30}Ag_{0.70}$ alloy strips with 450 µm thickness that develop surface cracks when dealloyed in 70 wt% HNO_3.[37] Cracking and film delamination are observed in 400 nm thick $Au_{0.24}Ag_{0.76}$ films when exposed to 32.5 wt% HNO_3.[38] Similarly, film cracking is extensive in free corrosion of $Au_{0.25}Ag_{0.75}$ alloy films up to 390 nm thickness when adhered to either silicon or glass.[24] In the constrained films, the cracking problem becomes less prevalent with decreasing film thickness. Blanket $Au_{0.30}Ag_{0.70}$ films within a thickness range of 45–75 nm lead to uniform crack-free NPG films upon selective etching in 70 wt% HNO_3 solution.[39] The nanoporous structure is homogeneous, but at this thickness range the porosity rapidly converges to a two-dimensional structure.

An effective method to reduce cracking in free corrosion is to use a low-concentration nitric acid solution which establishes a less severe environment. Sun and Balk[37] have shown that two-step dealloying in 35 wt% HNO_3 (70 h) followed by a second dealloying step in 70 wt% HNO_3 (10 h) leads to minimal volume shrinkage and no cracking in bulk NPG samples.

In summary, free corrosion is a fast, simple, and effective dealloying method for unconstrained alloy samples. It is also preferable in cases where the alloy cannot be connected to circuitry to allow electrochemical dealloying. This technique allows a simple design for microstructures with NPG coatings[31,40] and proves to be effective in dealloying of very thin Au–Ag leaves suspended in the electrolyte.[35,41] However, free corrosion allows limited process control and leads to cracking in constrained thin films, which makes it difficult to include as a process in fabrication of microscale structures. The electrochemical methods presented in the next section constitute a more controllable and hence viable alternative to microscale fabrication of NPG structures.

5.4 Dealloying by Using Electrochemical Cells

Electrochemical dealloying is a robust alternative to free corrosion in NPG fabrication. The selective removal of the less noble components is facilitated by introducing electric potential to the system. A conventional three-electrode electrochemical cell consists of an anode, cathode, reference electrode and electrolyte (*cf.* Figure 5.4). The electric potential is regulated by a potentiostat connected to the cell. In dealloying, the alloy serves as the anode of the electrochemical cell, and a positive potential is applied to induce selective dissolution of the alloying element, leaving behind a porous gold structure.

The most commonly used electrolyte in dealloying of AuAg alloy is aqueous perchloric acid ($HClO_4$) solution. Concentrations between 0.7 M and 1 M[22,27,42] are reported to give satisfactory results. Neutral silver nitrate solution is an alternative that leads to a small pore size.[43]

Electrochemical dealloying requires a more sophisticated set-up than free corrosion, and it provides superior process control. The potentiostat allows a precise and rapid control of the cell potential. The applied potential determines the rate of the two main mechanisms that shape the nanoporous surface: Ag dissolution and Au surface diffusion.

The Ag dissolution rate and the potential have an intricate relationship. The porous structure forms at a positive potential above the open circuit potential. First, low coordinated Ag atoms on the ledges dissolve, followed by the surface atoms on the terrace sites. Two important considerations are[9]: (1) the required potential to remove an Ag atom from an AuAg lattice is substantially higher than removing an atom from pure Ag; (2) following the selective dissolution of Ag, the uncoordinated Au atoms diffuse on the surface. This surface activity is not observed in pure Au but is specific to selectively corroded Au alloys. This is associated with the formation of a 'rough' surface as very mobile vacancies cluster on the surface. The driving force for Au atom mobility and vacancy

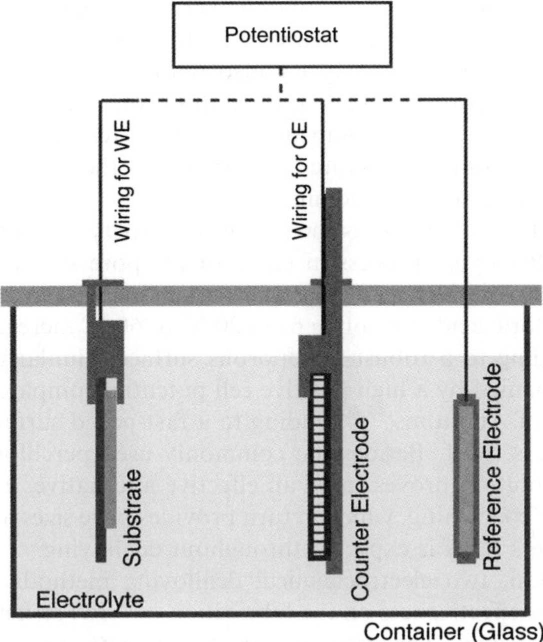

Figure 5.4   Three-electrode electrochemical set up.

cluster mobility is the overall reduction in surface free energy. At a certain positive potential value, Ag dissolution rate increases rapidly, and selective dissolution takes place many atomic layers below the surface leading to a three-dimensional nanoporous structure. This voltage value is called the *critical potential, E_c*.

An elegant description of the critical potential is that 'it marks the transition from a passivated alloy surface to sustained formation of bicontinuous porous structure'.[15] The critical potential values for some of the most commonly used alloys $Au_{0.20}Ag_{0.80}$, $Au_{0.25}Ag_{0.75}$, and $Au_{0.30}Ag_{0.70}$ are 0.80 V, 0.94 V, and 1.01 V (*vs.* normal hydrogen electrode), respectively. Slightly below the critical potential, selective dissolution occurs in the first few nanometers of the alloy surface; the anodic current decays rapidly, and the surface is passivated by the diffusing Au atoms and remains microscopically flat.[14] Slightly above E_c, the initial current decays slowly to a constant value until the accessible Ag is depleted throughout the alloy. The steady-state current and the time to reach depletion depend on the value of the potential. The initial Ag dissolution rate increases abruptly as the cell potential is set to much higher values than the critical potential. The dissolution rate may be as high as 400 mA cm^{-2} for an initial potential of 1.2 V (*vs.* Ag/AgCl electrode).[22]

A high potential leads to rapid dissolution as well as fast Au surface diffusion, leading to a NPG film with potentially low residual Ag content, but it may induce severe conditions with regard to the stress state in the film. Once the surface is sufficiently reactive, a monolayer of oxygen is adsorbed on the surface. The porous structure involves a high number of low-coordinated Au atoms that are susceptible to oxygen adsorption.[29] Oxygen stabilizes the low-coordinated Au atoms, quenches surface diffusion, and thus inhibits further roughening.[44] This is a critical impediment to the stress relief during dealloying and leads to transgranular fracture,[42] especially for the case of constrained thin alloy films, where cracking is common.

Any parameter that enhances the surface diffusivity of Au has an immediate effect on the dealloying process in terms of the pore size as well as cracking tendency. Senior and Newman[42] showed that an increase in temperature of the aqueous perchloric acid electrolyte from 20 °C to 60 °C increases the kinetics of dealloying leading to a robust nanoporous surface. Similarly, the acidic electrolyte accompanied by a high positive cell potential compared to E_c increases the diffusivity of Au atoms,[45,46] leading to a fast-paced surface reconstruction and a film stress relief. Besides the commonly used perchloric acid solution, silver nitrate solution proves to be an effective alternative. The neutral environment lessens coarsening, which in turn provides pore sizes as small as 3 nm,[43] but limited stress relief is expected throughout dealloying.

In this section, two electrochemical dealloying methods are presented in detail, and their effectiveness on the fabrication of NPG structures for different purposes is described. The processes are the *potentiostatic* and *galvanostatic* methods, which differ mainly in terms of which variable is controlled during potential application throughout the formation of the NPG structure. The general set-up requirements for the two methods are the same. Potentiostatic dealloying is based on controlling the applied potential to follow a pre-determined temporal history. On the other hand, galvanostatic dealloying is based upon regulating the applied potential to control the anodic current, which, depending on the electrolyte, is a direct measure of the rate at which Ag is dissolved from the precursor alloy.

5.4.1   Potentiostatic Dealloying

The philosophy in potentiostatic dealloying is to apply a predetermined potential history to the electrochemical cell until the anodic current falls below a threshold value, indicating the depletion of accessible Ag in the alloy. In potentiostatic dealloying, the electric potential (*i.e.* voltage) is controlled throughout the dealloying process, and the electric current adopts a value determined by the potential and the electrochemical circuit.[22,42,43,47] The complexity in assessing the effectiveness of the process arises from the dual effect of the potential on the Ag dissolution and the Au diffusion.

The most basic potentiostatic dealloying involves holding the cell potential at a constant value above the critical dealloying potential. This method is referred to as *stepped potential* application.[22,47] Applying a potential value around the critical potential results in a quasi-steady current history that corresponds to a slowly decreasing Ag removal rate throughout dealloying.[42] The resulting NPG film is generally free of cracks and surface damage, but the dealloying time is as long as several hours.

At higher potential values, the total process time may reduce down to several minutes. Also, the residual Ag amount is minimized. However, for high potential values, at which the monolayer of surface oxide forms, surface diffusion of Au is inhibited, leading to a higher risk of film cracking.[42] Furthermore, the Ag dissolution rate is very high initially, leading to high film stress in the earlier stages on dealloying.

The occurrence of cracks in constrained blanket NPG thin films of 250 nm thickness is significantly reduced with a potentiostatic method by application of the potential as a *ramp function* rather than a step function[17,22,47] using a three-electrode electrochemical cell. In this case, the Ag dissolution rate is non-uniform throughout dealloying (*cf.* Figure 5.7(a)); however, the maximum Ag dissolution rate is significantly less. In an example fabrication case, 250 nm thick $Au_{0.32}Ag_{0.68}$ blanket films are dealloyed by applying potential as either a ramp or step function. When the potential is increased by 1 V min^{-1} to a maximum value of 1.2 V, the peak Ag dissolution rate is 13 mA cm^{-2}, much less than 400 mA cm^{-2} that is observed in the early stage of step potential application.[22] The gradual potential increase avoids cracks and surface damage observed when a high potential is applied as a step function.

Potentiostatic dealloying is an effective method to fabricate unconstrained NPG leaves. Application of constant potential provides crack-free NPG leaves. The potential value determines the dealloying time, pore size, and residual silver content. Some examples from the literature are summarized in Table 5.1. For constrained alloy films, limiting the Ag dissolution rate by gradual potential increase is an effective method to avoid cracks.

5.4.2 Galvanostatic Dealloying

Galvanostatic dealloying is based on regulation of the potential in an electrochemical cell to control the anodic current to follow a prescribed temporal history. The areal current density is then determined by normalizing the anodic current by the surface area of the precursor AuAg alloy. Then, an electrolyte (*e.g.* aqueous perchloric acid) and a range of potentials are chosen so that the dissolution of Ag from the alloy is the only reaction that occurs within the electrochemical cell. Under those circumstances, the areal current density is a direct measure of the temporal rate, at which Ag is removed from solution on a

Table 5.1 Some process parameters for potentiostatic dealloying of AuAg alloy in 1 M $HClO_4$ solution.

Alloy	Potential	Time	Residual Ag	Pore size
Leaf, $Au_{0.25}Ag_{0.75}$, 120 nm[20]	600 mV (*vs.* Ag/AgCl)	8–10 h	4–10 at.%	20 nm
Leaf, $Au_{0.25}Ag_{0.75}$, 120 nm[20]	850 mV (*vs.* Ag/AgCl)	8–10 h	4–10 at.%	4 nm
Leaf, $Au_{0.25}Ag_{0.75}$, 120 nm[36]	600 mV (*vs.* Ag/AgCl)	90 min	4 at.%	5–30 nm
Bulk, $Au_{0.24}Ag_{0.76}$[59]	1000 mV (*vs.* SCE)	12 h	Low	3 nm

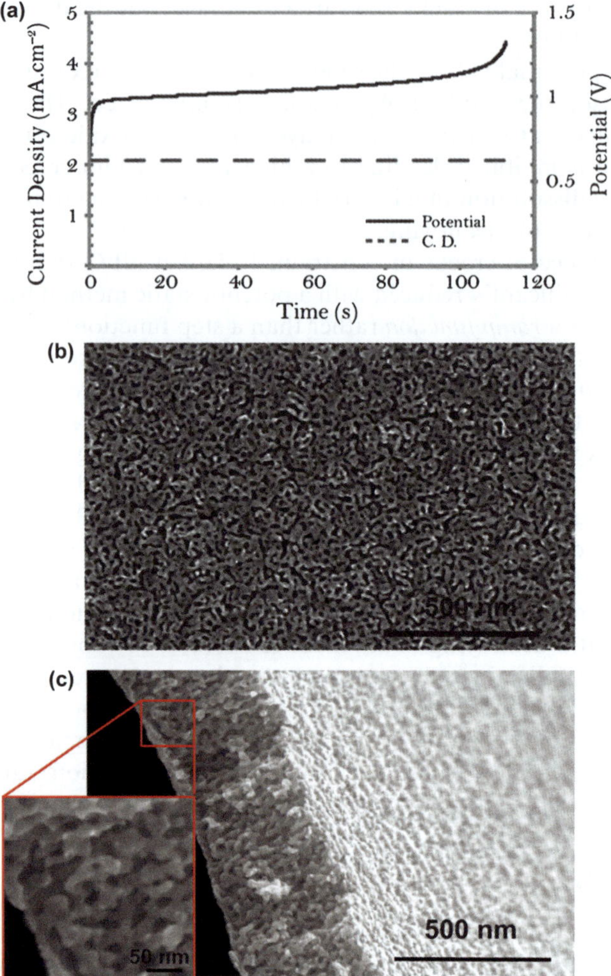

Figure 5.5 Galvanostatic dealloying of a 250 nm thick $Au_{0.30}Ag_{0.70}$ film: (a) areal current density and potential history; (b) top; and (c) cross-sectional views of the NPG (reprinted from Okman and Kysar,[17] Copyright 2011, with permission from Elsevier).

per-unit area basis. Thus, it is possible to control directly the rate of Ag dissolution. Figure 5.5 shows a typical potential history associated with galvanostatic dealloying of a 250 nm thick $Au_{0.30}Ag_{0.70}$ film at the prescribed areal current density of $2\,mA\,cm^{-2}$. To initiate the current density at the desired level, the cell potential begins at a value above the critical potential. The potential then continues to increase slowly to maintain the silver dissolution rate. The continuous increase in potential is due to higher energy requirements to part an Ag atom from the underlying atomic layers. As dealloying proceeds to the final stage, the cell potential undergoes a steep increase, indicating depletion of accessible Ag. The process is terminated when the potential reaches a maximum prescribed value, defined as the *cutoff potential*.

This method is useful for fabrication of crack-free uniform NPG thin films constrained to stiff substrates, such as silicon, where the precursor alloy encounters noticeable film stress variation in dealloying due to the restrictive boundary conditions. Fabricating crack-free NPG thin films on silicon is essential to make microdevices with functional NPG layers.

In terms of the evolution of the film stress, the dealloying procedure involves two competing mechanisms: the rate of stress buildup due to removal of Ag from a precursor alloy constrained to a substrate, and the rate of stress relief due to surface diffusion of the Au that is not removed. Experimental studies on unconstrained leaves and bulk material reveal that Ag removal from an unconstrained precursor alloy is accompanied by a significant shrinkage of up to 30% of the total volume.[20] Shrinkage can be prevented by tailoring the dealloying parameters to reduce the Ag removal rate.[37] Precursor alloy films constrained to a substrate cannot reduce their in-plane dimensions significantly, so an internal tensile stress is induced in the film.[42] Severe cracking is thus commonly observed on blanket thin films; however, it has been observed that cracking can be reduced by limiting the rate of Ag dissolution. The evolution of film stress in dealloying can thus be manipulated by regulating the Ag dissolution rate. On the other hand, as the tensile stress in a constrained film increases during the surface reconstruction, it becomes energetically favorable for the remaining Au to diffuse along the surface to effect a reduction in film stress via a Coble creep mechanism,[48] especially for precursor alloys with nanometer (20–100 nm) scale grains obtained via standard film deposition techniques. The diffusion of Au from the reconstructed surfaces into the grain boundaries in the ligaments leads to a diminution of tensile stress. Further, the surface diffusivity of a metal can increase significantly[45,46] in an electrolytic solution upon application of an electric potential below which an oxide surface layer is deposited. Hence, the stress-relief mechanism becomes more active at the applied potentials employed during dealloying.

In constrained NPG thin films, there is a competition between the rate of stress increase caused by Ag dissolution and the rate of stress decrease caused by Au diffusion, which controls whether film cracking will occur. Galvanostatic dealloying is an ideal method to obtain crack-free films because the rate of Ag dissolution can be controlled to be far below that attained during potentiostatic dealloying, while at the same time the electric potential is sufficiently high to significantly increase the surface diffusion.

The average film stress in a composite NPG-AuAg alloy film measured with the multi-beam stress sensor (MOSS) process[49,50] during dealloying is shown in Figure 5.6. The values indicate the change of film stress relative to the initial stress state in the alloy. They represent an average film stress value for a composite film of AuAg and NPG. The precursor material is 240 nm thick $Au_{0.32}Ag_{0.68}$ film sputter-deposited on a silicon substrate with underlying adhesion layers of Cu (7 nm) and Au (30 nm). The results from galvanostatic dealloying are shown in Figure 5.6(a) where the areal current density and potential cutoff values are prescribed to be 0.2 mA cm^{-2} and 1.2 V, respectively. The maximum rate of stress increase occurs at the initial stage of dealloying. During the course of dealloying, the potential increases from 0.95 V to 1.2 V to maintain the constant Ag removal rate. The peak increase in film stress is 120 MPa, which occurs relatively early in the dealloying process. The stress increase is likely to be governed by the Ag removal accompanied by a modest stress relief because the alloy has not yet developed its extensive network of surface. Subsequently, the stress decreases uniformly with time, which suggests that the diffusive stress-relief mechanism is predominant. The film from this dealloying process is crack-free.

Figure 5.6(b) shows the results of potentiostatic dealloying of a nominally identical film in which the potential increases linearly with time from open circuit potential ($\sim$0.2 V) to 1.25 V. The initial areal current density is negligibly small until the applied potential is about 1 V. At this point, the areal current density quickly achieves a value in excess of 12 mA cm^{-2}, which is a factor of 60 greater than that prescribed during the galvanostatic dealloying. The film stress then achieves a value of up to 230 MPa greater than the initial film stress. In addition, it is evident that the increase in film stress lags the increase in areal current density by about 8 seconds in this case. As before, this stress increase is likely due to the dissolution of silver with little concurrent surface diffusion. As the potential attains its maximum value, the film stress drops dramatically, indicating a stress relief due to Au surface diffusion, which again suggests that the diffusive stress relief mechanism, dominates. This may occur when the development of the porous structure increases total surface area and also decreases diffusion distances to the nearest grain boundary in a ligament. The film from this dealloying process has minor cracks. Thus, galvanostatic dealloying provides a methodology to maintain the peak stress induced in the film below the fracture threshold, whereas the potentiostatic method affords much less control over the stress.

A set of experiments shown in Figure 5.7 compares the galvanostatic and potentiostatic dealloying methods for significantly thicker films, in this case with precursor alloy films of thickness 1300 nm and composition 32 at.% Au. Potentiostatic dealloying is used to dealloy a sample with ramped potential rate of 1 V min^{-1} with a maximum potential of 1.2 V, which is then held constant for about 600 s until the Ag dissolution rate decreases below 1 mA cm^{-2} (*cf.* Figure 5.7(b)). The peak value of Ag dissolution rate is 50 mA cm^{-2}. Cracks appear to initiate from the free surface of the NPG films and propagate toward the substrate. Despite severe cracking, the film does not delaminate. Occasional nanoscale cracks are observed within the islands separated by large-sized cracks. On the other hand, galvanostatic dealloying at a constant current density of 3 mA cm^{-2} is used on a second nominally identical sample with a cutoff potential of 1.2 V. As shown in Figure 5.7(a), no through-thickness cracks are observed over an area of many hundreds of square micrometers, although the dealloying is completed in a shorter time.

The post-dealloying compositions of the NPG films are determined by energy-dispersive X-ray spectroscopy (EDS) by Princeton-Gamma Technologies Inc. using a Hitachi 4700 SEM. The underlying Si substrate and Cr

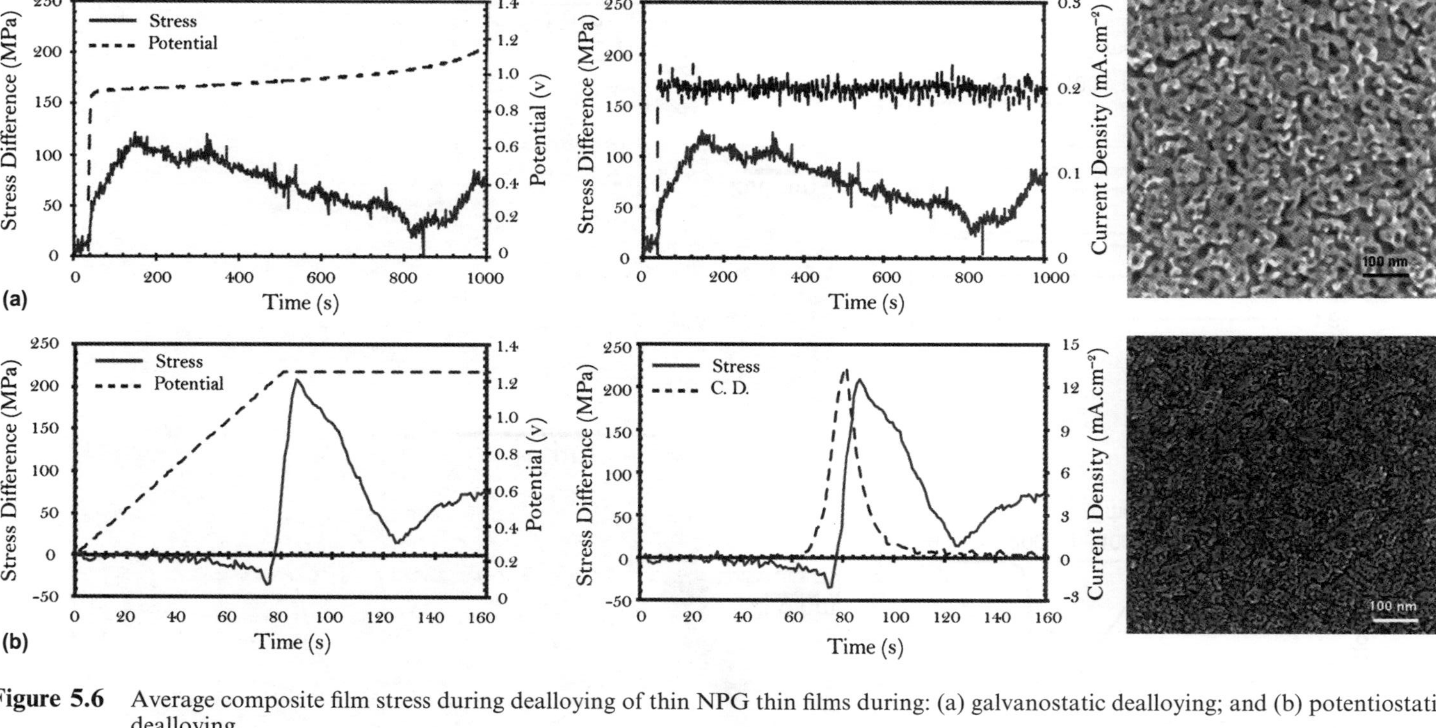

Figure 5.6 Average composite film stress during dealloying of thin NPG thin films during: (a) galvanostatic dealloying; and (b) potentiostatic dealloying.

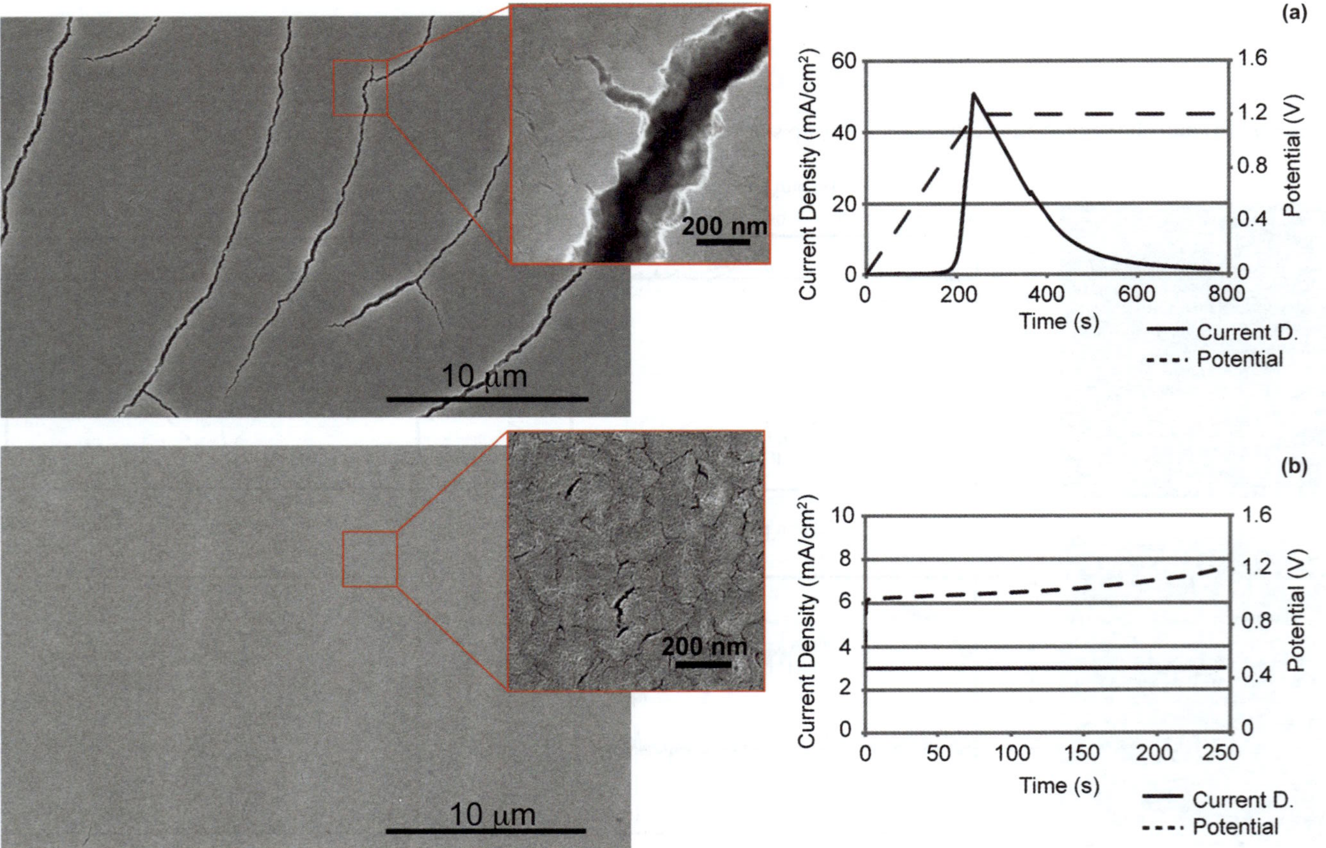

Figure 5.7 Potential and current history and resulting NPG structure obtained by dealloying of 1300 nm thick $Au_{0.32}Ag_{0.68}$ film on Si wafer: (a) potentiostatic method with ramped potential; and (b) galvanostatic method with constant current density of 3 mA cm^{-2} (reprinted from[17], Copyright 2011, with permission from Elsevier).

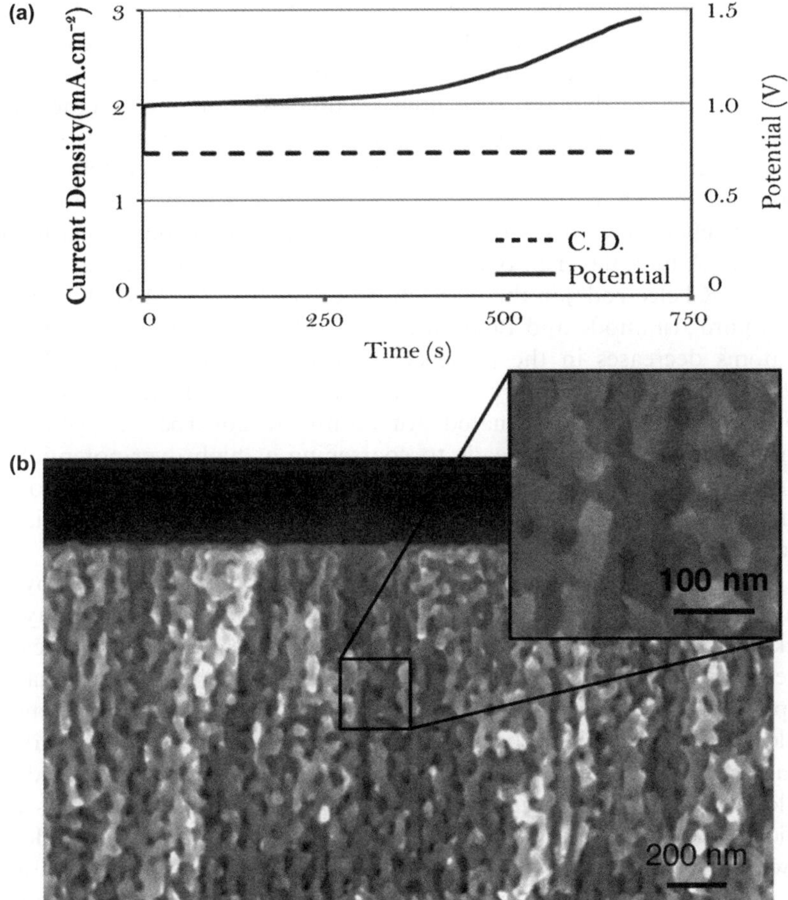

Figure 5.8 Galvanostatic dealloying of a 1300 nm thick $Au_{0.32}Ag_{0.68}$ precursor film on Si: (a) potential history for current density of 1.5 mA cm^{-2}; (b) cross-sectional view of crack-free blanket NPG film.

adhesion layers are not detected in the analysis; thus the results are indicative of the composition of the NPG film. For the 1300 nm thick films prepared by the potentiostatic method in Figure 5.7(a), the residual Ag in the NPG films is 20 at.%, whereas the NPG film prepared by galvanostatic dealloying in Figure 5.7(b) has a residual content of 14 at.% Ag. To achieve a lower residual Ag content for the 1300 nm NPG films, a nominally identical sample is dealloyed using galvanostatic dealloying at a lower current density of 1.5 mA cm^{-2} and a higher cutoff potential of 1.45 V. The potential and current density histories are shown in Figure 5.8(a). Figure 5.8(b) shows the cross-section of the resulting NPG film that demonstrates a uniform three-dimensional porous structure. The EDS measurements show the residual Ag content to be 1.5 at.%. It is also evident in comparing the insets of Figures 5.7(b) and 5.8(b) that a lower current density and higher cutoff potential lead to significantly larger ligament and pore sizes. In addition, the film shown in Figure 5.8(b) is free of superficial surface damage. The high residual Ag content together with the small pore sizes observed in thick films in Figure 5.7(b) suggests that mass transport through the pores of the thick film may play the limiting role in removal of the Ag.

Surface chemistry affects the evolution of the nanoporous structure at high potentials. A monolayer of oxide is adsorbed on the Au surface around 1.26 V (*vs.* Ag/AgCl electrode) in the three-electrode cell with 0.7 M $HClO_4$ electrolyte with a pure Au anode and Pt counter electrode. The surface diffusivity of the Au atoms decreases in the presence of oxide monolayer and stabilizes the existing porous Au structure. Newman *et al.*[12] and Biener *et al.*[44] confirm stabilization of low coordinated Au atoms by adsorbed oxygen. Electro-chemical annealing, which leads to coarsening at high over-potential values, is quenched if the potential exceeds the value at which the electrosorption occurs.[44] This is an impediment to the use of high dealloying potentials, since the enhanced Ag dissolution and insufficient stress relaxation by Au diffusion lead to cracking. Therefore, stepped or ramped potentiostatic dealloying per-formed at a potential near the onset of surface oxide formation may lead to cracking. In contrast, in galvanostatic dealloying the current density can be set to keep the potential below the surface oxidation threshold for the majority of the process time. It increases only due to a paucity of Ag near the end of the dealloying process. The galvanostatic method provides a more suitable deal-loying regime that results in a lower residual Ag content and yet reduces the dealloying time at potentials above the oxidation threshold, thus allowing fabrication of crack-free films with relatively large thickness. In addition, it allows a uniform Ag dissolution rate and a relatively smaller variation of the overall films stress at high over-potential values. Finally, galvanostatic deal-loying is amenable to the incorporation of the NPG films onto silicon sub-strates using scalable standard clean-room processes, which is the first step towards MEMS devices with functional NPG coatings.

5.5 Fabrication of Micropatterned NPG Features

Microfabrication technologies facilitate production of self-supporting NPG microscale structures as well as incorporation of NPG thin films into microdevices. The fabrication should provide NPG structures that are free of cracks, robustly attached to the underlying substrate, and precisely dimensioned.

Microsystems structures comprising NPG may be built to perform certain functions as well as to serve as specimens to test its properties. Examples of the latter include NPG microbeams, uniform thin films and micropillars that have been used in nanoindentation tests to characterize the mechanical response of NPG.[40,51,52] The mechanical response of NPG can be a function of surface charging and chemi-adsorption.[25,27,53] When deposited onto generic silicon wafers or onto silicon cantilevers, the NPG coatings can serve as the functional layer of cantilever sensors or actuators.[54] Another potential application of NPG is the ability to detect neural activity due to the low impedance of the nanoporous structure.[30] Testing and implementation of the NPG as functional devices require accurately shaped network NPG elements.

It is important to establish the compatibility of the standard microfabrication techniques with the specific NPG fabrication methods and the material itself. The main steps of microfabrication consist briefly of deposition of a precursor alloy film, shaping the structure and dealloying. In this section, techniques of making microstructures with NPG are discussed.

5.5.1 Incorporation of Thin Film

In fabrication of silicon devices, there are two simple choices when incorporating NPG film onto a silicon substrate: either the final precursor alloy or the NPG film is adhered directly on to the substrate, or the alloy film is grown on silicon substrate using film-deposition techniques and dealloyed afterwards.

The manual adhesion of thin precursor alloy leaves or of NPG has a limited number of examples in the literature. In fabricating suspended NPG beams, Lee *et al.*[40,55] used 100 nm thick $Au_{0.37}Ag_{0.63}$ leafs and applied them manually onto an epoxy-coated silicon wafer.

The second main approach in film incorporation is to deposit the precursor alloy film onto the silicon prior to dealloying. The thin alloy film and the metallic adhesion layers are grown on the silicon using standard film deposition techniques, such as magnetron sputtering or thermal evaporation. The constrained film or structure is then dealloyed. Homogeneity within the film and adequate adhesion are critically important to obtain intact nanoporous structures. Typically, prior to the deposition of the precursor film, an adhesion layer is deposited directly onto the silicon. The adhesion layer can be either a Ta[24] or Cr layer with a thickness of 5–10 nm.[17,23] This is followed by an Au layer that serves as a barrier to isolate the underlying adhesion layer from the electrolyte to prevent delamination during dealloying.[22,23]

One method to prepare the precursor alloy is to deposit alternating Au and Ag layers using thermal evaporation in a high vacuum chamber.[22,47] The thickness of the metal layers is 7–20 nm, depending on the desired final alloy composition (*cf.* Figure 5.9(a)). A Cr adhesion layer is deposited thermally prior to the alternating layers; however, the protective Au layer is omitted to maintain the desired alloy composition. The multilayer film is then annealed in an Ar atmosphere at temperatures varying between 310 and 400 °C. The alloying elements are completely mixed after 3 h. The upper limit of the annealing temperature is based on the phase diagram of Si–Cr, Cr–Au, and Cr–Ag systems, such that Cr is not expected to diffuse a significant distance into either the underlying Si or AuAg alloy.[47] During dealloying, the absence of the additional Au layer increases the film-delamination risk, since Cr may be

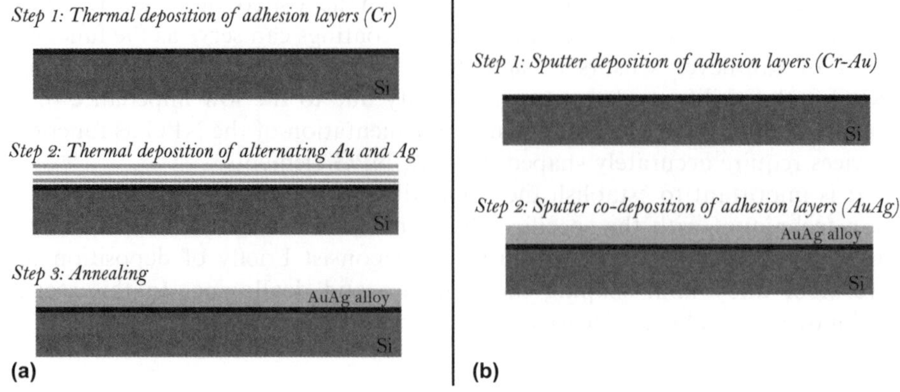

Figure 5.9 Preparation of AuAg alloy film by: (a) thermal deposition; (b) sputter deposition.

exposed to the electrolyte at the latter stages of dealloying. Applying a thin layer of masking aid around the periphery of each specimen reduces this problem.

Magnetron sputtering is a better alternative for depositing alloy films. In sputtering, accelerated argon ions erode material from Au and Ag targets, and the ejecta are deposited on the desired substrate in the vacuum chamber. The alloy components are sputter-deposited simultaneously to form a homogeneous alloy, and so no additional annealing step is needed. Sputtered films have small grain sizes relative to the film thickness. The grain size of the precursor alloy is about 40 nm for the 250 nm thick films and about 150 nm for the 1300 nm thick films, respectively. An isolating Au layer is deposited between the alloy and the metallic adhesion layer, without altering the alloy composition. Due to the non-directional nature of this deposition method, both the surface and the sides of the sample are coated with Au, adequately isolating the underlying Cr layer from the electrolyte. Sputter-deposited films, therefore, experience minimal delamination problems compared to thermally deposited films.

5.5.2 Fabrication of Microscale Structures

The most basic NPG microcomponents are precisely defined regions of NPG with known physical dimensions and are adhered to a substrate, such as silicon. In the simplest design the NPG patterns are fabricated on the conductive adhesion layer. A sample process flow in Table 5.2 describes the fabrications steps. The silicon substrate is coated with Cr and Au layers, and the sample is removed from the vacuum chamber. Then, the specimens are spin-coated with LOR series resist followed by a photoresist; the resist bilayer improves the integrity of the patterns during the lift-off. The photoresist is

Table 5.2 Process flow for fabrication of NPG islands on silicon surfaces.

Step	No.	Description
Cleaning silicon wafer	1	Clean in sonicator in acetone for 5 min, isopropanol rinse, bake on hot plate at 150 °C between 5 and 10 min
Adhesion layers	2	Sputter deposit Cr at 6 mTorr at 0.2 nm s^{-1}
	3	Sputter deposit Au at 6 mTorr at 0.2 nm s^{-1}
Photolithography	4	Spin coat of positive photoresist/e-beam resist
	5	Expose resist through the photomask/by electron beam
	6	Develop
AuAg alloy	7	Co-deposit using Au and Ag targets at 6 mTorr
	8	Lift off photoresist
Cleaning	9	Clean sample with a suitable solvent
Dealloying	10	Dealloy metal alloy in an electrochemical cell

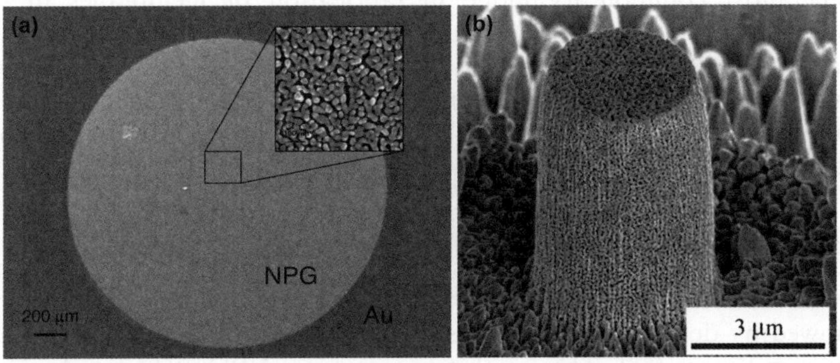

Figure 5.10 NPG structures: (a) a circular area of NPG on Au surface; and (b) a NPG micropillar (adapted with permission from Biener *et al.*[52] Copyright 2006 American Chemical Society).

then exposed in a mask aligner. After developing the resist, care should be taken to remove fully the photoresist residue to ensure perfect adhesion of the alloy. The samples are placed in the sputter-deposition system and coated with the alloy of the desired composition using simultaneous sputter deposition of the Au and Ag. The resist is then removed leaving behind precisely patterned AuAg precursor regions on an Au-coated surface. The alloy is then dealloyed, preferably by electrochemical dealloying. The photolithography methods used to pattern the surface provide dimensional accuracy of the order of several micrometers. The known surface area of the precursor allows the areal current density to be calculated in galvanostatic dealloying (*cf.* Figure 5.10(a)).

Micrometer-sized NPG pillars are useful structures for micro-compression tests.[51,52] The NPG (*cf.* Figure 5.10(b)) is fabricated from a sheet of $Au_{0.25}Ag_{0.75}$ alloy that is electrochemically dealloyed. The microcolumns are then milled into the NPG part with a focused ion beam. Fabricating regular columns with vertical side walls is challenging due to the finite diameter and tails of the ion beam, the angular dependence of the sputter yield, ion reflection, and redeposition. The optimum results are obtained by a two-step milling process,[56] in which a large annular volume is extracted around the microcolumn, and then the diameter of the column is incrementally reduced to the desired size. The final columns retain the porous structure of the parent porous film, except at the side walls, where there are vertical striations due to milling.

Micro- and macro-size suspended membranes are commonly used in applications, where large deflections are required either for testing or to meet service requirements. Fabrication of NPG membranes from a constrained AuAg alloy is challenging, due to the brittleness of the material. However, standard clean-room fabrication techniques are adequate in fabrication of composite membranes involving a NPG layer.

The most common technique in fabrication of membranes in silicon devices is wet etching of silicon through a masking layer.[57] Silicon wafers coated with silicon nitride (Si_3N_4) on both sides are useful candidates for this purpose. A sample fabrication scheme is given in Table 5.3. The Si_3N_4 coating is not damaged during the etching of silicon; the coating on the bottom layer is used as an etch mask, while the layer on the top surface acts as a suspended platform for the deposition of the NPG thin film. The minimum size of the window fabricated by the method depends on the wafer thickness due to the anisotropic etch rate of silicon in KOH solution. Metallic adhesion and the alloy layers are deposited on the membrane, followed by dealloying. A composite membrane with $6\,mm^2$ surface area is shown in Figure 5.11. The membrane comprises NPG (250 nm), Au (30 nm), Cr (7 nm), and Si_3N_4 (150 nm). The NPG film is deposited on a limited area around the membrane.

Table 5.3 Process flow for fabrication of a composite NPG membrane.

Step	No.	Description
Cleaning	1	Clean in sonicator in acetone for 5 min, isopropanol rinse, bake on hot plate at 150 °C between 5 and 10 min
Photolithography	2	Spin coat positive resist
	3	Expose resist through the photomask (Back side)
	4	Develop to remove the photoresist in the window area
Si_3N_4 removal	5	Remove Si_3N_4 by reactive ion etching (RIE)
	6	Lift off photoresist
Wet etch	7	Etch silicon in KOH
	8	Clean wafer with nanostrip (sulfuric acid and hydrogen peroxide) solution (highly corrosive environment)
Film deposition	9	Sputter deposit adhesion layers Cr and Au (Front side)
	10	Sputter deposit AuAg alloy
Dealloying	11	Dealloy metal alloy in an electrochemical cell

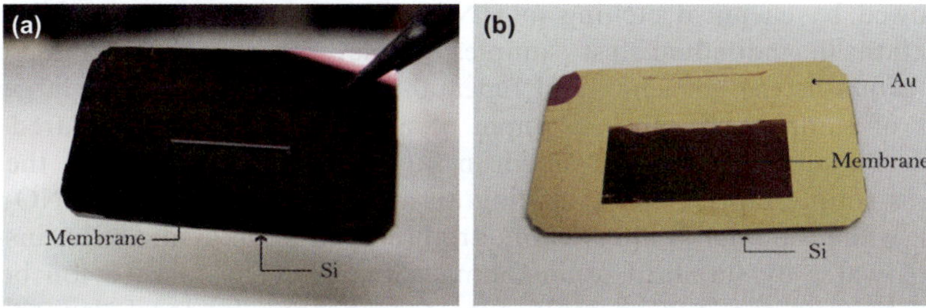

Figure 5.11 Composite membrane with Si$_3$N$_4$, Cr, Au, NPG (250 nm) layers: (a) back side; and (b) front side of the membrane.

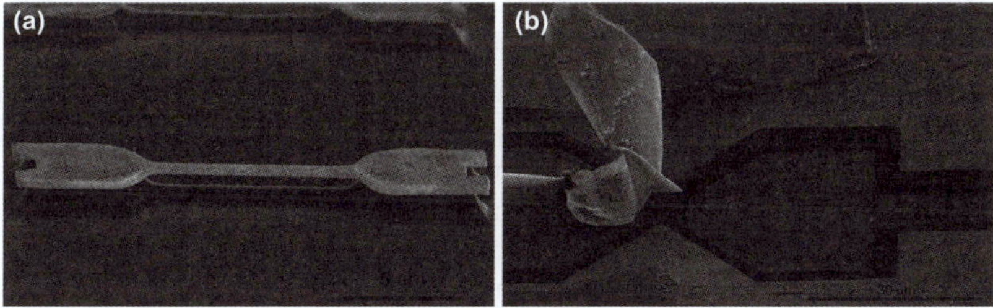

Figure 5.12 Manual incorporation of NPG onto silicon structures: (a) arrays of suspended NPG films precursor alloy leaves adhered to silicon by epoxy (courtesy of Dr Dongyun Lee, Pusan National University); and (b) an intermediate step in 'welding' of NPG leaf on silicon anchors (courtesy of Mehmet Yilmaz, Columbia University).

The NPG membranes can be fabricated by transferring a suspended NPG leaf onto a substrate. Zeis *et al.*[58] fabricated Pt-coated NPG/Nafion membrane electrode assemblies, to be used as a component of proton-exchange membrane fuel cells, by transferring a floating NPG leaf on a mica disk and attaching the film on Nafion at high temperature and pressure. In a similar process, Seker *et al.*[41] transferred NPG leaf suspended in water onto a photoresist coated Si wafer. The substrate is then coated with PDMS of about 40 µm thickness and cured. The composite PDMS–NPG membrane is then easily detached from silicon by removing the photoresist; the composite membrane can then be attached to a glass sample with circular holes to make suspended composite membranes.

Suspended microbeams are used for mechanical testing of the NPG as well as microsensors and actuators that deflect due to surface charging[54] or chemi-adsorption.[53] For constructing suspended NPG beams as shown in Figure 5.12(a), Lee *et al.*[40] suggested manually attaching a thin AuAg leaf on an epoxy-coated silicon wafer, patterning with e-beam lithography, followed by chemical etching in HNO_3. This method is successful in pre-paration in individual test samples; however, the manual intervention compromises the efficiency and repeatability of the process for larger batches of samples. Another example of affixing an already-prepared NPG onto two silicon anchors is shown in Figure 5.12(b). In this case, the NPG film is simply fabricated by etching an alloy leaf in a 70 wt% HNO_3 solution, while the microscale anchors are prepared upon a silicon-on-insulator (SOI) wafer by standard clean-room techniques. The focused ion beam is used to weld the NPG sample onto the anchors, by local deposition of platinum. Prior to welding, a small piece of NPG film is picked with a microprobe and carried to the desired location on the silicon structure. Figure 5.12(b) shows incorporation in an intermediate step for a case in which the NPG leaf is bent due to random exposure to the focused ion beam. The technique requires significant manual intervention and has a low success ratio.

Microbeams can be manufactured entirely by standard clean-room methods, without the need for manual intervention. Figure 5.13 illustrates the fabrication steps of a double-clamped beam coated with NPG. The SOI wafers are coated with adhesion and alloy layers, and dog-bone-shaped patterns are transferred on to the sample by photolithography. A protective Ni coating is applied on the patterns. The metal and silicon layers around the patterns are then removed by inductively coupled plasma etching. The structure is subsequently exposed to aqueous HF solution to etch the SiO_2 and suspend the composite beam consisting of a layer of NPG on the silicon handle layer of the SOI wafer. The exposure time is set carefully to avoid removing the SiO_2 anchors of the microbeams. The protective Ni layer is chemically removed before dealloying. In Figure 5.13, the surface of the beam is shown after dealloying in nitric acid. As described earlier, film cracking is a common problem in chemical dealloying. For the manufacture of MEMS devices, galvanostatic dealloying yields a superior film quality.

Developing robust NPG fabrication techniques requires a thorough understanding of the complex mechanisms of corrosion, surface reconstruction, and stress evolution. In microdevice fabrication, it is critically important to keep the stress level below the values at which cracks form. Regulating the Ag dissolution rate as well as enhancing Au diffusion during electrochemical dealloying is effective to avoid cracking. Also, precursor films with a minimum amount of impurities and microvoids are essential to obtain crack-free NPG films. Standard methods for photolithography as well as for wet and dry etching can be easily incor-porated into the process flow to fabricate NPG. During processing, an important process limitation is to avoid exposing NPG to thermal or electrochemical annealing that would alter the porous structure and lead to coarsening of the nanoporous structure. Current standard clean-room techniques are applicable to a wide range of architectures involving NPG components.

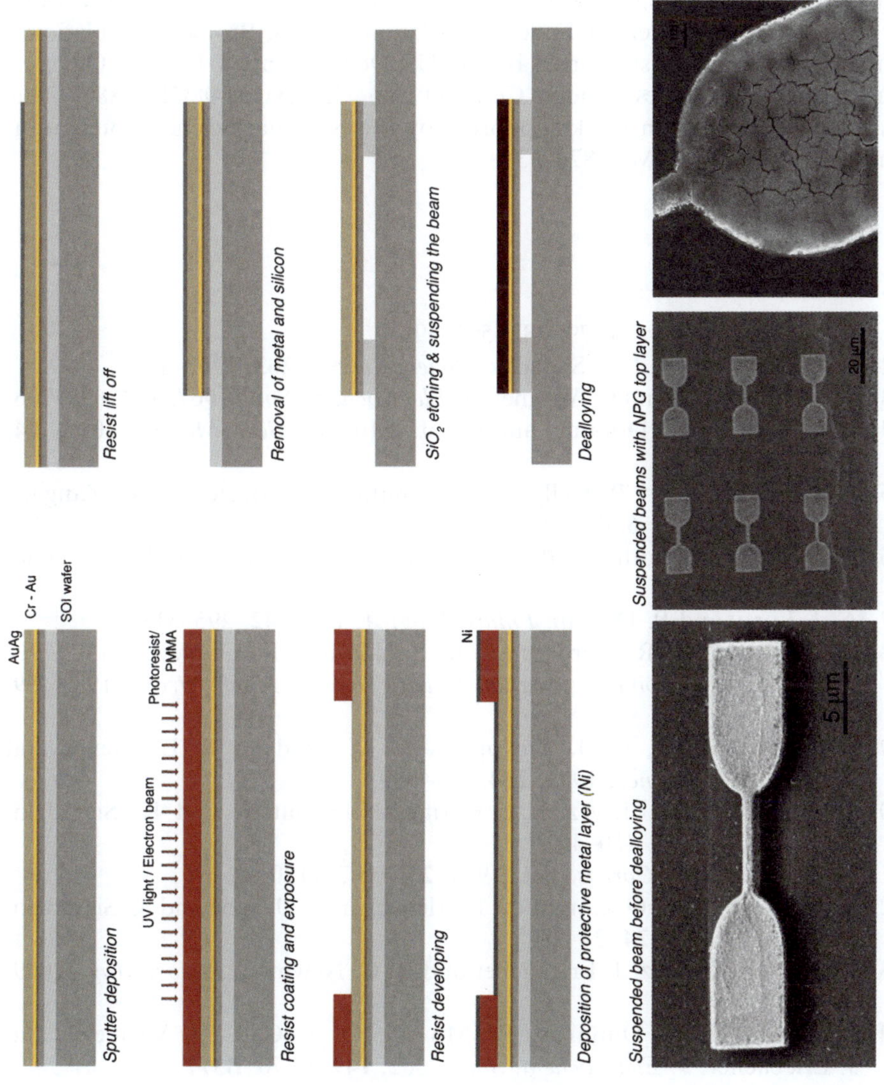

Figure 5.13 Fabrication steps for suspended beams coated with NPG. SEM images show the structure of the beams.

Acknowledgments

The authors would like to thank Prof. Eric Chason and Dr Nitin Jadhav, for providing the experimental facilities for film stress measurement using the MOSS and sharing their invaluable insight in the this field; Prof. Alan West, for enlightening discussions in electrochemistry; and Ms Emine Eda Kuran, for her assistance in microfabrication. This research was carried out (in whole or in part) at the Center for Functional Nanomaterials, Brookhaven National Laboratory, which is supported by the US Department of Energy, Office of Basic Energy Sciences, under Contract No. DE-AC02-98CH10886. This material is based upon work supported by the National Science Foundation under Grant No. CMMI-0826093.

References

1. A. J. Forty, *Nature*, 1979, **282**, 597–598.
2. R. C. Newman and K. Sieradzki, *Science*, 1994, **263**, 1708–1709.
3. H. W. Pickering and C. Wagner, *J. Electrochem. Soc.*, 1967, **114**, 698–706.
4. R. Morrish, K. Dorame and A. J. Muscat, *Scr. Mater.*, 2011, **64**, 856–859.
5. E. Rouya, M. Reed, R. Kelly, H. Bart-Smith, M. R. Begley and G. Zangari, *ECS Trans.*, 2007, **6**, 41–50.
6. J. Snyder, P. Asanithi, A. B. Dalton and J. Erlebacher, *Adv. Mater.*, 2008, **20**, 4883–4886.
7. A. J. Forty and P. Durkin, *Philos. Mag. A*, 1980, **42**, 295–318.
8. K. Sieradzki, R. R. Corderman, K. Shukla and R. C. Newman, *Philos. Mag. A Phys. Condens. Matter Struct. Defects Mech. Prop.*, 1989, **59**, 713–746.
9. I. C. Oppenheim, D. J. Trevor, C. E. D. Chidsey, P. L. Trevor and K. Sieradzki, *Science*, 1991, **254**, 687–689.
10. J. Erlebacher, M. J. Aziz, A. Karma, N. Dimitrov and K. Sieradzki, *Nature*, 2001, **410**, 450–453.
11. H. W. Pickering, *Corros. Sci.*, 1983, **23**, 1107–1120.
12. R. C. Newman, S. G. Corcoran, J. Erlebacher, M. J. Aziz and K. Sieradzki, *MRS Bull.*, 1999, **24**, 24–28.
13. D. M. Artymowicz, J. Erlebacher and R. C. Newman, *Philos. Mag.*, 2009, **89**, 1663–1693.
14. K. Sieradzki, N. Dimitrov, D. Movrin, C. McCall, N. Vasiljevic and J. Erlebacher, *J. Electrochem. Soc.*, 2002, **149**, B370–B377.
15. A. Dursun, D. B. Pugh and S. G. Corcoran, *J. Electrochem. Soc.*, 2005, **152**, B65–B72.
16. S. V. Petegem, S. Brandstetter, R. Maass, A. M. Hodge, B. S. El-Dasher, J. Biener, B. Schmitt, C. Borca and H. V. Swygenhoven, *Nano Lett.*, 2009, **9**, 1158–1163.
17. O. Okman and J. W. Kysar, *J. Alloys Compd.*, 2011, **509**, 6374–6381.

18. J. Biener, A. Wittstock, L. A. Zepeda-Ruiz, M. M. Biener, V. Zielasek, D. Kramer, R. N. Viswanath, J. Weissmueller, M. Baeumer and A. V. Hamza, *Nat. Mater.*, 2009, **8**, 47–51.
19. L. Zhang, L. Chen, H. Liu, Y. Hou, A. Hirata, T. Fujita and M. Chen, *The J. Phys. Chem. C*, 2011, **115**, 19583–19587.
20. S. Parida, D. Kramer, C. A. Volkert, H. Roesner, J. Erlebacher and J. Weissmueller, *Phys. Rev. Lett.*, 2006, **97**, 035504.
21. E. Seker, M. L. Reed and M. R. Begley, *Scr. Mater.*, 2009, **60**, 435–438.
22. O. Okman, D. Lee and J. W. Kysar, *Scr. Mater.*, 2010, **63**, 1005–1008.
23. M. C. Dixon, T. A. Daniel, M. Hieda, D. M. Smilgies, M. H. V. Chan and D. L. Allara, *Langmuir*, 2007, **23**, 2414–2422.
24. Y. Sun and T. J. Balk, *Metallurg. Mater. Trans. A Phys. Metallurg. Mater. Sci.*, 2008, **39A**, 2656–2665.
25. A. Wittstock, J. Biener and M. Baeumer, *Phys. Chem. Chem. Phys.*, 2010, **12**, 12919–12930.
26. J. Weissmuller, R. N. Viswanath, D. Kramer, P. Zimmer, R. Wurschum and H. Gleiter, *Science*, 2003, **300**, 312–315.
27. H. Jin, S. Parida, D. Kramer and J. Weissmueller, *Surf. Sci.*, 2008, **602**, 3588–3594.
28. A. Wittstock, J. Biener and M. Baeumer, *ECS Trans.*, 2010, **28**, 1–13.
29. T. A. Baker, X. Liu and C. M. Friend, *Phys. Chem. Chem. Phys.*, 2011, **13**, 34–46.
30. E. Seker, Y. Berdichevsky, M. R. Begley, M. L. Reed, K. J. Staley and M. L. Yarmush, *Nanotechnology*, 2010, **21**, 125504.
31. E. Seker, J. T. Gaskins, H. Bart-Smith, J. Zhu, M. L. Reed, G. Zangari, R. Kelly and M. R. Begley, *Acta Mater.*, 2008, **56**, 324–332.
32. L. B. Freund and S. Suresh, *Thin Film Materials: Stress, Defect Formation, and Surface Evolution*, Cambridge University Press, Cambridge, 2003.
33. A. J. Forty, *Gold Bull.*, 1981, **14**, 25–35.
34. X. Lu, E. Bischoff, R. Spolenak and T. J. Balk, *Scr. Mater.*, 2007, **56**, 557–560.
35. A. Mathur and J. Erlebacher, *Appl. Phys. Lett.*, 2007, **90**, 061910.
36. H. Roesner, S. Parida, D. Kramer, C. A. Volkert and J. Weissmueller, *Adv. Eng. Mater.*, 2007, **9**, 535–541.
37. Y. Sun and T. J. Balk, *Scr. Mater.*, 2008, **58**, 727–730.
38. X. Lu, T. J. Balk, R. Spolenak and E. Arzt, *Thin Solid Films*, 2007, **515**, 7122–7126.
39. Y. Sun, K. P. Kucera, S. A. Burger and T. J. Balk, *Scr. Mater.*, 2008, **58**, 1018–1021.
40. D. Lee, X. Wei, X. Chen, M. Zhao, S. C. Jun, J. Hone, E. G. Herbert, W. C. Oliver and J. W. Kysar, *Scr. Mater.*, 2007, **56**, 437–440.
41. E. Seker, M. Reed, M. Utz and M. R. Begley, *Appl. Phys. Lett.*, 2008, **92**.
42. N. A. Senior and R. C. Newman, *Nanotechnology*, 2006, **17**, 2311–2316.
43. J. Snyder, K. Livi and J. Erlebacher, *J. Electrochem. Soc.*, 2008, **155**, C464–C473.

44. J. Biener, M. M. Biener, T. Nowitzki, A. V. Hamza, C. M. Friend, V. Zielasek and M. Baeumer, *Chemphyschem*, 2006, **7**, 1906–1908.
45. J. M. Doña and J. Gonzalezvelasco, *Surf. Sci.*, 1992, **274**, 205–214.
46. S. A. Policastro, J. C. Carnahan, G. Zangari, H. Bart-Smith, E. Seker, M. R. Begley, M. L. Reed, P. F. Reynolds and R. G. Kelly, *J. Electrochem. Soc.*, 2010, **157**, C328–C337.
47. M. Kim, W. Ha, J. Anh, H. Kim, S. Park and D. Lee, *J. Alloys Compd.*, 2009, **484**, 28–32.
48. D. Josell and F. Spaepen, *Acta Metallurg. Mater.*, 1993, **41**, 3007–3015.
49. J. A. Floro, E. Chason, S. R. Lee, R. D. Twesten, R. Q. Hwang and L. B. Freund, *J. Electron. Mater.*, 1997, **26**, 969–979.
50. E. Chason and B. W. Sheldon, *Surf. Eng.*, 2003, **19**, 387–391.
51. C. A. Volkert, E. T. Lilleodden, D. Kramer and J. Weissmueller, *Appl. Phys. Lett.*, 2006, **89**, 061920.
52. J. Biener, A. M. Hodge, J. R. Hayes, C. A. Volkert, L. A. Zepeda-Ruiz, A. V. Hamza and F. F. Abraham, *Nano Lett.*, 2006, **6**, 2379–2382.
53. M. M. Biener, J. Biener and C. M. Friend, *Langmuir*, 2005, **21**, 1668–1671.
54. D. Kramer, R. N. Viswanath and J. Weissmuller, *Nano Lett.*, 2004, **4**, 793–796.
55. D. Lee, M. Zhao, X. Wei, X. Chen, S. C. Jun, J. Hone, E. G. Herbert, W. C. Oliver and J. W. Kysar, *Appl. Phys. Lett.*, 2006, **89**, 111916.
56. C. A. Volkert and E. T. Lilleodden, *Philos. Mag.*, 2006, **86**, 5567–5579.
57. S. D. Senturia, *Microsystem Design*, Kluwer Academic Publishers, Boston, 2000.
58. R. Zeis, A. Mathur, G. Fritz, J. Lee and J. Erlebacher, *J. Power Sources*, 2007, **165**, 65–72.
59. R. Li and K. Sieradzki, *Phys. Rev. Lett.*, 1992, **68**, 1168–1171.

Optical Properties and Applications of Nanoporous Metals

X. Y. LANG*[a,b] AND M. W. CHEN*[a]

[a] WPI Advanced Institute for Materials Research, Tohoku University, Sendai 980-8577, Japan; [b] Key Laboratory of Automobile Materials (Jilin University), Ministry of Education, School of Materials Science and Engineering, Jilin University, Changchun 130022, PR China
*Email: xylang@jlu.edu.cn or mwchen@wpi-aimr.tohoku.ac.jp

6.1 Introduction

Materials with feature sizes smaller than 100 nm have fascinating physico-chemical properties, bridging the gap between isolated atoms and bulk counterparts, which have been recent subjects of extensive theoretical[1,2] and experimental studies.[3–5] In principle, a large fraction of under-coordinated atoms at surfaces and confinement of electrons by nano-scale sizes are expected to be responsible for the occurrence of novel physical and chemical properties in nanostructures.[1–5] This has spurred the development of a large number of methodologies to synthesize novel nanostructured materials with controllable electronic, optical, and magnetic properties.[6–8] Among all kinds of inorganic nanosolids, metallic nanostructures deserve special attention, due to a wide range of intriguing physical and chemical properties and diverse applications, of key importance to our society, such as in catalysis, electronics, and optical systems.[1–10]

RSC Nanoscience & Nanotechnology No. 22
Nanoporous Gold: From an Ancient Technology to a High-Tech Material
Edited by Arne Wittstock, Jürgen Biener, Jonah Erlebacher and Marcus Bäumer
© Royal Society of Chemistry 2012
Published by the Royal Society of Chemistry, www.rsc.org

The optical properties of noble and transition metals are governed primarily by coherent oscillations of conduction electrons on metal surfaces excited by electromagnetic radiation at a metal–dielectric interface, known as surface plasmons (SPs).[4] This branch of research has attracted much attention, due to its potential applications in miniaturized optical devices,[11,12] sensors,[13–16] and photonic circuits as well as medical imaging, diagnostics, and therapeutics.[17,18] There are two types of surface plasmon resonance (SPR) depending on the metallic structures. One is propagating surface plasmon polaritons (SPPs) occurring on the thin metal films excited using grating or prism couplers. SPPs can propagate tens to hundreds of micrometers along the metal surface with an associated electric field that decays exponentially from the surface (normal to the dielectric–metal interface).[19–21] Without grating or prism couplers, surface polaritons cannot be excited directly with light on a smooth metal interface, and surface polaritons propagating on a smooth surface cannot be transformed into light in the adjacent medium.[19] The other is non-propagating localized SPRs (LSPRs) that are excited on metallic nanoparticles with different shapes, and around nanoholes or voids in thin metal films.[22–24] LSPRs of metallic nanostructures confined to very small regions are always associated with strong electromagnetic fields in the vicinity of the metal surfaces, contributing to numerous phenomena, such as surface-enhanced spectroscopy.[15] The properties of LSPR, such as wavelength and bandwidth, are determined by a set of physical parameters that include the size, shape, and composition of metallic nanostructures as well as the dielectric conditions of the surface media.[21] For example, the increase in nanostructure sizes gives rise to increasing amounts of absorption and scattering, accompanied by changes in the LSPR wavelength and the broadening of plasmon bands due to the damping of LSPR. The shape of nanostructures is another key factor that strongly affects the LSPR, and optimizing geometric morphology has been the ultimate goal of many studies.[11–18,22–24] Typically, this can be achieved by designing nanostructures that have an inhomogeneous configuration, such as core/shell particles,[25,26] or that have sharp geometrical features such as nanoprism,[27,28] nanocubes,[29–31] and star-shaped particles,[32,33] as well as particle pairs or aggregations,[34–36] by one of two approaches, commonly labeled 'top-down' and 'bottom-up'.[37] The top-down technique involves using various lithographic techniques to pattern nanostructures, whereas bottom-up methods exploit the interactions of atoms, molecules, or more complex mesoscale objects, in conjunction with the controlling influences of process kinetics, to assemble nanostructures either on substrates or in solutions.[15] The exquisite synthetic control using the present nanotechnology[6,7,38,39] in combination with advances in theory[40] and the emergence of quantitative electromagnetic modeling tools, such as the finite difference time domain[41], the discrete dipole approximation (DDA),[42,43] Green's function approach,[44] multiple multipole,[45] boundary element methods,[46,47] and the hybridization model,[48,49] has provided a better understanding of the optical properties of isolated and electromagnetically coupled nanostructures of various sizes, shapes, and/or spacing.[50] Even though this chapter will focus on optical properties of nanoporous metals, here we would like to start from the brief introduction of simple-shaped nanostructured metals' SP properties before moving to three-dimensional (3D), bicontinuous nanoporous metals.

6.2 Theoretical Consideration: Optical Properties of Metal Nanostructures

Figure 6.1 schematically illustrates the creation of an SP oscillation in a spherical metal particle with the complex dielectric constant, $\varepsilon(\omega) = \varepsilon_1(\omega) + i\varepsilon_2(\omega)$, in the dielectric medium with a dielectric constant, ε_m. The electric field of an incoming light wave with a angular frequency, ω, induces a polarization of conduction electrons with respect to the much heavier ionic core of a spherical particle. The net charge difference occurring at the nanoparticle boundaries (surfaces) creates the dipolar oscillation with the same phase, which in turn acts as a restoring force. Although all electrons are oscillating with respect to positive-ion background, the surface plays a very important role for the observation of SPR, because it alters the boundary condition of the polarizability of metals. This enables nanoparticles with different sizes and shapes to exhibit unique and intense colors when the frequency of the electromagnetic field becomes resonant with the coherent electron motion.[51,52] This phenomenon was explained theoretically by the pioneering work of Mie in 1908, in which the general solution of Maxwell's equations for the optical properties of a spherical particle was first given on the basis of the assumption that the particle and the surrounding medium are homogeneous and describable by their bulk optical dielectric functions.[40] Solutions have subsequently been developed for other particles with different shapes.[53–56] These analyses are greatly simplified when the particles are much smaller than the wavelength of light, since only the lowest order of the Mie theory (dipolar absorption)

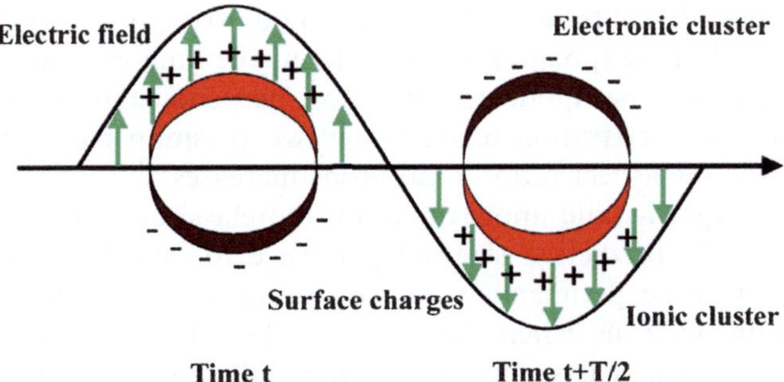

Figure 6.1 Schematic illustration of the excitation of the dipole surface plasmon oscillation. The electric field of an incoming light induces the polarization of conduction electrons with respect to the ionic core with the period T. A net charge difference is only felt at the nanoparticle boundaries (surface).

contributes to the extinction cross-section (σ_{ext}). The Mie theory then reduces to the following relationship (quasi-static or dipolar approximation):

$$\sigma_{ext} \propto \left[\frac{\varepsilon_2(\varpi)}{[\varepsilon_1(\omega) + \chi\varepsilon_m]^2 + \varepsilon_2(\omega)^2} \right], \tag{6.1}$$

where $\chi = (1 - P_i)/P_i$ is a form factor that describes the nanoparticle's aspect ratio R, with P_i denoting the depolarization factors along the three axes x, y, and z, and $P_x = \frac{1-e^2}{e^2}\left[\frac{1}{2e}\ln\left(\frac{1+e}{1-e}\right) - 1\right]$, $P_y = P_z = (1 - P_x)/2$, $e = \left[1 - \left(\frac{1}{R^2}\right)^2\right]^{1/2}$.[57]

$\chi = 2$ for a sphere and increases with the aspect ratio of the nanoparticle. ε_1 and ε_2 are the real and imaginary parts of the dielectric function of the metallic nanostructure, respectively, and $\varepsilon_1(\omega) = 1 - [\omega_p^2\tau^2/(1 + \omega^2\tau^2)]$, $\varepsilon_2(\omega) = \omega_p^2\tau/[\omega(1 + \omega^2\tau^2)]$ according to the Drude–Sommerfeld free-electron theory.[58] Here, $\tau = l_b/v_f$ is the electron relaxation time in bulk metals, with l_b being its mean free path (MFP) and v_f the Fermi velocity of electron, and $\omega_p = (4e^2\pi N/m_0)^{1/2}$ is the bulk plasmon resonance frequency with N being the density of conductor electrons, and e and m_0 denoting the charge and the effective optical mass, respectively. For metals at near-infrared frequencies $\omega \gg 1/\tau$, $\varepsilon_1(\omega) = 1 - \omega_p^2/\omega^2$, $\varepsilon_2(\omega, T, D) = \omega_p^2/\omega^3\tau$ for further simplicity.[58]

Usually, ε_m is taken as a real constant, and the extinction coefficient has the maximum value at a resonance frequency where

$$\varepsilon_1(\omega) = -\chi\varepsilon_m. \tag{6.2}$$

Apparently, the position of this resonance red shifts with an increase in magnitude of the dielectric constant of the medium surrounding the nanostructure due to the buildup of polarization charges on the dielectric side of the interface, which is responsible for the weakening of the total restoring force. Eqn (6.2) also shows that, for non-spherical metallic nanoparticles, the LSPRs are dependent on the shapes due to the difference of surface charge distributions on the nanoparticle surface.[37–39,57,59] For example, the plasmon resonance of metallic nanorods splits into two peaks: (1) a strongly red-shifted long axis or longitudinal mode (polarization parallel to the long axis) and (2) a slightly blue-shifted transverse mode (polarization perpendicular to the long axis). Furthermore, the separation between the two plasmon bands becomes more remarkable as the aspect ratio of nanorods increases.[60,61]

Within the quasi-static approximation, the classical condition of SP resonance [eqn (6.2)] predicts a size independence for small particles, in strong contradiction with experimental results on nanoparticles smaller than the MFP of conduction electrons, where the size-dependent shift and damping of SP are observed.[4,62–65] This casts into doubt the validity of the method to account for the optical properties of small nanoparticles using a bulk dielectric function in Mie's theory, and suggests the modification of dielectric function.[62–67] Generally, for nanoparticles smaller than the ordinary MFP (in the order of 30–50 nm for silver and gold), the surface scattering caused by phonons, impurities, and lattice defects becomes important. Kreibig suggested a simple picture where the dielectric function can be decomposed into a contribution from the inter-band transitions and a free electron part, and the latter can be modified within the Drude free-electron model to account for the enhanced electron-surface scattering as a function of the particle diameter.[4,62–64] The classical consideration leads to the effective relaxation time τ_{eff}, $\tau_{eff}^{-1} = \tau^{-1} + \tau_s^{-1}$ with $\tau_s^{-1} = 2v_f/D_{eff}$ denoting the surface scattering rate, where D_{eff} is the effective size of the particle, equal to the diameter in the case of a sphere with diffuse surface scattering.[4,62,68] If the inter-band transitions are neglected, and only the free electron movement is taken into account, then $\varepsilon_1(\omega) = 1 - [\omega_p^2/(1/\tau_{eff}^2 + \omega^2)]$. In Mie's theory, the resonance condition [eqn (6.2)] gives rise to the resonance frequency,[5]

$$\omega_{sp} = \left\{ \left[\omega_p^2/(1 + \chi\varepsilon_m)\right] - v_f(D_{eff} + 2l_b)/l_b D_{eff} \right\}^{1/2}, \tag{6.3}$$

for which the prediction is in agreement with experimental results of small metal particles. As for very small nanoparticles with a size below 2 nm, the assumption of a delocalized free electron gas is no longer valid, and it is better to treat them as molecular clusters because of the discrete electronic states.[4,60,64]

In contrast with the aforementioned intrinsic size effect, the extrinsic size effect occurs in larger metallic nanoparticles (beyond the Rayleigh approximation, $D_{eff} > 30$ nm), for which the extinction cross-section is also dependent on higher-order multipole (quadrupole and octopole) modes arising from inhomogeneous polarizations, in addition to dipole oscillation, within the full Mie equation.[57,69] The introduction of these higher oscillation modes gives rise to a reduction in the depolarization field at the particle center generated by the surrounding polarized matter because of a retardation effect, where the conduction electrons do not all move in phase.[57,69] This explicitly enables the dipolar resonance to shift to a longer wavelength as well as a substantial broadening of bands with increasing particle sizes. The total plasmon band can also be considered as the superposition of all contributing multipole oscillations peaking at different energies, associated with the dephasing of the coherent oscillation.[70–74] This pure dephasing process primarily contributes to the total plasmon damping in contrast with the population relaxations consisting of the radiative and non-radiative dampings. Depending on the location of the hole within conduction bands or in the lower-lying bands, the radiative loss takes place *via* the transformation of particle plasmons into photons, while the non-radiative damping arises from the intra-band excitation within the conduction bands and the inter-band excitation between other bands (*e.g.* the d bands for noble metals) and the conduction bands.[73,75] If one assumes a simple two-level model for the plasmon absorption, the total dephasing time T_2 is given by

$$1/T_2 = 1/(2T_1) + 1/T_2^*, \tag{6.4}$$

where T_1 describes the population relaxation time involving both radiative and non-radiative processes, and T_2^* is the pure dephasing time. Often T_2^* (on the order of 10^{-15} s) is much shorter than the energy relaxation T_1 (10^{-13} s) and thus determines the value of T_2. The narrow plasmon bandwidth corresponds to slow loss of coherent electron motion, which generally benefits the optical applications of metal nanostructures, such as surface-enhanced spectroscopy. The electron dephasing time in the order of a few femtoseconds correlates with the homogeneous bandwidth of SP (Γ) and can be computed using $1/T_2 = \pi c \Gamma$, where c is the speed of light, strongly suggesting that the main relaxation process involves electron–electron collisions.[70–74] A narrow distribution in size and shape of a nanostructure is essential for determining the value of T_2 accurately. The scattering process at the surface of a nanoparticle is thought to contribute to the total damping for nanoparticles smaller than the free-electron scattering length. The depolarization field and additional damping mechanisms for large and small particles can be seen as lowest-order corrections to the quasi-static theory leading to a decrease in the total enhancement of the excitation field.

In addition to the isolated nanoparticles, the assembly or aggregation of two or more particles with a very small separation distance expands the range of available LSPR modes.[34–36] Because of the hybridization effect between the plasmon modes of neighboring particles,[48,49] high- and low-energy modes with different electromagnetic distributions can be produced.[76,77] The electromagnetic coupling between LSPRs of these assembled particles leads to a strong electromagnetic field enhancement within the gap between the particles, namely the so-called 'hot spots',[78,79] which has been used in the pursuit of single-molecule surface-enhanced Raman scattering[80–82] and surface-enhanced fluorescence (SEF)[83] in the systems of aggregated particles.

6.3 Microstructure and Optical Properties of Nanoporous Metals

Although the LSPRs confined in small regions are associated with strong electromagnetic fields, the interaction between light and the LSPR modes of particles is only a weak function of the angle of the light with respect to the particles.[39] In contrast, the SPP mode of a planar film may have a very specific and strong dependence on the angle if an appropriate wavelength-scale periodicity is built into a nanostructure.[19,21] The hybridization effect of SP modes has stimulated great interest in the fabrication of nanostructure supporting both localized/propagating models simultaneously and thus allowing high field enhancements and good directional control. Among the interesting mixed localized/propagating systems,[84,85] dealloyed nanoporous metals (NPMs) recently have attracted special attention because of their unique quasi-periodic, 3D nanostructure.[86]

NPMs, such as gold,[87–90] silver,[91] copper,[92] platinum,[93,94] palladium,[95] and nickel,[96] *etc.* can be fabricated by an electrochemical method called dealloying during which remaining more noble components of selective etching form a 3D nanoporous structure through an interface-diffusion-controlled self-assembly process while the less noble components selectively dissolve from alloy precursors.[97–99] Because of the proper optical properties, the dealloyed NPMs exhibit a remarkable color change in comparison with the original solid alloys, as shown in Figure 6.2(a) and (b), where the color of the representative example of the $Ag_{65}Au_{35}$ sheet changes from a shiny luster to dark dullness accompanying the formation of nanoporous gold (NPG) by etching in a concentrated HNO_3 solution. The top view [Figure 6.2(c) and (d)] and cross-section [Figure 6.2(e)] scanning electron microscope (SEM) images illustrate the unique bicontinuous nanostructure consisting of quasi-periodic metal ligaments and nanopore channels.[87–97] In addition, the SEM micrographs [Figure 6.2(c) and (d)] also demonstrate that the NPG films dealloyed at different conditions have different characteristic lengths. In this article, the characteristic length is defined by the equivalent diameter of nanopores or the gold ligaments that are measured by a rotational fast Fourier transform method.[100] The changeable structural feature suggests that the etching process allows us to tailor the characteristic lengths of NPMs from several nanometers to micrometers by adjusting the dealloying conditions, such as the dealloying time and temperature.[101–103] Although the characteristic lengths vary with dealloying conditions, it is interesting to note that the morphologies and sizes of nanopore channels and ligaments are statistically identical in each sample with a different characteristic length. This is further verified by 3D electron tomography [Figure 6.2(f)].[104] However, these observations are different from that of

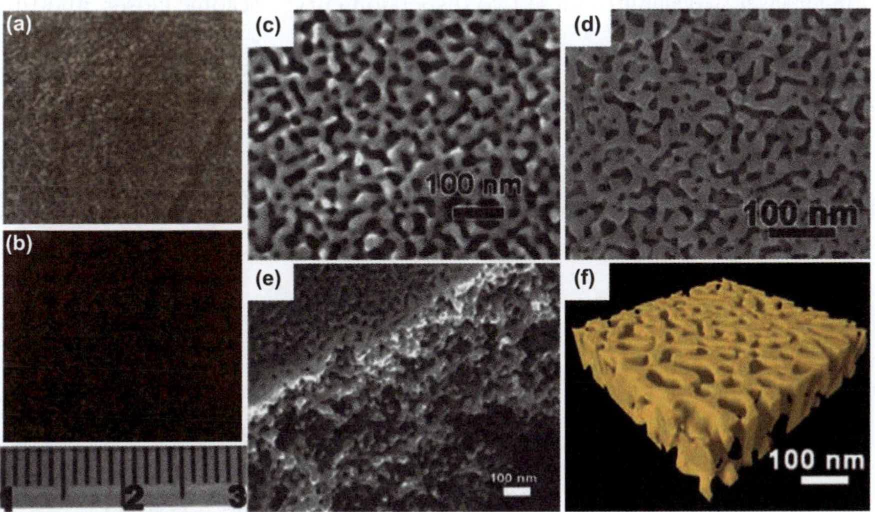

Figure 6.2 Optical images of $Ag_{65}Au_{35}$ sheets (a) before (b) after free etching in concentrated HNO_3. (c, d) Top-view SEM micrographs of $Ag_{65}Au_{35}$ sheets chemically dealloyed for 30 and 120 min, respectively. (e) Cross-section image of sample (d). (f) 3D electron tomographic image of the sample (c).

Rösner *et al.*[105] In their electron tomographic images, the NPG prepared by electrochemically dealloying $Ag_{80}Au_{20}$ shows a very open nanoporous structure, and the mean pore size ($\sim 28 \pm 8$ nm) is about twice that of the ligament size ($\sim 16 \pm 5$ nm). The remarkable dissimilarity is probably caused by the different dealloying approaches. Free-chemical etching may allow the nanoporosity to have sufficient time for structural relaxation by surface diffusion for a more stable nanostructure. In contrast, the applied potential (~ 600 mV) for electrochemical dealloying in the study of Rösner *et al.* may lead to the formation of gold hydroxide that inhibits the coarsening of gold ligaments and the relaxation of the porous structure.[106]

The unique bicontinuous nanostructure enables NPMs to possess structural features of both planar metal films and nanostructured metals, which exhibit propagating and localized SPR excitations under the light irradiations. The properties of propagating SPR of NPG films were first characterized by Yu *et al.*, employing a classic experimental setup,[107] the so-called Kretschmann configuration, where a probe laser reflects off the backside of the metal film through a high-index glass prism. The reflectivity of NPG as a function of incident angle with irradiation at different wavelengths (632.8, 780, 820, and 1152 nm, respectively) clearly shows a strong dependence of the propagating SPR on the excitation wavelength, *i.e.* the longer the laser wavelength, the sharper the propagating SPR dip, and the smaller the dip angle (Figure 6.3). Propagating SPR is dependent on the extent to which the dispersion-relation (between angular frequency and wave vector) curve at the NPG/dielectric interface asymptotically approaches that of the free phonon in air.[19] At a longer laser wavelength, these two dispersion curves become closer, and thus a

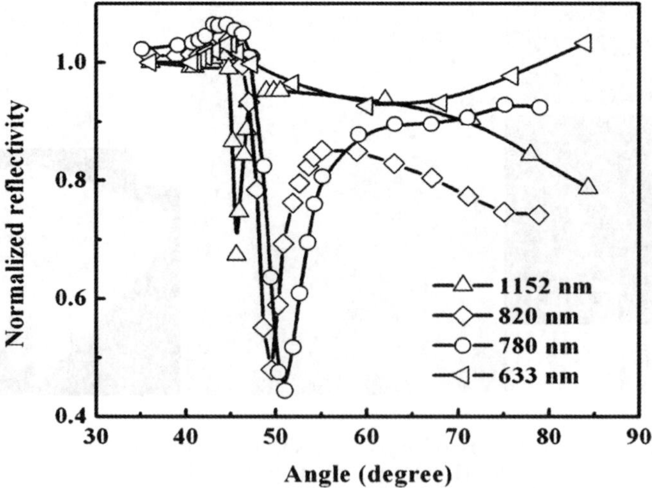

Figure 6.3   Excitation of propagating SPR on NPG film with the thickness of 100 nm in air using a right-angled BK7 glass prism.[107]

smaller wave-vector amplification factor for the high-refractive-index prism is needed to fulfil the SPR excitation conditions. This argument was further clarified by the reflectivity measurements in which the SPR minimum was absent when the sample was irradiated with an s-polarized laser.[107] While under the fixed excitation wavelength, the resonance angle exhibits a significant shift, depending on the dielectric constants of surrounding media. This property can be used to detect specific targets for chemical or biological sensing.[19,107] Note that the reflectivity band is also influenced by the roughness of NPG film due to the damping of propagating SP mode, which may affect the analytical sensitivity measurements. Such perturbations may be described by the consideration of the forward scattering and directional backward scattering of non-radiative SPs due to roughness.[107]

In contrast, the localized SPR of noble metal nanostructures, such as nanoparticles with different shapes and nanoholes in planar metal films, can be easily excited by visible light when their feature sizes are much smaller than the wavelength of light. Usually, localized SPR can be characterized using a UV-visible (UV-Vis) spectrometer wherein the absorption, scattering, or extinction spectrum as a function of wavelength at a fixed angle is collected, and the anomalous peaks ascribed to the localized SPs can be detected. Typical UV-Vis extinction spectra of as-prepared NPG films with different characteristic lengths in air, collected using a UV-Vis spectrometer with a standard component of a ϕ60 mm integrated sphere, are shown in Figure 6.4(a), demonstrating that there are two peaks in each spectrum, *i.e.* the lower peaks located at $\sim$490 nm, and the higher peaks red shifted to $\sim$512, 516, 522, and

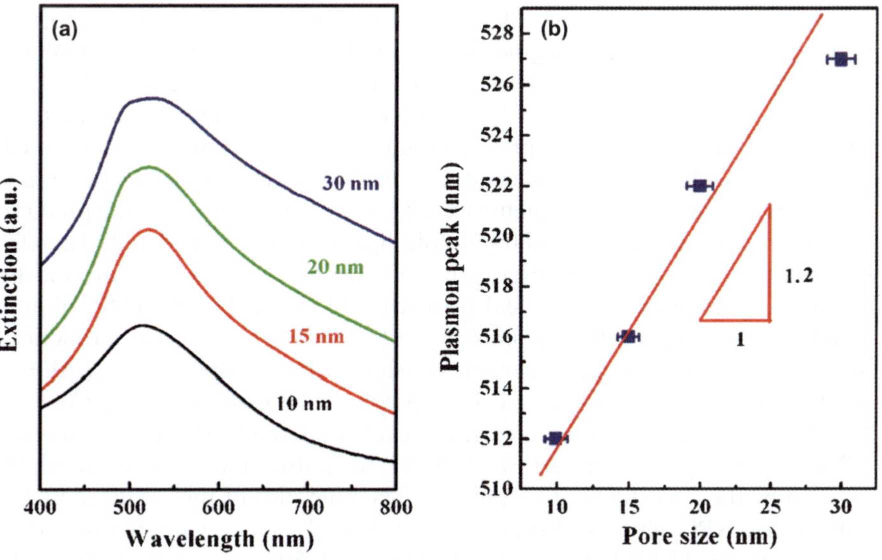

Figure 6.4   (a) UV-vis spectra of as-prepared NPG films with different pore sizes in air. (b) Plasmon peaks of NPG in air as a function of pore size.

527 nm for NPG films with nanopore sizes of $D \approx 10$, 15, 20, and 30 nm, respectively. The former corresponds to the bulk plasmon edge of gold, independent of nanopore sizes and dielectric surroundings, while the latter is ascribed to the localized SPR resulting from the collective oscillation of electrons confined in nano-scaled metal ligaments and thus shows remarkable dependence on the characteristic lengths of NPG.[108] Figure 6.4(b) shows the localized plasmon resonance peaks of NPG films as a linear function of characteristic length (D) with the slope of $\Delta\lambda_{res}/\Delta D \approx 1.2$, quantitatively consistent with the experimental and theoretical results of individual nanoholes in planar gold films.[47,109,110] Analogous results can also be observed from nanoporous silver (NPS), for which the peak appearing at 317 nm arises from the bulk plasmon edge of silver, and the localized SPR peaks show red shift as the characteristic length increases to 100–200 nm.[91]

There are two factors responsible for the band shift of the localized SPR with decreasing characteristic lengths of NPG films: reduced diameter of gold ligament and shrunk nanopores. The former can be explained by modified resonance condition in the Mie theory [eqn (6.3)]. However, how the shrunk nanopores affect the localized SPR remains unknown, since reducing ligament sizes by changing dealloying conditions simultaneously leads to a decrease in nanopores. To tackle this problem, we investigated the localized SPR of gold-plated NPG that has an opposite relation between nanopores and ligaments, *i.e.* shrunk nanopores corresponding to increasing ligaments. The localized SPR peaks of the gold-plated NPG films blue shift as the pore size (d) decreases, and the ligament size (D) increases concurrently.[111] The gold-plated NPG with the novel nanoporous structure consisting of very small nanopore channels and large gold ligaments was fabricated by electroless plating of gold onto as-dealloyed NPG frameworks. Figure 6.5(a) shows the representative top-view SEM image of the NPG with gold plating for 80 min, for which the morphology remarkably differs from that of the as-dealloyed one [Figure 6.5(b)]. The UV-vis extinction spectra of gold-plated NPG films with different characteristic lengths that are controlled by plating conditions [Figure 6.5(c)] are presented in Figure 6.5(d), showing the blue shifts of the localized SPRs with the reducing pore sizes and increasing ligament sizes. This result is contrary to the prediction by Mie's theory if only the enlarged gold ligaments are considered. Therefore, the nanopores play a more important role, rather than the sizes of gold ligaments, in determining the optical properties of NPG. Figure 6.5(e) plots the wavelength at a maximum transmission of the SPR band (λ_{max}) as a function of the pore sizes, exhibiting a linear relationship with the slope of $\Delta\lambda_{max}/\Delta d = 2.1$, much larger than that observed in as-prepared NPG films ($\Delta\lambda_{max}/\Delta d = 1.2$). The difference in the slopes in turn indicates that the SPPs of gold ligaments influence the optical properties of NPG in addition to the localized SPR induced by the curvatures of nanopores. The blue shift of the localized SPRs of gold-plated NPG films is also accompanied by the narrowing of full width at half-height (Γ) of SPR bands, or $\Gamma \propto 1/D$ [Figure 6.5(f)]. As mentioned in Section 6.2, in the two-level model the plasmon bandwidth is associated with the dephasing of coherent electron oscillation *via* $\Gamma = 1/\pi c T$.[70–75] For the NPG films,

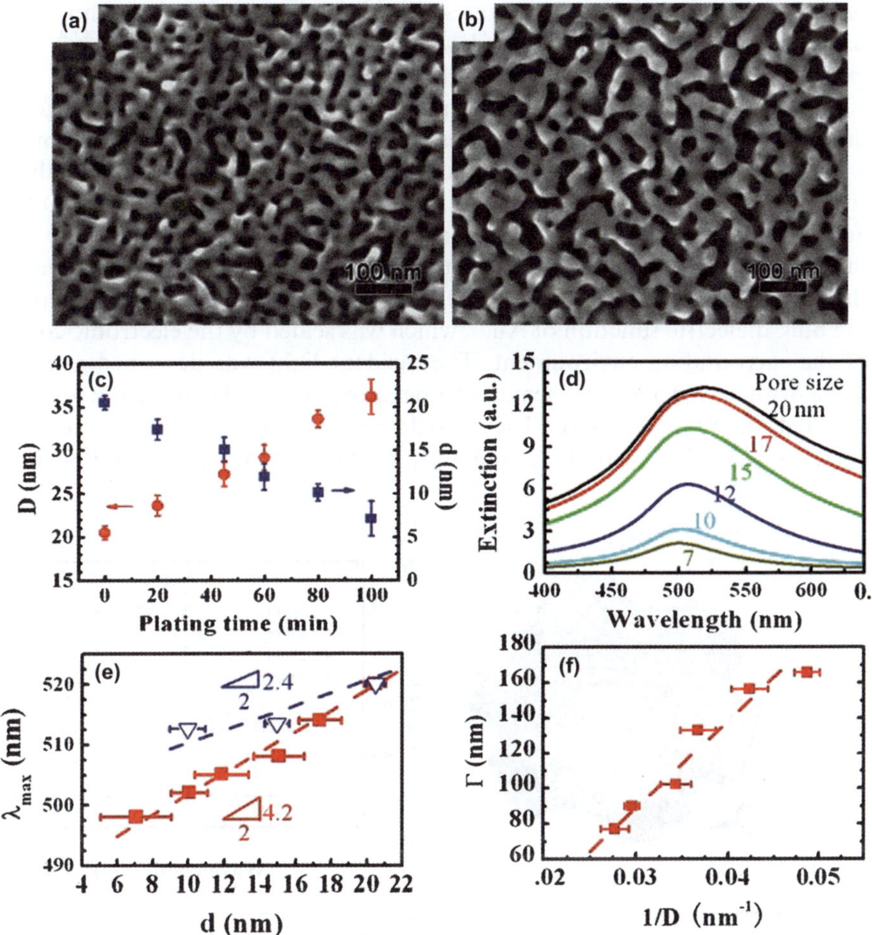

Figure 6.5 Representative top-view SEM micrographs of (a) NPG gold-plated for 100 min and (b) as-prepared NPG with $d \approx D \approx 20.5$ nm. (c) Relationship between d or D of gold-plated NPG and plating time t. (d) UV-vis extinction spectra of gold-plated NPG films with different pore size d/D ratios. (e) Resonance wavelengths of as-prepared (∇) and gold-plated ($\blacksquare$) NPG films as a function of pore size d. (f) Dependence of linewidth Γ of gold-plated NPG on the inverse D.[111]

the electron scattering is controlled by the internal surface of NPG.[112] Thus, the dephasing time of coherent electron oscillation increases with increasing D, namely $T \propto D$, which gives rise to the increasing local-field enhancement, and thus the better performance in surface-enhanced spectroscopy.

As shown in Figure 6.4(a), while the characteristic length decreases from a larger size, the localized SPR peak blue shifts to overlap with the bulk plasmon peak, which makes it difficult to clearly identify the detailed information on the localized SPR of nanoporous metals with a small pore size. To circumvent this problem, we carried out computer simulations using a DDA method on the basis of the DDSCAT 7.0 code developed by Draine and Flatau.[42,43] As an example, we show the simulated SP properties of NPG.[103] In view of the complicated structure of NPG, a simplified quasi-three-dimensional target with dipolar elements more than 10^5 representing the key structural features of NPG with the identical pore and ligament sizes was introduced into the qualitative simulation [Figure 6.6(a)]. The requirement of $|m|dk < 0.5$ for the DDA was adequately satisfied, where m is the complex refractive index of the target materials, d is the inter-dipole separation, and $k \equiv 2\pi/\lambda$ with λ the wavelength in the vacuum. The dielectric properties of the nanostructure were simulated by the bulk dielectric function of Au,[58] which was scaled by the electronic constant of the surrounding environment. The incident light was described as a plane wave with a specific wavelength propagating along the direction of x axis denoted in Figure 6.6(a). The calculated spectra of extinction efficiency ($C_{ext} = C_{abs} + C_{sca}$, with C_{abs} and C_{sca} being the absorption and scattering

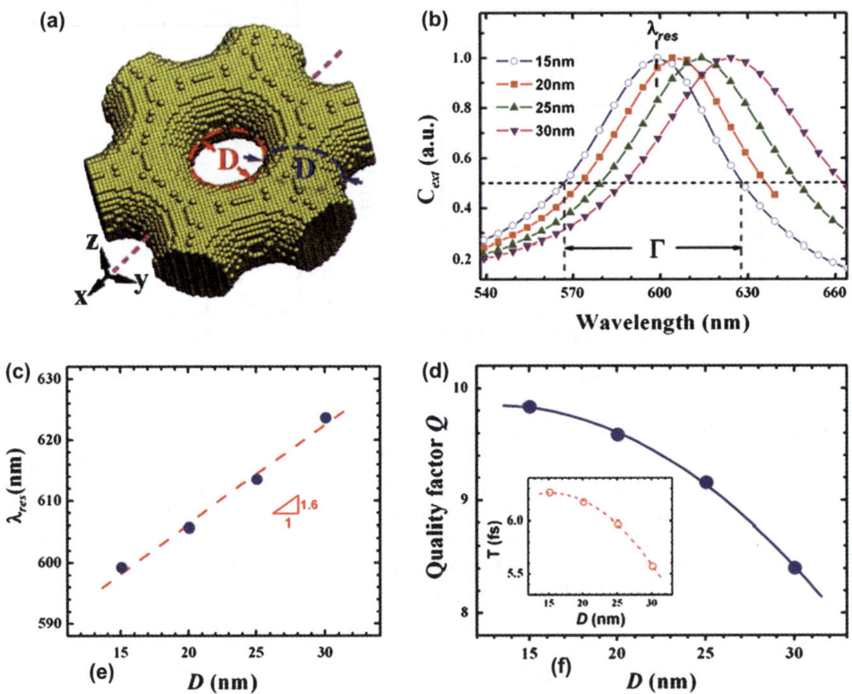

Figure 6.6 (a) Schematic nanostructure used for DDA simulations, wherein D denotes the characteristic length, *i.e.* the equivalent diameters of metal ligaments and nanopore channels. (b) Typical extinction spectra of target nanostructure with different characteristic lengths (D). (c) Resonance plasmon peaks (λ_{res}) of target nanostructure (a) as a linear function of D with the slope of 1.6. (d) Relationship between quality factor $Q = E_{res}/\Gamma$ and D. Inset: dephasing time (T_2) as a function of D.

cross-sections, respectively) for the target nanostructure [Figure 6.6(a)] with different characteristic lengths (D) are presented in Figure 6.6(b), showing a single SP band for each spectrum and the red-shift of SP peak with increasing D. In the calculation, the intensity is normalized to the maximum extinction for each nanostructure with D. Figure 6.6(c) illustrates the relationship between the resonance wavelengths and the sizes of the nanostructured targets, exhibiting the linear function with the slope of $\Delta\lambda_{res}/\Delta D \approx 1.6$, fairly consistent with the experimental result (~ 1.2) of as-prepared NPG [Figure 6.4(b)]. As the nanopore size increases, the red shift of the SPR peak is also accompanied by the broadening of the full width at the half-height of SP bands because of the strong plasmon damping, *i.e.* a rapid dephasing of the optical polarization associated with the electron oscillation.[70–75] The dephasing time, characterizing the speed of the dephasing of the nanostructure, can be calculated from the homogeneous linewidth Γ *via* $T_2 = 2\hbar/\Gamma$ with $\hbar = h/(2\pi)$, where h is the Plank constant. The plot of T_2 against D is presented in the inset of Figure 6.6(d), demonstrating that rapider dephasing occurs in a nanostructure with a larger pore size. The dephasing time of $T \sim 6\,fs$ is consistent with that of nanoparticles[74] but larger than that experimentally obtained from gold-plated NPG films. This is probably because the overlapping of the bulk plasmon band and localized SPR band gives rise to the smaller estimation. Figure 6.6(d) shows another important quantity, namely, the quality factor of the resonance, $Q = E_{res}/\Gamma = \pi c T_2/\lambda_{res}$, for the calculated nanostructured target. The quality factor is the enhancement of the oscillation amplitude of a driven oscillating system with respect to the driven amplitude, *i.e.* the near-field enhancement in the case of nanostructure plasmons. It is clear from this plot that the factor Q drops with increasing D because the decreasing T_2 is accompanied by the red shift of λ_{res}. This trend of Q against D implies that the NPG with a large D exhibits a relatively weaker enhancement in surface-enhanced spectroscopy, such as surface-enhanced Raman scattering (SERS), which is usually proportional to Q^4.[70–75]

In addition to the nanopore size dependence of the localized SPR, the classical resonance theory [eqn (6.2)] also predicts that the plasmon peaks of NPG films red-shift with increasing ε_m. This phenomenon is observed from the NPG films coated with a layer of human serum albumin (HSA) [Figure 6.7(a)]. The larger slope of $\Delta\lambda_{max}/\Delta d = 2.7$ [Figure 6.7(b)], compared to 1.2 in air, is caused by the large dielectric constant of the surrounding medium, while the bulk plasmon band again is independent of the dielectric constant. The SPR sensitivity of the NPG films was tested in the media with different refractive indexes $n = \sqrt{\varepsilon_m}$. Figure 6.7(c) shows the typical extinction spectra of the NPG film with the characteristic length of $\sim 50\,nm$ in water, ethanol, and toluene, demonstrating a substantial red shift with increasing refractive index n, analogous to that observed from the localized SPR in other nanostructures.[15] The sensitivity of the localized SPR of NPG, characterized by the slope of $\Delta\lambda_{LSP}/\Delta n$, exhibits a strong dependence on the characteristic length. The data yield $\Delta\lambda_{LSP}/\Delta n = 274.8\,nm/$refractive index unit (RIU) and $227.6\,nm/$RIU for the NPG films with the characteristic lengths of 50 and 30 nm, respectively,

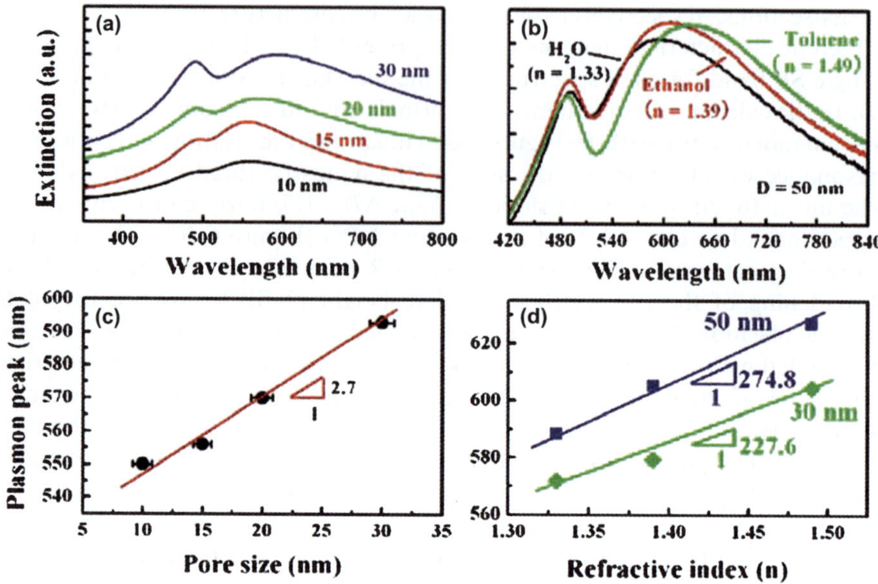

Figure 6.7 (a) UV-vis extinction spectra and (b) pore-size dependence of plasmon wavelengths of as-prepared NPG films coated with HSA layer. (c) Typical extinction spectra of as-prepared NPG film with the characteristic length of ∼50 nm in different refractive index environments. (d) Dependence of localized surface plasmon resonance peaks of NPG films with characteristic lengths of 50 and 30 nm on the refractive index (n).

which are much larger than $\Delta\lambda_{\mathrm{LSP}}/\Delta n \approx 100$ nm of the planar Au films decorated with nanometric holes made by colloidal lithography films.[109,110]

6.4 Applications of Plasmonic Nanoporous Metals

6.4.1 Biosensing with Plasmonic Nanosensors

Since propagating and localized SPR spectroscopy is a surface-sensitive technique, it can provide quantitative information about the surface-binding events of plasmonic substrates by measuring the refractive index of surrounding media. The localized SPR spectral shift ($\Delta\lambda_{\mathrm{LSP}}$) of nanostructured metals in response to changes in refractive index is approximately described as:[51]

$$\Delta\lambda_{\mathrm{LSP}} = m\Delta n\left(1 - e^{-2h/l}\right), \tag{6.5}$$

where m is the sensitivity factor (in nm per RIU) $\Delta n = n_{\mathrm{absorbate}} - n_{\mathrm{medium}}$ with $n_{\mathrm{absorbate}}$ and n_{medium} being the refractive indices (in RIU) of the adsorbate and medium surrounding the nanoparticle, respectively; h is the effective thickness of the adsorbate layer (in nm); and l is the electromagnetic field decay length (nm). Obviously, there are two methods to improve the effectiveness of applications of SPR spectroscopy in biosensing. One is to optimize the nanostructure characteristics, such as size, shape, and composition, to increase the values of m and l. The other is to enlarge the dielectric constant difference between nanostructured metals and adsorbed molecules. The former has led to recent renovation in NPMs with fascinating nanostructures, for example, NPG with nanostructures consisting of large pore channels and large gold ligaments,[102] very small pore channels and large gold ligaments,[111] or multimodal nanoporosity.[113] Core-shell Au–Ag porous films have been developed that exhibit dramatic improvements in optical properties and performance as well.[114]

Additionally, the use of bare NPMs (with very small $n_{absorbate}$ values) as biosensors has large advantages over that of nanoparticles covered by molecules, or embedded in a solid matrix with large $n_{absorbate}$ values to prevent them from agglomeration and precipitation.[6,8,74] Yu *et al.* first reported the applicability of NPG films as plasmonic substrates to detect an avidin/biotinylated anti-avidin antibody (BAA) double-layer system[107,115] and demonstrated that the peak shift of NPG with large pore sizes is larger than that of NPG with small pore sizes, similar to the results shown in Figure 6.4(a) and 6.7(a). They also investigated the sensitivity of localized SPR of NPG films, with a characteristic length of ~15 nm, to the environmental refractive index change by using an aqueous solution of glycerol as well as the layer-by-layer system based on charged dendrimers {cationic $[G_4(NH^+Et_2Cl^-)_{96}]$ and anionic $[G_4(CH\text{-}COO^-Na^+)_{96}]$}. The refractive index is changed by controlling the concentrations of glycerol ranging from 0 to 60% or the layer numbers of the layer-by-layer system.[107,115] They found that higher glycerol concentrations or the increase in layer number of the layer-by-layer systems gives rise to an obvious increase in the peak intensity and red shift of the peak wavelength, different from the observation that the wavelength and intensity of plasmonic peaks remain constant for flat/dense gold.[116] These results suggest that the localized SPR signal is directly linked to the accessibility of the interior of the NPG nanoporosity.[107,115] Meanwhile, Yu *et al.* also investigated the application of the propagating SPR of NPG for biorecognition events in a well-studied biotin–streptavidin binding system.[107] Upon the formation of each biological layer on an NPG membrane, significant shifts of the resonance angle were observed. The increased value of Δn can also be achieved by using large molecules and resonant labels.[117–119] Additionally, the SPR signal is enhanced by labeling the target analytes with dielectric or plasmonic nanoparticles. These labels increase the refractive index at the metal surface and can couple electromagnetically to the flat metal film in the case of plasmonic nanoparticles, leading to large SPR peak shifts.

6.4.2 Surface-Enhanced Raman Scattering

Raman spectroscopy, which is based on the inelastic scattering of photons by chemical entities, has been successfully utilized for detecting and identifying organic and biological molecules.[120,121] In general, normal Raman scattering is extremely weak, due to the small cross-section, typically in the range of 10^{-30} to 10^{-25} cm^2, which limits its applications, particularly in detecting molecules in dilute solutions. Amplified by nanostructured metals, the intensity of the SERS can be dramatically improved by a factor of up to 10 orders of magnitude. This phenomenon was independently discovered by Jeanmarie and Van Duyne[122] and Albrecht and Creighton[123] in 1977, and has recently attracted great attention as result of rapid developments in nanoscience and nanotechnology. It is now well understood that the creation of SPs in metallic nanostructures is mainly responsible for the enhancement of Raman scattering through the associated electric field.[4,13,124,125] Great enhancements in the Raman intensity of one molecule in the vicinity of metallic nanostructure suggests the possibility that SERS can be used as an invaluable tool for detection and identification of extremely minute quantities of target molecules, which is sufficient to study analytes at the single-molecule level.[126–132] Although SERS effects can be achieved simply by exploiting the electromagnetic resonance properties of roughened surfaces or nanoparticles of noble or transition metals,[13,17,128–133] the use of the SERS effect in a quantitative chemical analysis is intricately dependent on the reproducibility to fabricate nanostructured surfaces with well-defined morphologies and uniformly high enhancement factors.[78,134–136] This has given rise to a wealth of research centered on new methods to create nanostructured materials with controllable nanoscale features for improving the SERS sensitivity to the levels required for sensing application. Such forms of SERS substrates span a vast geometrical range encompassing colloids, templated colloidal crystal films, electrochemically roughened electrodes, deposited metal island films, nanohole arrays, and lithographically defined thin films. We direct the interested reader to a number of review articles on this subject,[13,38,126,137,138] while here we focus on the SERS effect of NPMs.

In contrast to SERS substrates mentioned above, NPMs, which are fabricated by the dealloying method, possess a unique bicontinuous nanostructure consisting of quasi-periodic metallic ligaments and nanopore channels.[102–104] The proper 3D nanoporous structure results in a coupling between localized and propagating SPR modes, thus enabling NPMs to have fascinate optical properties and promising applications in SERS. Analogous to other metallic nanostructures, the SERS effect of NPMs has been attributed to both electromagnetic and chemical enhancement mechanisms.[4,13,124] The chemical mechanism usually offer an SERS enhancement with a factor of 10^2 to 10^3, depending on the interaction between the orbitals in metal and adsorbed molecules, where charge transfer probably occurs for a lower-energy resonant Raman mechanism.[139] Apparently, the large enhancement factor of $\sim 10^{10}$ cannot be achieved solely by the chemical effect even with the consideration of large internal surfaces of NPMs. The electromagnetic mechanism elucidates that the excitation of SP in metallic nanostructures generates a strong local electric field that strongly affects the optical properties of the adsorbate and is the main factor in creating the phenomenon of the giant SERS effect. Therefore, the ability to enhance the Raman signal *via* SERS is fundamentally linked to the precise nanoporosity as well as the choice of the metal itself, which includes but is not limited to Ag, Au, and Cu.

6.4.2.1 Nanoporous Gold

Among these NPMs, free-standing NPG films are of special interest as attractive SERS-active substrates by virtue of the tenability of nanoporosity, as well as the excellent chemical stability and biocompatibility properties of gold.[101,103,108,140] More recently, Qian *et al.* have systematically investigated the SERS effect of NPG with topologically and morphologically equivalent gold ligaments and nanopore channels, for which sizes ranging from ~ 5 nm to ~ 700 nm are tailored by the combination of low-temperature dealloying and annealing.[101] The simple free etching and annealing procedure does not affect the porosity and geometry of NPG, as shown in Figure 6.8(a)–(d) in which NPG with different characteristic lengths can be achieved by dealloying or annealing. Figure 6.8(e) shows SERS spectra for rhodamine 6G (R6G) molecules adsorbed on NPG with different pore sizes, showing that Raman scattering intensities of R6G molecules spontaneously increase with decreasing nanopore size. This demonstrates that the SERS enhancement of NPG is susceptible to the characteristic lengths of nanoporosity. The strongest SERS enhancement of NPG is achieved from the sample with an ultra-fine nanopore size of ~ 5 nm. The integrated intensity of Raman band for R6G at $1650\,\mathrm{cm}^{-1}$ from the 5 nm sample is about two orders of magnitude stronger than those from 700 nm NPG [Figure 6.8(g)]. The similar dependence of SERS enhancement on the nanopore size has also been observed from other organic molecules, such as crystal violet (CV), with different electronic vibration [Figure 6.8(f)].

The continuous increase in the SERS enhancements with decreasing pore size is contrary to the maximum SERS enhancement at a pore size of ~ 250 nm observed by Kucheyev *et al.* [Figure 6.9(a)].[140] The anomalous enhancement from large pore size most likely arises from the rough surfaces of gold ligaments [Figure 6.9(b) and (c)], which produce the so-called antenna effect for high SERS enhancement.[101,141] As shown in Figure 6.9, the Raman intensity of the porous sample with a smooth gold surface is almost 15 times weaker than that of 500 nm porous gold with a large number of surface irregularities. This is further verified by the SERS measurements of R6G molecules adsorbed on the 60 nm NPG decorated without and with Au nanoparticles [Figure 6.10(a)];[141] the surface morphologies of these are shown in Figure 6.10(b) and (c), respectively. The size of the deposited gold nanoparticles, 5–20 nm, is fairly comparable to the diameter of the pimples on the rough ligament surface (5–20 nm) in the experiment carried out by Kucheyev *et al.* Large SERS enhancement factors and size effects in the optical properties of NPG prepared by dealloying have also been observed by other groups. For example, Dixon *et al.* studied the SERS response of supported thin films of NPG and reported enhancement factors of up to 10^4. The observed SERS enhancement showed a complex dependence on the film thickness and dealloying time, which both affect the morphology of the material.[108]

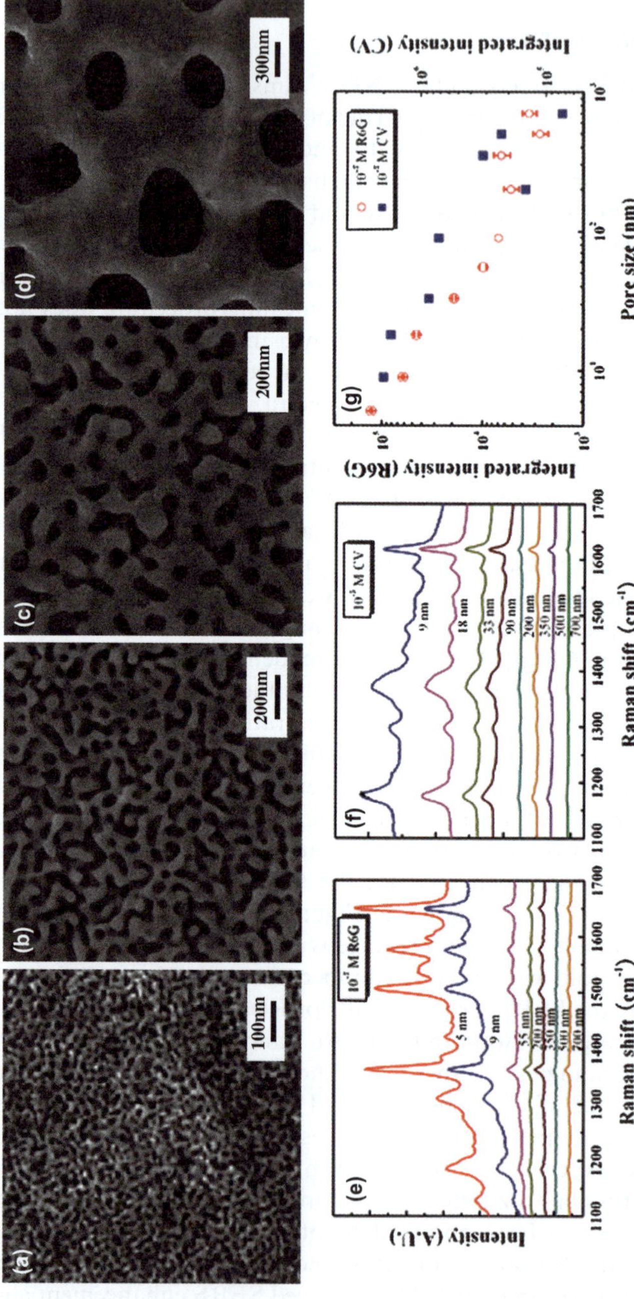

Figure 6.8 Representative SEM micrographs of nanoporous gold with various nanopore sizes: (a) nanoporous gold film after 5 min dealloying at room temperature; (b) dealloyed for 48 h at room temperature; (c) thoroughly rinsed nanoporous gold annealed at 200 °C for 2 h (d) at 500 °C for 2 h. SERS spectra of nanoporous gold with different pore sizes for (e) a 10^{-7} mol L^{-1} R6G aqueous solution and (f) 10^{-5} mol L^{-1} CV methanol solution. (g) Pore-size dependence of the integrated intensities of R6G and CV at Raman band of 1650 and 1175 cm^{-1}, respectively. Laser excitation: 514.5 nm for R6G and 632.8 nm for CV.[101]

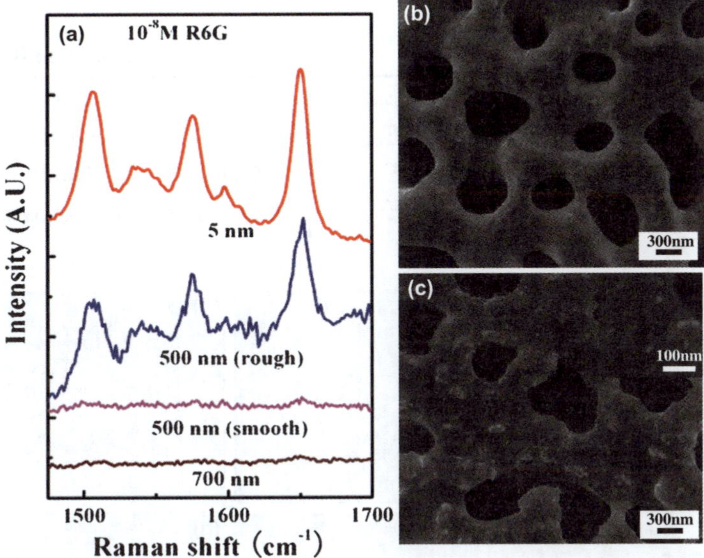

Figure 6.9 (a) SERS spectra of nanoporous gold with different ligament surface roughness for a 10^{-8} mol L^{-1} R6G aqueous solution. SEM images of 500 nm nanoporous gold with (b) smooth and (c) rough surfaces.[101]

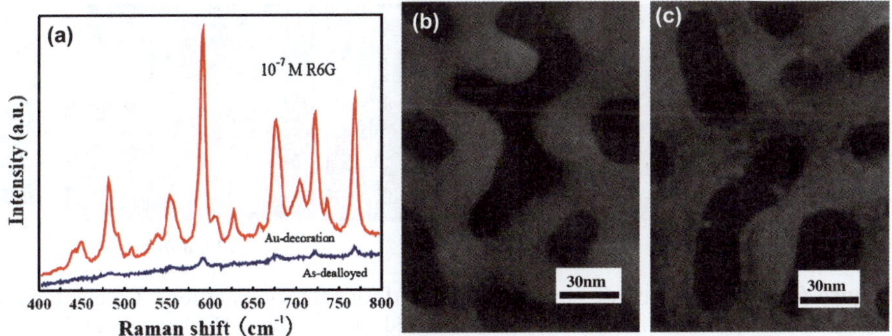

Figure 6.10 SEM micrographs of as-prepared nanoporous gold (a) without and with (b) the deposited gold nanoparticles. (c) Corresponding SERS spectra of 10^{-7} mol L^{-1} R6G molecules from the samples (a) and (b).[141]

To explore the origins of electromagnetic enhancement of NPG for the SERS effect, Lang *et al.* have systematically investigated the electromagnetic distribution in NPG using DDA simulation.[103] In view of the complicated structure of NPG, a simplified 2D nanoporous structure with identical pore and ligament sizes [Figure 6.11(a)] is introduced to qualitatively simulate the electromagnetic fields of NPG with different characteristic lengths. Under the plane wave with a wavelength of 514 nm propagating along the direction normal to

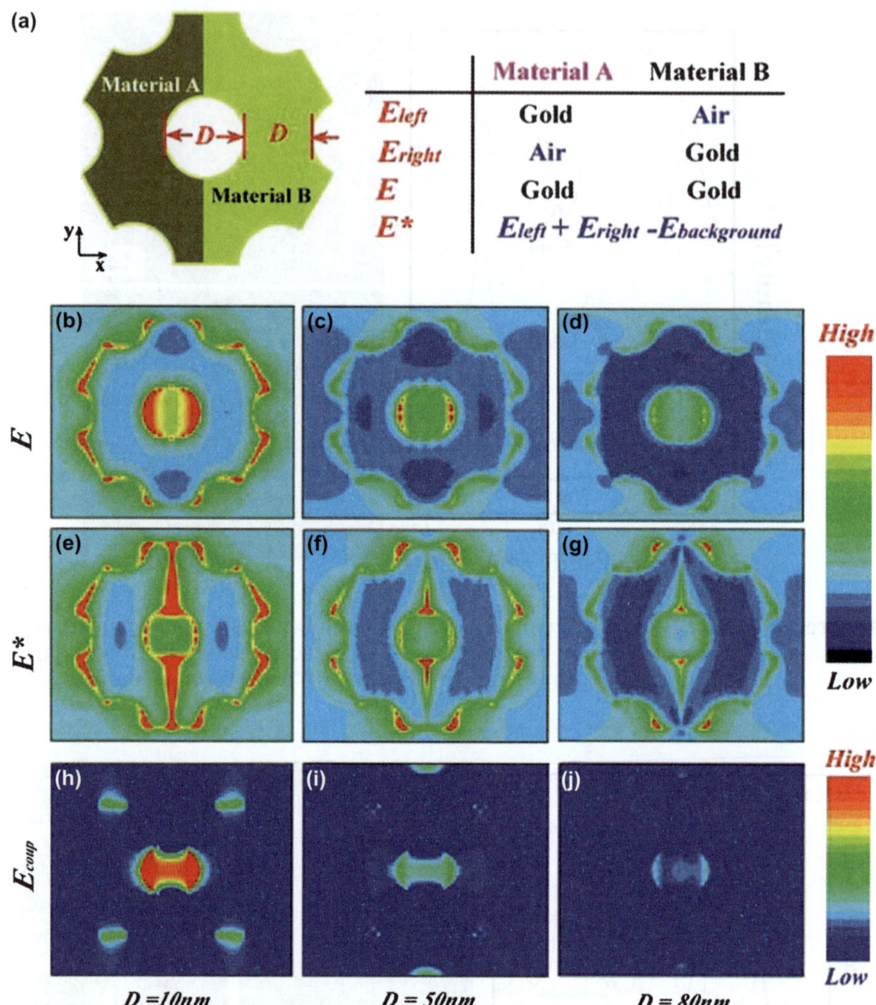

Figure 6.11 (a) Schematic nanostructure used for DDA simulations. Calculated total electric field (E) (b–d), localized electric field (E*) (e–g), and electromagnetic coupling electric field (E_{coup}) distributions (h–j) on the top surface of the NPG with $D = 10$, 50, and 80 nm, respectively. Incident light with a wavelength of 514 nm was polarized in the x direction.[103]

the top surface of NPG films, the total electric field distributions $[E(D)/E_0]$, calculated on the basis of the nanostructure shown in Figure 6.11(a), are illustrated in Figure 6.11(b)–(d) for NPG with $D = 10$, 50, and 80 nm, respectively, thus illustrating that $E(D)/E_0$ increases with decreasing characteristic lengths. This is qualitatively consistent with the experimental results. Here, $E(D)$ and E_0 denote the average electric fields inside the NPG films and the externally applied electric field, respectively. Figure 6.11(e)–(g) present the electric field distribution only contributed from gold ligaments $[E^*(D)/E_0]$, which is the linear sum of electric field distribution of the left [$A =$ gold and $B =$ air, $E_{\text{left}}(D)$] and right [$A =$ air and $B =$ gold, $E_{\text{right}}(D)$] parts of the nanostructure [Figure 6.11(a)]. Note that although the $E^*(D)/E_0$ located at the boundary between the left and the right parts is higher than that in other spots in the nanostructure because of the boundary effect, it does not influence to any great degree the evaluation of electric field distribution in the nanoporous channels. The difference between $E(D)/E_0$ and $E^*(D)/E_0$, *i.e.* $E_{\text{coup}}(D)/E_0 = E(D)/E_0 - E^*(D)/E_0$,[81] as a function of characteristic length is shown in Figure 6.11(h)–(j). The increase in electric field distributions of $E_{\text{coup}}(D)/E_0$ with decreasing pore size indicates that the electromagnetic coupling between neighboring ligaments becomes increasingly important as the pore size decreases to a few nanometers.[47,109] This theoretical result is also confirmed by the experimental investigations of SERS effect on the ruptured NPG with different protrusion densities.[141]

Therefore, the measured SERS intensity of NPG films [$I_{\text{SERS}}(D)$] consists of the contributions of $I_{\text{SERS,lig}}(D)$ from the localized electric field of individual gold ligaments and the electromagnetic coupling between neighboring ligaments [$I_{\text{SERS,coup}}(D)$],[81] which are determined by the sizes of ligament and pore sizes, respectively.[103] Consequently, $I_{\text{SERS,coup}}(D) = I_{\text{SERS}}(D) - I_{\text{SERS,lig}}(D)$. The relative SERS intensity induced by the EM coupling [$m_{\text{SERS,coup}}(D) = I_{\text{SERS,coup}}(D)/I_{\text{SERS}}$] is shown in Figure 6.12. Similarly to the optical properties of NPG influenced by a evanescent field,[107] $m_{\text{SERS,coup}}(D)$ increases dramatically with decreasing characteristic length by an exponential function, which can be fitted well by the scaling behavior of distance decay of plasmon coupling between metal nanoparticles,[76,142,143] $m_{\text{SERS,coup}}(D) = A\exp(-D/l)$, where A is a constant, and l denotes the decay length. From the fitting curve in Figure 6.12, $l = 6$ nm, which corresponds approximately to the decay length of the electromagnetic coupling between Au nanoparticles.[76,142,144,145] This qualitative agreement with other experimental results of assembled nanoparticles suggests that the electromagnetic

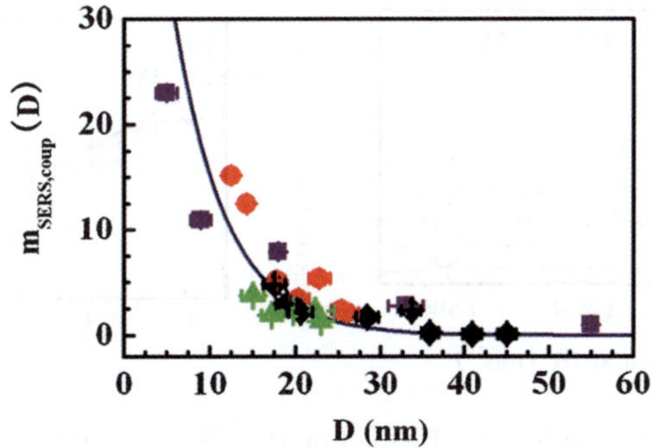

Figure 6.12   Relative SERS intensity induced by electromagnetic coupling $m_{\text{SERS,coup}}$ (D) as a function of the pore size of NPG films. The solid line is described by the exponential function $m_{\text{SERS,coup}}(D) = A\exp(-D/l)$ with $A = 81$ and $l = 6$ nm.[103]

coupling between neighboring ligaments gives rise to a large number of 'hot spots' in the quasi-periodic nanostructure when the nanopore size becomes small and thus results in reliable and uniform SERS spectra with a high degree of enhancement.[101,103,141]

Smaller nanopores yield stronger SERS enhancements because large curvatures and short interligament distances can enhance the localized electromagnetic fields. However, for NPG fabricated by conventional chemical dealloying, a small nanopore size is always accompanied by a small gold ligament, which gives rise to strong plasmon damping due to electron-surface scattering and thus weakens the near-field SPR coupling and SERS effect.[70] To improve further the SERS enhancement of NPG by optimizing the nanoporosity, Lang *et al.* tailored the nanoporosity by the combination of dealloying and electroless plating.[111] The NPG fabricated by this facile method has a tunable porosity with large gold ligaments and very small pores, which gives rise to much more significant improvements in SERS effect as well as structure rigidity and stability than the conventional NPG. Figure 6.13(a) and (b) present the SERS spectra of R6G and CV molecules adsorbed on gold-plated NPG films with different D/d ratios, respectively, demonstrating that as D decreases and d increases, the SERS enhancement is remarkably improved by an approximately exponential relationship with decreasing D and increasing d

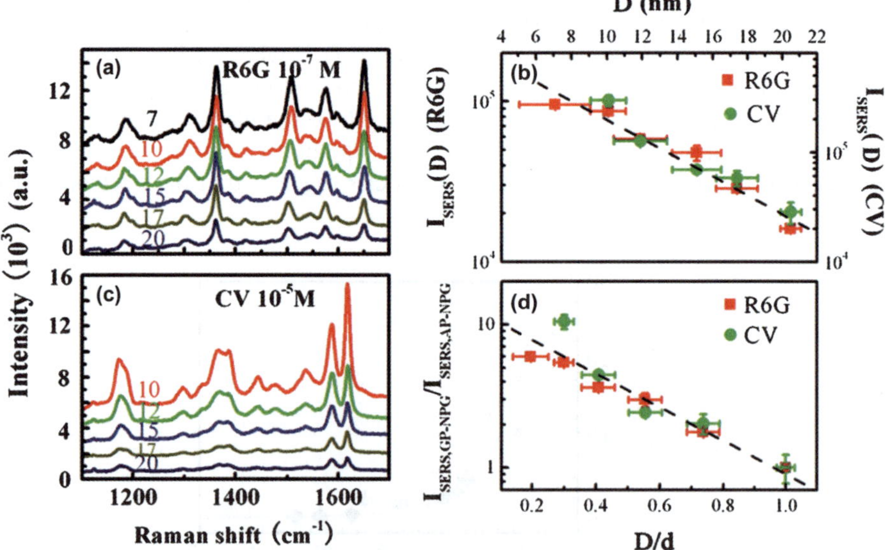

Figure 6.13   SERS spectra of (a) R6G and (b) CV molecules adsorbed on gold-plated NPG films with different D/d ratios. Laser excitation: 514.5 nm for both molecules. (c) Nanopore-size dependence of integrated SERS intensity of Raman bands of R6G and CV located at 1650 and 1175 cm^{-1}, respectively. (d) Normalized SERS enhancement of gold-plated NPG films as a function of the D/d ratios.[111]

[Figure 6.13(c)]. The detection limits for R6G and CV molecules are $\sim 10^{-10}$ and $10^{-8}\,\mathrm{mol\,l^{-1}}$, respectively, much better than the as-dealloyed NPG with identical nanopore and ligament sizes. Figure 6.13(d) illustrates the relative Raman intensity ($I_{\mathrm{SERS,GP\text{-}NPG}}/I_{\mathrm{SERS,AP\text{-}NPG}}$) as an exponential function of the ratio of D/d, where $I_{\mathrm{SERS,GP\text{-}NPG}}$ and $I_{\mathrm{SERS,AP\text{-}NPG}}$ denote the SERS intensities of gold-plated and as-prepared NPG, respectively. As shown in this plot, the SERS enhancements of gold-plated NPG are strongly dependent on not only D but also the ratio of D/d, very analogous to the general behavior of SERS enhancements in nanoparticle arrays, which is intrinsically dependent on the gap/diameter ratio due to the enhanced near-field coupling between nanoparticles. This is also verified by the near-field calculation by DDA, in which the targets with different D and d shown in Figure 6.14(a) are introduced for the calculations of distributions of electric fields. As shown in these plots, for the SERS effect of gold-plated NPG films, in addition to the localized electric field from small nanopores, the weakening of plasmon damping with increasing ligament sizes obviously enhances the near-field coupling between neighboring ligaments, which gives rise to the observed SERS improvements with the decrease in D/d ratios.

In addition to the features of morphology, the SERS enhancement of NPG is also dependent on the measurement temperature, which further verifies the electromagnetic field mechanism for the SERS effect of NPG.[103] Figure 6.15(a)–(c) presents the SERS spectra of R6G molecules adsorbed on NPG films with $D\approx d\approx 15$, 26, and 41 nm at temperatures ranging from ~ 80 to $300\,\mathrm{K}$, respectively. As shown in these plots, no peak shifts can be observed, which indicates that the bond nature of the R6G molecules does not change during the course of temperature variation, excluding the effect of strong chemical enhancement.[146–149] The integrated SERS intensities [$I_{\mathrm{SERS}}(T,D)$] as a function of temperature are plotted in Figure 6.15(d), from which we can see that for each NPG with the given characteristic length, the SERS intensity increases with decreasing temperature. According to the electromagnetic enhancement mechanism for SERS effect based on the quasi-static approximation,[13,139] when the SP resonance condition [$\varepsilon_1(\omega) = -2\varepsilon_{\mathrm{m}}$] is satisfied, the observed temperature dependence of a larger enhancement at lower temperatures was attributed to the fact that the imaginary part of the dielectric constant is an increasing function of temperature, i.e. $I_{\mathrm{SERS}}(T,D) \propto \{3\varepsilon_{\mathrm{m}}\omega^3/[\omega_{\mathrm{p}}^2\omega_{\mathrm{c}}(T,D)]\}^4$, or $I_{\mathrm{SERS}}(T,D) \propto 1/[\omega_{\mathrm{c}}(T,D)]^4$. Here, $\omega_{\mathrm{c}}(T,D)$ is the temperature and characteristic length-dependent collision frequency of electrons, which consists of two parts that arise from the phonon–electron scattering [$\omega_{\mathrm{c}}(T)$][148,150–152] and the electron mean free path effect [$\omega_{\mathrm{c}}(D)$][4,153] in the light of Matthiessen's rule, namely, $\omega_{\mathrm{c}}(T,D) = \omega_{\mathrm{c}}(T) + \omega_{\mathrm{c}}(D)$. Based on the simple Debye model for phonon spectrum, $\omega_{\mathrm{c}}(T)$ is derived by Holstein in the limit of $E_{\mathrm{F}} \gg \hbar\omega \gg k_{\mathrm{B}}T$ and $k_{\mathrm{B}}\theta_{\mathrm{D}}$,[148,150–152] $\omega_{\mathrm{c}}(T) = \omega_0\,2/5 + 4\left[(T/\Theta_{\mathrm{D}})^5\int_0^{\theta_D/T} z^4\,\mathrm{d}z/(e^z - 1)\right]$, where E_{F} is the Fermi energy, k_{B} is Boltzmann's constant, and Θ_{D} is the Debye temperature. The 2/5 term in this equation is due to spontaneous emission of a phonon by a photon-excited electron, while the temperature-dependent term corresponds to phonon adsorption

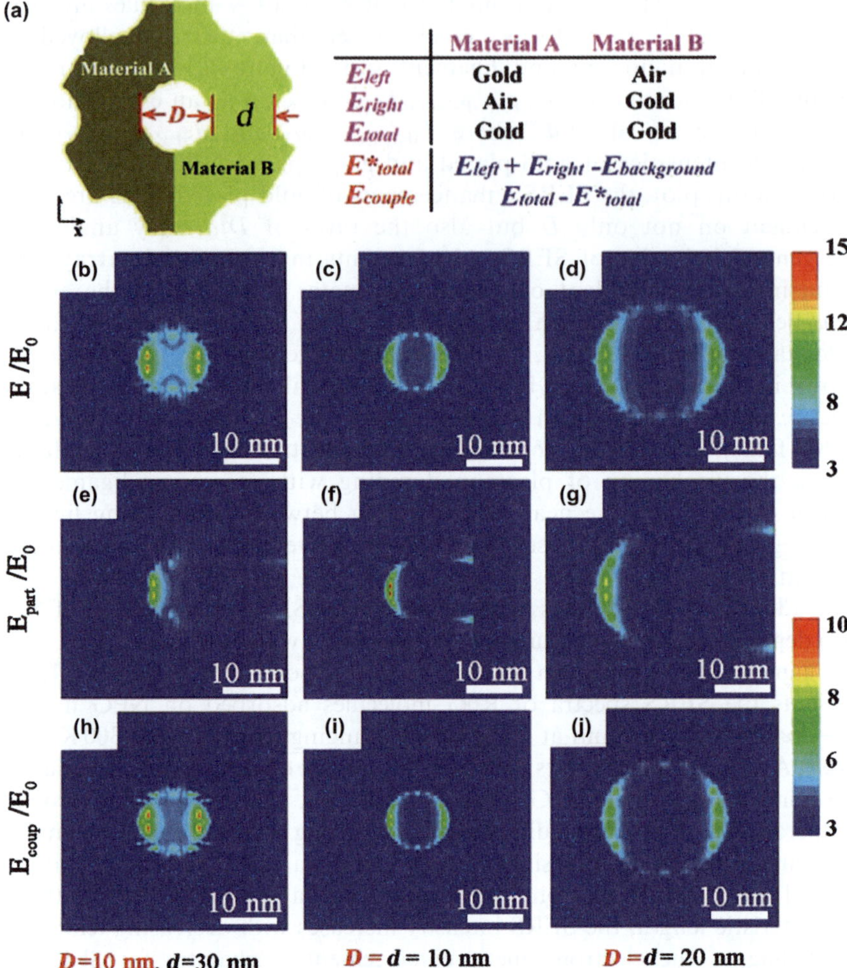

	Material A	Material B
E_{left}	Gold	Air
E_{right}	Air	Gold
E_{total}	Gold	Gold
E^*_{total}	$E_{left} + E_{right} - E_{background}$	
E_{couple}	$E_{total} - E^*_{total}$	

Figure 6.14 (a) Schematic nanostructure used for DDA simulations. Calculated electric-field distributions: (b) total electric field (E/E_0), (c) electric field of half hole (E_{part}/E_0), (d) electric field contributed from electromagnetic coupling (E_{coup}/E_0) for $D = 10$ nm and $d = 30$ nm; the distributions of (e) E/E_0, (f) E_{part}/E_0 and (g) E_{coup}/E_0 for $D = d = 10$ nm; and the distributions of (h) E/E_0, (i) E_{part}/E_0 and (j) E_{coup}/E_0 for $D = d = 20$ nm. The incident light of 514 nm was polarized in the x direction.[111]

and stimulated-emission processes. ω_0 is a constant that can be determined independently from the known value of the dc conductivity (σ_0). Since, for the dc case, $\hbar\omega \ll k_B T$ and $k_B\theta_D$, the Holstein expression becomes $\omega_c(T,\omega \to 0) = \omega_p^2/(4\pi\sigma_0) = \omega_0\left\{4(T/\Theta_D)^5 \int_0^{\theta_D/T} z^5 \, dz/[(e^z - 1)(1 - e^{-z})]\right\}$.[148,150–152] On the other hand, for an NPG with D smaller than the electronic mean free path l_b of bulk Au, the effective mean free path l_{eff} is dominated by collisions with the surfaces of gold

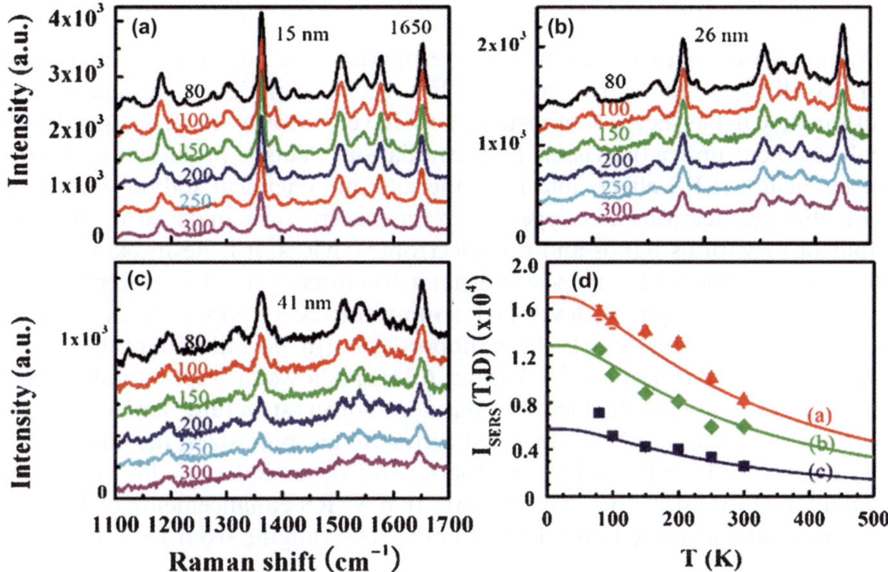

Figure 6.15 SERS spectra of R6G (10^{-7} mol L^{-1}) adsorbed on NPG films with characteristic lengths of (a) 15, (b) 26, and (c) 41 nm at temperatures of 80, 100, 150, 200, 250, and 300 K, respectively. (d) Characteristic length and temperature dependence of the integrated intensity of the 1650 cm^{-1} Raman band of R6G molecules on NPG according to the plot: (a) 15 (▲), (b) 26 (♦), and (c) 41 nm (■). The solid lines denote the predictions of $I_{SERS}(T,D)$ in terms of eqn (6.2) with $I_{SERS}(T_0,D) =$ 8178, 5958, and 2595 for 15, 26, and 41 nm NPG films, respectively, $\omega_0 = 2.0477 \times 10^{13}$ rad s^{-1}, $\Theta_D = 170$ K, $v_F = 1.4 \times 10^6$ m s^{-1}, $l_b = 47.5$ nm, $T_0 = 300$ K. [103]

ligaments. As a result, $\omega_c(D)$ can be given by $\omega_c(D) = 8v_F/(3l_{eff})$.[4,154] When taking the room temperature T_0 as a reference point, the scaling relationship $I_{SERS}(T,D)/I_{SERS}(T_0,D) = [\omega_c(T_0,D)/\omega_c(T,D)]^4$ quantitatively describes the experimental results, as shown in Figure 6.15(d). The good consistence indicates that the gradual increase in $I_{SERS}(T,D)$ with decreasing T results from the reduced effect of electron-phonon scattering. While T drops to ~ 30 K, the $I_{SERS}(T,D)$ approaches a constant, similar to the inverse of temperature dependence of the electronic resistivity of bulk and nanostructured Au,[154,155] which implies the underlying correlation between SERS effect and the natural properties of the electron transport in NPG. The effective mean free path decreases with D as a result of the inner surface scattering of NPG films, giving rise to the decrease in $\varepsilon_2(\omega,T,D)$ and thus the decrease in EM enhancements. This makes it reasonable that the degree of increment of $I_{SERS}(T,D)$ with decreasing T is strongly dependent on the characteristic length of NPG films. For instance, the $I_{SERS}(T = 80$ K$,D)$ is larger than the $I_{SERS}(T = 300$ K$,D)$ by factors of ~ 1.9, 2.0, and 2.7 for $D \approx 15$, 26, and 41 nm, respectively.

6.4.2.2   *Nanoporous Copper*

In comparison with the extensive investigations on gold nanostructures, the SERS effect of nanostructured Cu has been much less investigated,[156,157] although Cu has been demonstrated to have a strong electromagnetic field enhancement under the visible light irradiation.[4] This is probably due to their insignificant SERS effects and poor thermal and chemical stability. Given the potential applications of copper as an inexpensive raw material in the development of various nano- and optoelectronic devices, it is desirable to explore the characteristics of any possible nanostructured Cu. More recently, Chen *et al.* have successfully developed nanoporous Cu (NPC) by dealloying a $Cu_{30}Mn_{70}$ alloy, for which the nanoporosity can be tailored by controlling the dealloying conditions.[92] Figure 6.16(a) shows a representative SEM micrograph, showing the bicontinuous nanoporous structure, similar to that of NPG. On the basis of NPC obtained with different pore sizes, the authors have systematically investigated the nanopore size dependence of the SERS effect [Figure 6.16(b)–(d)]. This demonstrates that SERS enhancements of R6G and CV molecules adsorbed on NPC with pore sizes ranging from 15 to 118 nm are

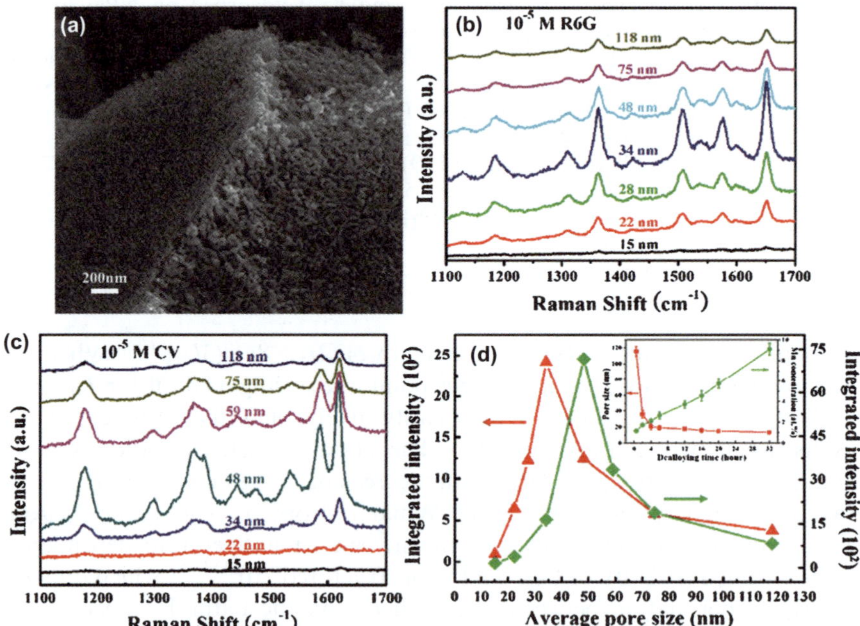

Figure 6.16   (a) Representative cross-section SEM image of nanoporous Cu produced by dealloying $Cu_{30}Mn_{70}$ ribbons in HCl solution. SERS spectra of nanoporous Cu with different pore size for (b) 10^{-5} mol l^{-1} R6G aqueous solution and (c) 10^{-5} mol l^{-1} CV methanol solution. Excitation: 514.5 nm laser. (d) Integrated intensities of R6G and CV at the Raman bands of 1650 and 1625 cm^{-1} as a function of pore size. Inset: relationship between nanopore size and residual Mn concentration with dealloying time.[92]

remarkably dependent on the characteristic lengths. The SERS intensities start to increase with pore size decreasing from 118 nm and reach their maxima at nanopore sizes of ~34 nm and ~48 nm for R6G and CV, respectively. While the pore size is further decreased, the SERS intensities begin to reduce evidently. The small difference in nanopore size corresponding to the maximum enhancements between R6G and CV is most likely caused by fluctuations in the characteristic lengths of NPC samples or possible variations in the electromagnetic enhancements for different organic molecules.[92]

As shown in Figure 6.16(d), the peak values of SERS enhancements for both R6G and CV molecules appear at the nanopore size of about 30 to 50 nm, which is different from that of NPG from which the strongest SERS effect is achieved at the smallest nanopore size of ~5 nm.[101] This is probably attributed to the fact that small nanopore sizes are usually fabricated by a short dealloying time or at low dealloying temperatures and thus contain a certain amount of residual dealloyed components, such as Mn in NPC. The SERS inactivity of element Mn may degrade the SERS enhancements of NPC with smaller nanopore sizes but a higher concentration of the residual Mn [inset of Figure 6.16(d)]. Thus, the maximum enhancement occurring at the nanoporous size of about 30–50 nm most likely results from the balance of the two effects, *i.e.* smaller pore sizes improve SERS enhancements, whereas a higher residual Mn in NPC decreases the enhancements.

6.4.2.3 Nanoporous Core-Shell Au–Ag, Cu–Ag Bimetals

It is well known that the hybridization effect arising from core-shell bimetallic nanostructures can effectively enhance the functionality in optical,[31,158,159] magnetic,[160] and catalytic[161,162] properties in comparison with their corresponding monometal counterparts.[25,26,163,164] This has stimulated recent developments of a variety of core-shell bimetallic nanostructures by many routes, such as electroless plating,[165] surface precipitation reaction,[166] surface seeding,[167] and self-assembly.[168] Of these, NPMs with various physicochemical properties have been exploited as a support substrate to synthesize 3D porous core-shell nanostructures.[114,169–171] The incorporation of secondary metals onto inner surface of nanoporous precursors significantly modifies the optical properties, and thus depresses or enhances the performance of porous core-shell bimetals for SERS enhancement, depending on whether or not the incorporated composition is active for SERS effect.[114,171] This is verified by the fact that the SERS enhancements of NPG and NPC with small pore sizes are improved and depressed by a certain amount of residual elements, such as Ag in NPG and Mn in NPC, respectively.[114,171]

It is well known that silver is used most commonly in SERS study because its *d–s* band gap is in the UV region and does not dampen the plasmon mode as strongly as for gold.[172] Recently, on the basis of the developed NPG films, Qian *et al.* have fabricated porous core-shell Au–Ag nanoarchitectures with the tunable thickness of Ag shell by a modified gas–liquid electroless plating method.[113,114] Figure 6.17(a) shows the representative transmission electron microscope (TEM)

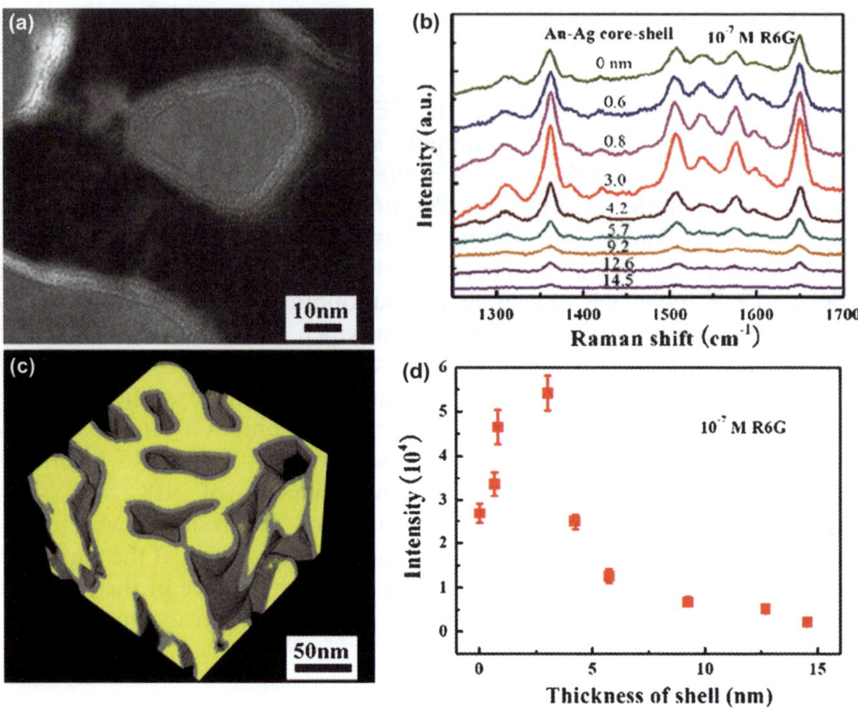

Figure 6.17   (a) Representative TEM micrograph and (b) 3D image of nanoporous core-shell Au–Ag structure. (c) SERS spectra and (d) their integrated intensities of R6G molecules adsorbed on nanoporous core-shell Au–Ag substrates with different thickness of Ag shell.[114]

microstructure, illustrating that the silver shell uniformly covers the internal surface of NPG. The core-shell structure can be clearly indentified by the contrast difference between bright silver shells and dark gold cores. The thickness of the shell ranges from 0.6 to 14.5 nm, controlled by the electroless plating time. Figure 6.17(b) presents the schematic diagram of the Ag@NPG nanoporous bimetals. The SERS spectra of R6G adsorbed onto the porous core-shell Au–Ag films with different shell thicknesses are plotted in Figure 6.17(c). In comparison with the NPG uncoated with Ag shell, an improvement by a factor of 2 in the Raman band intensities can be observed as the thickness of the silver shell increases from 0 to 3 nm. While the thickness of the Ag layer further increases to 14.5 nm, the SERS intensities decrease to about 1/20 of that of Ag@NPG with an Ag shell thickness of 3 nm, which is even lower than that of as-prepared NPG [Figure 6.17(d)]. The dramatic dependence of SERS intensities of Ag@NPG on the thickness of the Ag shell is attributed to the competition effect between chemical and electromagnetic enhancements. On the one hand, when a small amount of Ag is deposited onto NPG film, the absorption band of Ag@NPG does not change remarkably in comparison with that of as-prepared NPG films. This does not give rise to the large difference in electromagnetic fields in the Ag@NPG and

NPG films excited by the laser with the same wavelength. Therefore, the improvement in SERS intensities with Ag thickness increasing from 0 to 3 nm results from the chemical enhancement because silver itself has a stronger SERS enhancement than gold in the visible region. The evaluation of the chemical enhancement factor of 1.8 according to the theoretical model is quantitatively consistent with the experimental results shown in Figure 6.17(d). On the other hand, the incorporation of much more silver into the inner surface of NPG results in a deviation in the plasmon band from ~510 nm resonant with the laser wavelength (514.5 nm) to ~350 nm non-resonant with respect to 514.5 nm excitation, thus strongly depressing the electromagnetic enhancement. This result suggests a primary role of the electromagnetic field for total SERS enhancement.

While considering that NPG itself is composed of one of the most expensive precious metals, Chen *et al.* were motivated to develop a nanoporous non-noble metal substrate (Cu)[92] that they can easily decorate with a thin layer of noble metals for optical applications. Although NPC exhibits the highest SERS enhancement factor of 1.85×10^5, it is much lower than that of Ag nanostructures. To further improve the performance of NPC, they have incorporated a silver layer into the NPG skeleton using a spontaneous displacement reaction during which Cu acts as a reduction agent and template.[171] Figure 6.18(a)

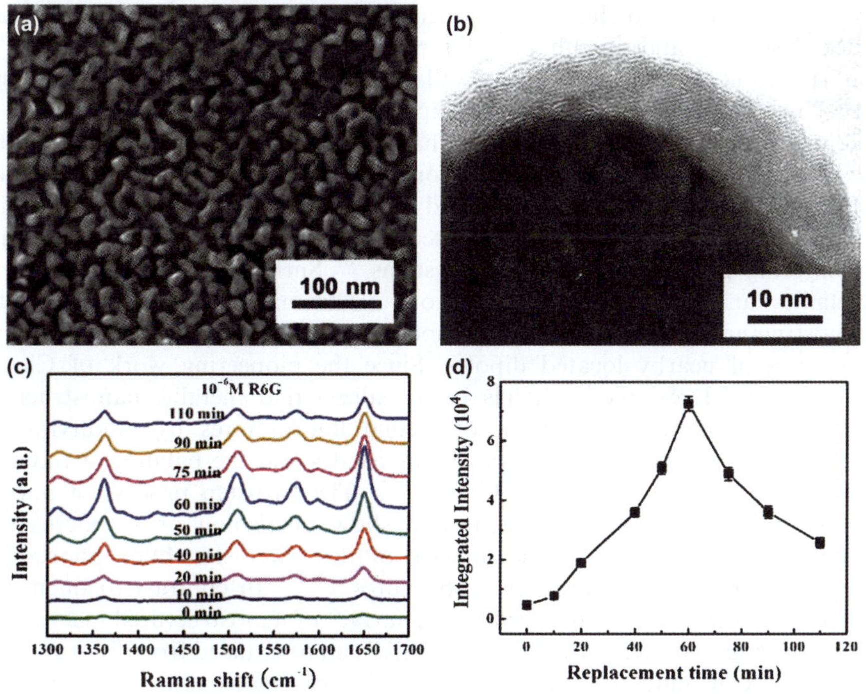

Figure 6.18 Representative (a) SEM and (b) TEM images of Ag@NPC plating for 60 min. (c) SERS spectra and (d) integrated intensity of 1650 cm^{-1} Raman band of R6G (10^{-6} mol L^{-1}) adsorbed on silver-decorated NPC with different displacement reaction times. Excitation: 514.5 nm laser.[171]

illustrates the representative top-view SEM image of Ag@NPC nanoporous composites fabricated in an $Ag(NH_3)_2NO_3$/PVP electrolyte for 60 min. The different contrast in the high-resolution TEM (HRTEM) micrograph clearly demonstrates the core-shell structure [Figure 6.18(b)] and the shell thickness of ~5–8 nm. Because the thickness of the Ag shell increases with increasing displacement time,[171] the SERS enhancement of the nanoporous Ag@NPC composite displays depends on the displacement time or the thickness of Ag shell, analogously to the observations in the case of Ag@NPG bimetals. As shown in Figure 6.18(c), significant improvements in the Raman bands can be observed with Ag decoration, and the SERS intensities of Ag@NPC with the reaction time up to 60 min can be improved by a factor of 16 in comparison with that of the as-prepared NPC [Figure 6.18(d)]. The dramatic improvement results from the synergistic effects between silver layers and the reduced nanopore size.

6.4.3   Nanoporous Plasmon-Enhanced Fluorescence

Since Purcell's first suggestion in 1946 that spontaneous emission could be modified by resonant coupling to the external electromagnetic environment,[173] a wealth of fundamental research and technological innovation have arisen based on this effect, including a number of SP-based fluorescence technologies, such as surface- and tip-enhanced fluorescence,[174–176] SP-enhanced LEDs,[11,12] and photonic crystal lasers,[177] as well as plasmon resonance energy transfer (PRET),[178] *etc.* As one of the principal detection/sensing methodologies in biological and chemical science, SEF has attracted great attention because the ability to increase the emission of fluorophores can dramatically improve the effectiveness of the fluorophore-based applications in fields,[174–176] including single-molecule detection,[179] DNA sequencing,[180] medical diagnostics, and monitoring processes in biological systems.[181] Surface plasmon excitation in metallic nanostructures generates strong electromagnetic fields close to the nanostructure surface, significantly modifying the non-radiative and radiative processes of nearby located dipoles. Since the pioneering work of Chance *et al.*[182] and Drexhage,[183] it has been realized that metallic nanostructures influence spontaneous emission of vicinal fluorophores by modifying the intensity of excitation incident on the molecules and the balance of radiative and non-radiative decay rates.[174–176,184,185] The two-step process of fluorescence enables fluorescent molecules to show quenching or an increase in fluorescence intensity due to the interplay of the excited-state fluorophores with free electrons in the metallic nanostructures.[186–188] In contrast to quenching from molecules directly adsorbed on the surface of nanostructured metals from the non-radiative energy transfer,[184,185] SEF can be realized from fluorophores, such as semiconductor nanocrystals,[189–192] dye molecules,[193–196] or light-harvesting complexes,[197] when they are placed at certain distances from metallic nanostructures without direct contact.[198]

The plasmon enhancement effect has spurred many efforts in fabricating metallic nanostructures with high electric fields or 'hot spots'. Among the many types of SEF substrates available today, free-standing nanoporous structures exhibit promising applications as SEF active-substrates due to their fascinating optical properties as mentioned earlier. Recently, Biteen *et al.* first observed photoluminescence (PL) enhancement by coupling nanocrystal Si with NPG under the light excitation at 488 nm.[192] The enhanced PL intensity was shown to depend on the NPG–Si nanocrystal separation distance. While the separation distance is varied between 0 and 20 nm, a maximum fourfold luminescence intensity enhancement was obtained, which is attributed to increasing absorption cross-section and effective quantum yield due to the near-field effect of NPG.[192]

The fluorescence enhancement of nanoporous metals relies not only on the distance between metal and dipole but also on the near-field intensity generated from the excitation of SP. More recently, Lang *et al.* demonstrated the underlying relationship on the investigation of the SEF effect of NPG films with different characteristic lengths. Figure 6.19(a) shows a representative scanning tunneling microscope image of NPG films, illustrating the nanostructure with the uniform distributions of nanopore channels and gold ligaments. Using an HSA monolayer ($\sim$3–8 nm) as the spacer layer,[199,200] the

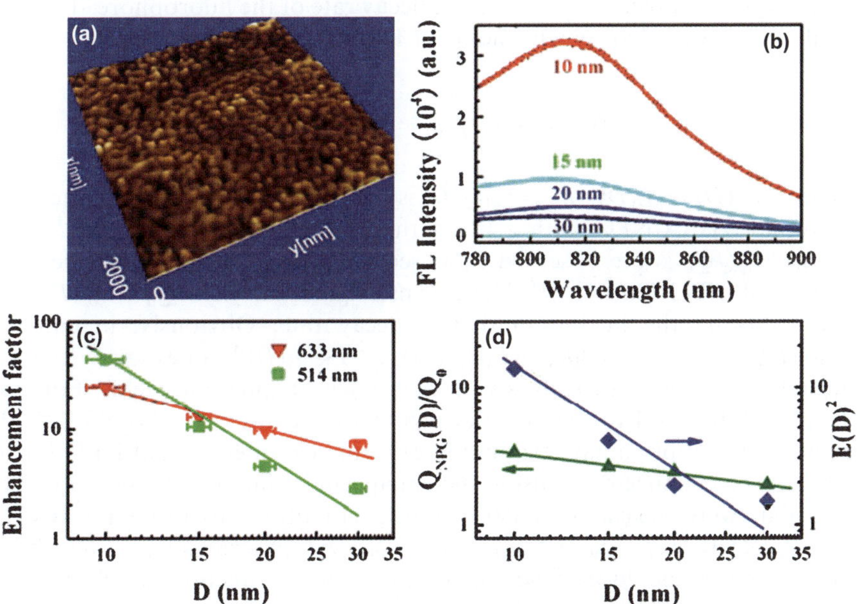

Figure 6.19 (a) Representative STM image of NPG. (b) Fluorescence emission spectra of ICG conjugated to HSA-coated NPG films with different characteristic lengths (D) under a laser excitation of 514.5 nm. (c) Fluorescence enhancement factor of ICG/HSA/NPG as a function of D under laser excitations of 514.5 nm and 632.8 nm. (d) Characteristic length dependence of near-field ($|E(D)|^2$) and quantum-yield enhancements ($Q_{\mathrm{NPG}}(D)/Q_0$) of ICG/HSA/NPG films under the illumination of 514.5 nm.

enhancement in fluorescence emission of indocynanine green (ICG) nearby NPG films is dramatically dependent on the characteristic lengths of NPG [Figure 6.19(b)]. Figure 6.19(c) shows the coverage corrected enhancement factor of SEF for ICG as a function of nanopore size (D), $m_{NPG}(D) = I_{NPG}(D)/I_0$, with $I_{NPG}(D) = I_f(D)/N(D)$ and $I_0 = I_{f0}/N_0$, where $I_f(D)$ and I_{f0} denote the fluorescence intensities measured from an area covered with a fluorophore/spacer layer on the NPG and glass sheet, respectively, and $N(D)$ and N_0 are the numbers of fluorescent molecules in the given beam area (S_0). Because of the coverage of the HSA monolayer, the thickness of NPG films covered by the fluorophore is assumed to be D. This gives rise to $N(D)/N_0 \approx [S_0 D R(D)/A_0]/(S_0/A_0) = DR(D) \approx 2.36$, 3.08, 3.62, 3.87 with a surface-to-volume ratio $R(D)$ of ≈ 0.236, 0.205, 0.181, 0.129 nm^2 nm^{-3} for NPG films with $D \approx 10$, 15, 20, 30 nm. Here, A_0 is the area of the fluorescent molecule.

According to the two-step process of fluorescence emission, the NPG excited by the incident light (I_{exc}) generates a strong electromagnetic field, $|E(D)|^2$, which not only increases the intensity of light absorbed by the molecules according to $I_{exc\text{-}NPG}(D) = |E(D)|^2 I_{exc}$,[174,185,187,201] but also improves the effective quantum yield of fluorophore, $Q_{NPG}(D)$, by increasing the effective radiative decay rate, $\Gamma(D)$, leading to $Q_{NPG}(D) = \Gamma(D)/[\Gamma(D) + k_{nr}]$,[174,185,201] where k_{nr} is the intrinsic non-radiative decay rate of the fluorophore. Therefore, the fluorescence enhancement factor of HSA/NPG films is given by

$$m_{NPG}(D) = I_{NPG}(D)/I_0 = |E(D)|^2 [Q_{NPG}(D)/Q_0], \tag{6.6}$$

where $I_{NPG}(D) = |E(D)|^2 I_{exc} \varepsilon Q_{NPG}(D)$ is the observed fluorescence intensity of fluorophore/HSA/NPG with ε being the absorptivity of molecules,[174,185,187] and $I_0 = I_{exc} \varepsilon Q_0$ is the observed fluorescence intensity of the molecules in the absence of NPG films.[185,187] The intrinsic quantum yield $Q_0 = \Gamma_0/(\Gamma_0 + k_{nr})$, with Γ_0 being the intrinsic radiative decay rate. Obviously, in eqn (6.6), $Q_{NPG}(D)/Q_0 = \{\Gamma_{NPG}(D)[k_{nr} + \Gamma_0]\}/\{\Gamma_0[k_{nr} + \Gamma_{NPG}(D)]\}$ does not give significant enhancements for dyes with high intrinsic quantum yields. For example, $Q_{NPG}(D)/Q_0 \approx 1$ for R6G with the intrinsic quantum yield of 95%.[202] This facilitates the approximate evaluation of local field intensities of NPG for ICG fluorescence enhancements using the enhancement factors of R6G/HSA/NPG films that were measured under identical conditions to those for ICG, *i.e.* $m_{NPG\text{-}R6G}(D) = I_{NPG\text{-}R6G}(D)/I_{0\text{-}R6G} \approx |E_{NPG}(D)|^2$ in terms of eqn (6.6). Figure 6.19(d) shows the local field enhancement as a function of characteristic lengths. In comparison with the small improvement in quantum yield of ICG, which is evaluated by the ratio of $m_{NPG\text{-}ICG}(D,d)/m_{NPG\text{-}R6G}(D,d)$ [Figure 6.19(d)], the enhanced absorption of fluorescent molecules due to the near-field enhancement of NPG plays a more important role in the dramatically improved fluorescence enhancement. This is verified by the fact that under the 632.8 nm laser irradiation, the SEF enhancement of ICG/HSA/NPG films shows a relatively weak dependence on D [Figure 6.19(c)] because the SP bands of ICG/HSA/NPG films shift away from the excitation wavelength of 632.8 nm with decreasing nanopore sizes [Figure 6.7(a)]. These results imply the simple relationship between the significant fluorescence enhancement and electric field intensity, which is intrinsically determined by the excitation of SP in nanoporous metals.

6.5 Concluding Remarks

Nanoporous metals exhibit localized and propagating SPR properties due to their unique three-dimensional, quasi-periodic nanoporosity. In this chapter, we have reviewed the fundamental SPR properties of nanoporous metals as well as their applications in surface-enhanced spectroscopy for biological and chemical sensing. Recent experiments have demonstrated that the spectral location of localized SPR and its sensitivity to changes of refractive index rely on the characteristic lengths of nanoporous metals. Moreover, the strong electromagnetic field generated by the resonance excitation of SP in nanoporous metals results from the confluence of the localized electromagnetic field and electromagnetic coupling between face-to-face ligaments, which are determined by the diameters of metal ligaments and porous channels, respectively. Understanding the optical properties of nanoporous metals is of great importance to develop new porous structures with improved optical performances. As new plasmonic materials, nanoporous metals with superior reproducibly, facile synthesis and excellent stability may open up new avenues for a wide range of applications in life sciences, environmental protection, and green energy innovation.

References

1. C. Q. Sun, *Prog. Solid State Chem.*, 2007, **35**, 1.
2. Q. Jiang and H. M. Lu, *Surf. Sci. Rep.*, 2008, **63**, 427.
3. W. P. Halperin, *Rev. Mod. Phys.*, 1986, **58**, 533.
4. U. Kreibig and M. Vollmer, *Optical Properties of Metal Cluster*, Springer, Berlin, 1995.
5. A. I. Gusev and A. A. Rempel, *Nanocrystalline Materials*, Cambridge International Science Publishing, Cambridge, 2004.
6. A. R. Tao, S. Habas and P. D. Yang, *Small*, 2008, **4**, 310.
7. A. R. Tao, J. X. Huang and P. D. Yang, *Acc. Chem. Res.*, 2008, **41**, 1662.
8. Y. N. Xia, Y. J. Xiong, B. Lim and S. E. Skrabalak, *Angew. Chem. Int. Ed.*, 2009, **48**, 60.
9. R. Zia, J. A. Schuller, A. Chandran and M. L. Brongersma, *Mater. Today*, 2006, **9**, 20.
10. S. Lal, S. Link and N. J. Halas, *Nat. Photonics*, 2007, **1**, 641.
11. J. Vučković, M. Lončar and A. Scherer, *IEEE J. Quantum Electron.*, 2000, **36**, 1131.
12. K. Okamoto, I. Niki, A. Shvartser, Y. Narukawa, T. Mukai and A. Scherer, *Nat. Mater.*, 2004, **3**, 601.

13. M. Moskovits, *Rev. Mod. Phys.*, 1985, **57**, 783.
14. J. N. Anker, W. P. Hall, O. Lyandres, N. C. Shan, J. Zhao and R. P. van Duyne, *Nat. Mater.*, 2008, **7**, 442.
15. M. E. Stewart, C. R. Anderton, L. B. Thompson, J. Maria, S. K. Gray, J. A. Rogers and R. G. Nuzzo, *Chem. Rev.*, 2008, **108**, 494.
16. Z. Q. Tian, B. Ren and D. Y. Wu, *J. Phys. Chem. B*, 2002, **106**, 9463.
17. P. K. Jain, X. H. Huang, I. H. El-Sayed and M. A. El-Sayed, *Acc. Chem. Res.*, 2008, **41**, 1578.
18. L. R. Hirsch, R. J. Stafford, J. A. Bankson, S. Sershen, B. Rivera, R. E. Price, J. D. Hazle, N. J. Halas and J. West, *Proc. Natl. Acad. Sci. U. S. A.*, 2003, **100**, 13549.
19. H. Raether, *Surface Plasmons on Smooth and Rough Surfaces and on Gratings*, Springer, Berlin, 1986.
20. A. V. Zayats and I. I. Smolyaninov, *J. Opt. A: Pure Appl. Opt.*, 2003, **5**, S16.
21. J. Homola, *Chem. Rev.*, 2008, **108**, 462.
22. S. Coyle, M. C. Netti, J. J. Baumberg, M. A. Ghanem, P. R. Birkin, P. N. Bartlett and D. M. Whittaker, *Phys. Rev. Lett.*, 2001, **87**, 176801.
23. Y. Sugawara, T. A. Kelf, J. J. Baumberg, M. E. Abdelsalam and P. N. Bartlett, *Phys. Rev. Lett.*, 2006, **97**, 266808.
24. R. M. Cole, J. J. Baumberg, F. J. Garcia de Abajo, S. Mahajan, M. Abdelsalam and P. N. Bartlett, *Nano Lett.*, 2007, **7**, 2094.
25. M. Sastry, A. Swami, S. Mandal and P. Selvakannan, *J. Mater. Chem.*, 2005, **15**, 3161.
26. J. P. Wilcoxon and B. L. Abrams, *Chem. Soc. Rev.*, 2006, **35**, 1162.
27. R. C. Jin, Y. W. Cao, C. A. Mirkin, K. L. Kelly, G. C. Schatz and J. G. Zheng, *Science*, 2001, **294**, 1901.
28. M. Rang, A. C. Jones, F. Zhou, Z. Y. Li, B. J. Wiley, Y. N. Xia and M. B. Raschke, *Nano Lett.*, 2008, **8**, 3357.
29. C. M. Cobley, D. J. Campbell and Y. N. Xia, *Adv. Mater.*, 2008, **20**, 748.
30. D. S. Kim, J. Heo, S. H. Ahn, S. W. Han, W. S. Yun and Z. H. Kim, *Nano Lett.*, 2009, **9**, 3619.
31. J. Y. Chen, B. Wiley, J. McLellan, Y. J. Xiong, Z. Y. Li and Y. N. Xia, *Nano Lett.*, 2005, **5**, 2058.
32. C. L. Nehl, H. W. Liao and J. H. Hafner, *Nano Lett.*, 2006, **6**, 683.
33. M. Yamamoto, Y. Kashiwagi, T. Sakata, H. Mori and M. Nakamoto, *Chem. Mater.*, 2005, **17**, 5392.
34. A. M. Funston, C. Novo, T. J. Davis and P. Mulvaney, *Nano Lett.*, 2009, **9**, 1651.
35. T. Shengai, Z. P. Li, T. Dadosh, Z. Y. Zhang, H. X. Xu and G. Haran, *Proc. Natl. Acad. Sci. U. S. A.*, 2008, **105**, 16448.
36. M. A. Mahmoud and M. A. El-Sayed, *Nano Lett.*, 2009, **9**, 3025.
37. M. Grzelczak, J. Pérez-Juste, P. Mulvaney and L. M. Liz-Marzán, *Chem. Soc. Rev.*, 2008, **37**, 1783.
38. S. Lal, N. K. Grady, J. Kundu, C. S. Levin, J. B. Lassiter and N. J. Halas, *Chem. Soc. Rev.*, 2008, **37**, 898.

39. W. A. Murray and W. L. Barnes, *Adv. Mater.*, 2007, **19**, 3771.
40. G. Mie, *Ann. Phys.*, 1908, **25**, 377.
41. E. K. Miller, *J. Electromagn. Waves Appl.*, 1994, **8**, 1125.
42. B. T. Braine and P. J. Flatau, *J. Opt. Soc. Am. A*, 1994, **11**, 1491.
43. *User Guide for the Discrete Dipole Approximation Code DDSCAT 7.0*, http://arxiv.org/abs/0809.0337.
44. J. Alegret, P. Johansson and M. Käll, *New J. Phys.*, 2008, **10**, 105004.
45. C. H. Hafner and R. Ballist, *Int. J. Comp. Elect. Elect. Eng.*, 1983, **2**, 1.
46. F. J. Garcĭa de Abajo and A. Howie, *Phys. Rev. Lett.*, 1998, **80**, 5180.
47. T. Rindzevicius, Y. Alaverdyan, B. Sepulveda, T. Pakizeh, M. Käll, R. Hillenbrand, J. Aizpurua and F. J. Garcĭa de Abajo, *J. Phys. Chem. C*, 2007, **111**, 1207.
48. E. Prodan, C. Radloff, N. J. Halas and P. Nordlander, *Science*, 2003, **302**, 419.
49. H. Wang, D. W. Brandl, P. Nordlander and N. J. Halas, *Acc. Chem. Res.*, 2007, **40**, 53.
50. S. K. Ghosh and T. Pal, *Chem. Rev.*, 2007, **107**, 4797.
51. K. A. Willets and R. P. Van Duyne, *Annu. Rev. Phys. Chem.*, 2007, **58**, 267.
52. V. Myroshnychenko, J. Rodríguez-Fernández, I. Pastoriza-Santos, A. M. Funston, C. Novo, P. Mulvaney, L. M. Liz-Marzán and F. J. García de Abajo, *Chem. Soc. Rev.*, 2008, **37**, 1972.
53. R. Gans, *Ann. Phys.*, 1915, **47**, 270.
54. C. F. Bohren and D. R. Huffman, *Absorption and Scattering of Light by Small Particles*, Wiley, New York, 1983.
55. N. Calander and M. Willander, *J. Appl. Phys.*, 2002, **92**, 4878.
56. S. Asano and G. Yamamoto, *Appl. Opt.*, 1975, **14**, 29.
57. G. C. Papavassiliou, *Prog. Solid State Chem.*, 1980, **12**, 185.
58. P. B. Johnson and R. W. Christy, *Phys. Rev. B*, 1972, **6**, 4370.
59. S. Link, M. B. Mohamed and M. A. El-Sayed, *J. Phys. Chem. B*, 1999, **103**, 3073.
60. S. Link and M. A. El-Sayed, *Annu. Rev. Phys. Chem.*, 2003, **54**, 331.
61. A. Brioude, X. C. Jiang and M. P. Pileni, *J. Phys. Chem. B*, 2005, **109**, 13138.
62. U. Kreibig and C. V. Fragstein, *Z. Phys.*, 1969, **224**, 307.
63. U. Kreibig and P. Zacharias, *Z. Phys.*, 1970, **231**, 128.
64. L. Genzel, T.P. Martin and U. Kreibig, *Z. Phys. B*, 1975, **21**, 339.
65. K. P. Charle, W. Schulze and B. Winter, *Z. Phys. D*, 1989, **12**, 471.
66. P. Apell and D. R. Penn, *Phys. Rev. Lett.*, 1983, **50**, 1316.
67. L. B. Scaffardi and J. O. Tocho, *Nanotechnology*, 2006, **17**, 1309.
68. J. Euler, *Z. Phys.*, 1954, **137**, 318.
69. J. A. Creighton and D. G. Eadon, *J. Chem. Soc., Faraday Trans.*, 1991, **87**, 3881.
70. S. Link and M. A. El-Sayed, *J. Phys. Chem. B*, 1999, **103**, 4212.
71. E. J. Heilweil and R. M. Hochstrasser, *J. Chem. Phys.*, 1985, **82**, 4762.

72. T. Klar, M. Perner, S. Grosse, G. von Plessen, W. Spirkl and J. Feldmann, *Phys. Rev. Lett.*, 1998, **80**, 4249.
73. C. Sönnichsen, T. Franzl, T. Wilk, G. von Plessen and J. Feldmann, *Phys. Rev. Lett.*, 2002, **88**, 077402.
74. S. Link and M. A. El-Sayed, *Int. Rev. Phys. Chem.*, 2000, **19**, 409.
75. A. Wokaun, J. P. Gordon and P. F. Liao, *Phys. Rev. Lett.*, 1982, **48**, 957.
76. P. K. Jain, W. Y. Huang and M. A. El-Sayed, *Nano Lett.*, 2007, **7**, 2080.
77. P. Pramod and K. G. Thomas, *Adv. Mater.*, 2008, **20**, 4300.
78. H. X. Xu, J. Aizpurua, M. Käll and P. Apell, *Phys. Rev. E*, 2000, **62**, 4318.
79. C. E. Talley, J. B. Jackson, C. Oubre, N. K. Grady, C. W. Hollars, S. M. Lane, T. R. Huser, P. Nordlander and N. J. Halas, *Nano Lett.*, 2005, **5**, 1569.
80. W. Y. Li, P. H. C. Camargo, X. M. Lu and Y. N. Xia, *Nano Lett.*, 2009, **9**, 485.
81. L. Gunnarsson, E. J. Bjerneld, H. Xu, S. Petronis, B. Kasemo and M. Käll, *Appl. Phys. Lett.*, 2001, **78**, 802.
82. K. A. Bosnick, J. Jiang and L. E. Brus, *J. Phys. Chem. B*, 2002, **106**, 8096.
83. J. Zhang, Y. Fu, M. H. Chowdhury and J. R. Lakowicz, *Nano Lett.*, 2007, **7**, 2101.
84. F. Le, N. Z. Lwin, J. M. Steele, M. Käll, N. J. Halas and P. Nordlander, *Nano Lett.*, 2005, **5**, 2009.
85. G. Lévêque and O. J. F. Martin, *Opt. Express*, 2006, **14**, 9971.
86. J. Biener, G. W. Nyce, A. M. Hodge, M. M. Biener, A. V. Hamza and S. A. Maier, *Adv. Mater.*, 2008, **20**, 1211.
87. A. J. Forty, *Nature*, 1979, **282**, 597.
88. Y. Ding, Y. J. Kim and J. Erlebacher, *Adv. Mater.*, 2004, **16**, 1897.
89. A. Dursun, D. V. Pugh and S. G. Corcoran, *J. Electrochem. Soc.*, 2003, **150**, B355.
90. J. F. Huang and I. W. Sun, *Adv. Funct. Mater.*, 2005, **15**, 989.
91. R. Morrish and A. J. Muscat, *Chem. Mater.*, 2009, **21**, 3865.
92. L. Y Chen, J. S. Yu, T. Fujita and M. W. Chen, *Adv. Funct. Mater.*, 2009, **19**, 1221.
93. D. V. Pugh, A. Dursun and S. G. Corcoran, *J. Mater. Res.*, 2003, **18**, 216.
94. J. C. Thorp, K. Sieradzki, L. Tang, P. A. Crozier, A. Misra, M. Nastasi, D. Mitlin and S. T. Picraux, *Appl. Phys. Lett.*, 2006, **88**, 033110.
95. J. S. Yu, Y. Ding, C. X. Xu, A. Inoue, T. Sakurai and M. W. Chen, *Chem. Mater.*, 2008, **20**, 4548.
96. L. Sun, C. L. Chien and P. C. Searson, *Chem. Mater.*, 2004, **16**, 3125.
97. R. C. Newman and K. Sieradzki, *MRS Bull.*, 1999, **24**, 12.
98. J. Erlebacher, M. J. Aziz, A. Karma, N. Dimitrov and K. Sieradzki, *Nature*, 2001, **410**, 450.
99. J. Erlebacher, *J. Electrochem. Soc.*, 2004, **151**, C614.
100. T. Fujita and M. W. Chen, *Jpn J. Appl. Phys.*, 2008, **47**, 1161.
101. L. H. Qian, X. Y. Yan, T. Fujita, A. Inoue and M. W. Chen, *Appl. Phys. Lett.*, 2007, **90**, 153120.
102. L. H. Qian and M. W. Chen, *Appl. Phys. Lett.*, 2007, **91**, 083105.

103. X. Y. Lang, P. F. Guan, L. Zhang, T. Fujita and M. W. Chen, *J. Phys. Chem. C*, 2009, **113**, 10956.

104. T. Fujita, L. H. Qian, K. Inoke, J. Erlebacher and M. W. Chen, *Appl. Phys. Lett.*, 2008, **92**, 251902.

105. H. Rösner, S. Parida, D. Kramer, C. A. Volkert and J. Weissmüller, *Adv. Eng. Mater.*, 2007, **9**, 535.

106. R. C. Newman, S. G. Corcoran, J. Erlebacher, M. J. Aziz and K. Sieradzki, *MRS Bull.*, 1999, **24**, 24.

107. F. Yu, S. Ahl, A. M. Caminade, J. P. Majoral, W. Knoll and J. Erlebacher, *Anal. Chem.*, 2006, **78**, 7346.

108. M. C. Dixon, T. A. Daniel, M. Hieda, D. M. Smilgies, M. H. W. Chan and D. L. Allara, *Langmuir*, 2007, **23**, 2414.

109. T. H. Park, N. Mirin, J. B. Lassiter, C. L. Nehl, N. J. Halas and P. Nordlander, *ACS Nano*, 2008, **2**, 25.

110. J. Prikulis, P. Hanarp, L. Olofsson, D. Sutherland and M. Käll, *Nano Lett.*, 2004, **4**, 1003.

111. X. Y. Lang, L. Y. Chen, P. F. Guan, T. Fujita and M. W. Chen, *Appl. Phys. Lett.*, 2009, **94**, 213109.

112. T. Fujita, H. Okada, K. Koyama, K. Watanabe, S. Maekawa and M. W. Chen, *Phys. Rev. Lett.*, 2008, **101**, 166601.

113. Y. Ding and J. Erlebacher, *J. Am. Chem. Soc.*, 2003, **125**, 7772.

114. L. H. Qian, Y. Ding, T. Fujita and M. W. Chen, *Langmuir*, 2008, **24**, 4426.

115. S. Ahl, P. J. Cameron, J. Liu, W. Knoll, J. Erlebacher and F. Yu, *Plasmonics*, 2008, **3**, 13.

116. D. H. Kim, P. Karan, P. Goring, J. Leclaire, A. M. Caminade, J. P. Majoral, U. Gosele, M. Steinhart and W. Knoll, *Small*, 2005, **1**, 99.

117. L. A. Lyon, M. D. Musick and M. J. Natan, *Anal. Chem.*, 1998, **70**, 5177.

118. J. S. Mitchell, Y. Wu, C. J. Cook and L. Main, *Anal. Biochem.*, 2005, **343**, 125.

119. L. He, M. D. Musick, S. R. Nicewarner, F. G. Slinas, S. J. Benkovic, M. J. Natan and C. D. Keating, *J. Am. Chem. Soc.*, 2000, **122**, 9071.

120. W. Suëtaka, *Surface Infrared and Raman Spectroscopy: Methods and Applications*, Plenum, New York, 1995.

121. J. K. Wang, C. S. Tsai, C. E. Lin and J. C. Lin, *J. Chem. Phys.*, 2000, **113**, 5041.

122. D. L. Jeanmarie and R. P. Van Duyne, *J. Electroanal. Chem.*, 1977, **84**, 1.

123. M. G. Albrecht and J. A. Creighton, *J. Am. Chem. Soc.*, 1977, **99**, 5215.

124. G. C. Schatz, *Acc. Chem. Res.*, 1984, **17**, 370.

125. A. Campion and P. Kambhampati, *Chem. Soc. Rev.*, 1998, **27**, 241.

126. K. Kneipp, H. Kneipp, I. Itzkan, R. R. Dasari and M. S. Feld, *Chem. Rev.*, 1999, **99**, 2957.

127. M. Moskovits, *Surface-Enhanced Raman Scattering: A Brief Perspective, Surface-Enhanced Raman Scattering: Physics and Applications, Topics in Applied Physics*, 2006, **103**, 1.

128. S. M. Nie and S. R. Emory, *Science*, 1997, **275**, 1102.

129. K. Kneipp, Y. Wang, H. Kneipp, L. T. Perelman and I. Itzkan, *Phys. Rev. Lett.*, 1997, **78**, 1667.
130. T. Dadosh, J. Sperling, G. W. Bryant, R. Breslow, T. Shegai, M. Dyshel, G. Haran and I. Bar-Joseph, *ACS Nano*, 2009, **3**, 1988.
131. N. P. W. Pieczonka and R. F. Aroca, *Chem. Soc. Rev.*, 2008, **37**, 946.
132. P. G. Etchegoin, E. C. Le Ru and M. Meyer, *J. Am. Chem. Soc.*, 2009, **131**, 2713.
133. H. Xu, E. J. Bjerneld, M. Käll and L. Börjesson, *Phys. Rev. Lett.*, 1999, **83**, 4357.
134. C. L. Hynes and R. P. Van Duyne, *J. Phys. Chem. B*, 2003, **107**, 7426.
135. G. L. Liu and L. P. Lee, *Appl. Phys. Lett.*, 2005, **87**, 074101.
136. H. H. Wang, C. Y. Liu, S. B. Wu, N. W. Liu, C. Y. Peng, T. H. Chan, C. F. Hsu, J. K. Wang and Y. L. Wang, *Adv. Mater.*, 2006, **18**, 491.
137. J. P. Camden, J. A. Dieringer, J. Zhao and R. P. Van Duyne, *Acc. Chem. Res.*, 2008, **41**, 1653.
138. M. J. Banholzerm, J. E. Millstone, L. D. Qin and C. A. Mirkin, *Chem. Soc. Rev.*, 2008, **37**, 885.
139. R. Aroca, *Surface-Enhanced Vibrational Spectroscopy*, Wiley, Chichester, 2006.
140. S. O. Kucheyev, J. R. Hayes, J. Biener, T. Huser, C. E. Talley and A. V. Hamza, *Appl. Phys. Lett.*, 2006, **89**, 053102.
141. L. H. Qian, A. Inoue and M. W. Chen, *Appl. Phys. Lett.*, 2008, **92**, 093113.
142. K. H. Su, Q. H. Wei, X. Zhang, J. J. Mock, D. R. Smith and S. Schultz, *Nano Lett.*, 2003, **3**, 1087.
143. P. K. Jain and M. A. El-Sayed, *Nano Lett.*, 2007, **7**, 2854.
144. H. X. Xu and M. Käll, *Phys. Rev. Lett.*, 2002, **89**, 246802.
145. A. J. Hallock, P. L. Redmond and L. E. Brus, *Proc. Natl. Acad. Sci. U. S. A.*, 2005, **102**, 1280.
146. Y. S. Pang, H. J. Hwang and M. S. Kim, *J. Phys. Chem. B*, 1998, **102**, 7203.
147. C. H. Kwon, D. W. Boo, H. J. Hwang and M. S. Kim, *J. Phys. Chem. B*, 1999, **103**, 9610.
148. P. T. Leung, M. H. Hider and E. J. Sanchez, *Phys. Rev. B*, 1996, **53**, 12659.
149. H. P. Chiang, P. T. Leung and W. S. Tse, *J. Phys. Chem. B*, 2000, **104**, 2348.
150. A. K. Sharma and B. D. Gupta, *Appl. Opt.*, 2006, **45**, 151.
151. R. Chang, P. T. Leung, S. H. Lin and W. S. Tse, *Phys. Rev. B*, 2000, **62**, 5168.
152. T. Holstein, *Phys. Rev.*, 1954, **96**, 535.
153. U. Kreibig, *J. Phys. F: Met. Phys.*, 1974, **4**, 999.
154. E. H. Sondheimer, *Adv. Phys.*, 1952, **1**, 1.
155. G. Kästle, H. G. Boyen, A. Schröder, A. Plettl and P. Ziemann, *Phys. Rev. B*, 2004, **70**, 165414.
156. J. S. Gao and Z. Q. Tian, *Chem. Phys. Lett.*, 1996, **262**, 151.

157. K. S. Shin, H. S. Lee, S. W. Joo and K. Kim, *J. Phys. Chem. C*, 2007, **111**, 15223.
158. S. W. Kim, S. Y. Kim and S. H. Park, *J. Am. Chem. Soc.*, 2009, **131**, 8380.
159. K. J. Major, C. De and S. O. Obare, *Plasmonics*, 2009, **4**, 61.
160. J. J. Schneider, *Adv. Mater.*, 2001, **13**, 529.
161. C. J. Zhong and M. M. Maye, *Adv. Mater.*, 2001, **13**, 1507.
162. B. Lim, J. G. Wang, P. H. C. Camargo, M. J. Jiang, M. J. Kim and Y. N. Xia, *Nano Lett.*, 2008, **8**, 2535.
163. A. C. Templeton, W. P. Wuelfing and R. W. Murray, *Acc. Chem. Rev.*, 2000, **33**, 27.
164. F. Caruso, *Adv. Mater.*, 2001, **13**, 11.
165. Y. Kobayashi, V. Salgueirino-Maceira and L. M. Liz-Marzan, *Chem. Mater.*, 2001, **13**, 1630.
166. M. Giersig and P. Mulvaney, *Langmuir*, 1996, **12**, 4329.
167. S. J. Oldenburg, J. B. Jackson, S. L. Westcott and N. J. Halas, *Appl. Phys. Lett.*, 1999, **75**, 2897.
168. F. Caruso, H. Lichtenfeld and H. Mohwald, *J. Am. Chem. Soc.*, 1998, **120**, 8523.
169. Y. Ding, M. W. Chen and J. Erlebache, *J. Am. Chem. Soc.*, 2004, **126**, 6876.
170. J. T. Zhang, H. Y. Ma, D. J. Zhang, P. P. Liu, F. Tian and Y. Ding, *Phys. Chem. Chem. Phys.*, 2008, **10**, 3250.
171. L. Y. Chen, L. Zhang, T. Fujita and M. W. Chen, *J. Phys. Chem. C*, 2009, **113**, 14195.
172. J. H. Hodak, I. Martini and G. V. Hartland, *J. Phys. Chem. B*, 1998, **102**, 6958.
173. E. M. Purcell, *Phys. Rev.*, 1946, **69**, 681.
174. J. R. Lakowicz, *Principles of Fluorescence Spectroscopy,* 3rd edn., Springer, New York, 2006.
175. J. R. Lakowicz, *Anal. Biochem.*, 2001, **198**, 1.
176. J. R. Lakowicz, *Anal. Biochem.*, 2005, **337**, 171.
177. O. Painter, R. K. Lee, A. Scherer, A. Yariv, J. D. O'Brien, P. D. Dapkus and I. Kim, *Science*, 1999, **284**, 1819.
178. G. L. Liu, Y. T. Long, Y. Choi, T. Kang and L. P. Lee, *Nat. Meth.*, 2007, **4**, 1015.
179. W. P. Ambrose, P. M. Goodwin, J. H. Jett, A. V. Orden, J. H. Werner and R. A. Keller, *Chem. Rev.*, 1999, **99**, 2929.
180. A. C. Pease, D. Solas, E. J. Sullivan, M. T. Cronin, C. P. Holmes and S. P. A. Fodor, *Proc. Natl. Acad. Sci. U. S. A.*, 1994, **91**, 5022.
181. E. M. C. Hillman and A. Moore, *Nat. Photonics*, 2007, **1**, 526.
182. R. R. Chance, A. Prock and R. Silbey, *J. Chem. Phys.*, 1974, **60**, 2744.
183. K. H. Drexhage, *J. Lumin.*, 1974, **1**, 693.
184. E. Dulkeith, A. C. Morteani, T. Niedereichholz, T. A. Klar, J. Feldmann, S. A. Levi, F. C. J. M. van Veggel, D. N. Reinhoudt, M. Möller and D. I. Gittins, *Phys. Rev. Lett.*, 2002, **89**, 203002.
185. P. Anger, P. Bharadwaj and L. Novotny, *Phys. Rev. Lett.*, 2006, **96**, 113002.

186. K. Ray, R. Badugu and J. R. Lakowicz, *Langmuir*, 2006, **22**, 8374.
187. R. Baradhan, N. K. Grady and N. J. Halas, *Small*, 2008, **4**, 1716.
188. S. Neretina, W. Qian, E. Dreaden, M. A. El-Sayed, R. A. Hughes, J. S. Preston and P. Mascher, *Nano Lett.*, 2008, **8**, 2410.
189. O. Kulakovich, N. Strekal, A. Yaroshevich, S. Maskevich, S. Gaponenko, I. Nabiev, U. Woggon and M. Artemyev, *Nano Lett.*, 2002, **2**, 1449.
190. K. Ray, R. Badugu and J. R. Lakowicz, *J. Am. Chem. Soc.*, 2006, **128**, 8998.
191. K. T. Shimizu, W. K. Woo, B. R. Fisher, H. J. Eisler and M. G. Bawendi, *Phys. Rev. Lett.*, 2002, **89**, 117401.
192. J. S. Biteen, D. Pacifici, N. S. Lewis and H. A. Atwater, *Nano Lett.*, 2005, **5**, 1768.
193. F. Tam, G. P. Goodrich, B. R. Johnson and N. J. Halas, *Nano Lett.*, 2007, **7**, 496.
194. R. Baradhan, N. K. Grady, J. R. Cole, A. Joshi and N. J. Halas, *ACS Nano*, 2009, **3**, 744.
195. S. H. Guo, S. J. Tsai, H. C. Kan, D. H. Tsai, M. R. Zachariah and R. J. Phaneuf, *Adv. Mater.*, 2008, **20**, 1424.
196. A. G. Brolo, S. C. Kwok, M. G. Moffitt, R. Gordon, J. Riordon and K. L. Kavanagh, *J. Am. Chem. Soc.*, 2005, **127**, 14936.
197. S. Mackowski, S. Wörmke, A. J. Maier, T. H. P. Brotosudarmo, H. Harutyunyan, A. Hartschuh, A. O. Govorov, H. Scheer and C. Bräuchle, *Nano Lett.*, 2008, **8**, 558.
198. E. M. Goldy, K. Drozdowicz-Tomsia, F. Xie, T. Shtoyko, E. Matveeva, I. Gryczynski and Z. Gryczynski, *J. Am. Chem. Soc.*, 2007, **129**, 12117.
199. C. D. Geddes, H. S. Cao, I. Gryczynski, Z. Gryczynski, J. Y. Fang and J. R. Lakowicz, *J. Phys. Chem. A*, 2003, **107**, 3443.
200. C. Röcker, M. Pötzl, F. Zhang, W. J. Parak and G. U. Nienhaus, *Nat. Nanotechnol.*, 2009, **4**, 577.
201. K. Sokolov, G. Chumanov and T. M. Cotton, *Anal. Chem.*, 1998, **70**, 3898.
202. G. Ritchie and E. Burstein, *Phys. Rev. B*, 1981, **24**, 4843.

Actuation with High-Surface-Area Materials

L.-H. SHAO,*[a] H.-J. JIN[b] AND J. WEISSMÜLLER[a,c]

[a] Institut für Werkstoffphysik und -Technologie, Technische Universität Hamburg-Harburg, Hamburg, Germany; [b] Shenyang National Laboratory for Materials Science, Institute of Metal Research, Chinese Academy of Sciences, Shenyang, PR China; [c] Institut für Werkstoffforschung, Werkstoffmechanik, Helmholtz-Zentrum Geesthacht, Geesthacht, Germany
*Email: lihua.shao@tu-harburg.de

7.1 Introduction

Materials that can reversibly change their dimensions upon the application of an external stimulus, such as an applied voltage, are used as actuators in many applications.[1] High-performance actuators require a combination of large actuation stroke, stiffness, and strength for the sake of high strain energy density and load. Ceramic piezoelectric actuation materials are stiff and strong, while their strain amplitude is restricted. The more recent polymer actuation materials explore another extreme of parameter space, with a much larger strain amplitude that invites comparison to muscles in living organisms.[2–4] However, the mechanical performance of polymers is restricted by their low stiffness and strength. Other actuator material classes include electrostrictive ceramics, shape memory alloys, and magnetostrictive metal actuators.

The suitability of an actuator material for a given application is not determined by stiffness and stroke alone. Important figures of merit include the magnitude of the required electric voltage or magnetic control field.

RSC Nanoscience & Nanotechnology No. 22
Nanoporous Gold: From an Ancient Technology to a High-Tech Material
Edited by Arne Wittstock, Jürgen Biener, Jonah Erlebacher and Marcus Bäumer
© Royal Society of Chemistry 2012
Published by the Royal Society of Chemistry, www.rsc.org

Since different scenarios favor different combinations of properties, there remains an interest in exploring new actuator systems that explore uncharted regions of the parameter space. Here, we inspect materials that exhibit extremely large volume- or mass-specific surface area and that can be made to expand or contract by controlling the interatomic bond strength at the surfaces and, thereby, the capillary forces. Typically, the control is achieved by manipulating space-charge regions at the surface or by inducing reversible adsorption/desorption processes. (Throughout this text, we use the term 'reversible' to describe changes that can be reversed in a cyclic process, even when there is hysteresis during the cycle; this does not imply reversibility in the sense of thermodynamics.) One way of setting up space-charge regions or of controlling adsorption processes is by using electrochemistry. Nanoporous metals and high-surface-area carbon materials exemplify that approach. The important general aspects of nanoporous metals have been reviewed in a recent issue of MRS Bulletin,[5–7] and an overview on actuation can be found in Ref. [8]. Here, we inspect actuation by solids with a large mass- or volume-specific surface area, including nanoporous metals and carbon nanostructures. For brevity, our emphasis is on exemplifying the relevant observations and approaches rather than providing a complete review.

7.2   Actuation Driven by Capillary Forces

7.2.1   General Phenomenology

Consider an experiment where a macroscopic sample of nanoporous metal is impregnated with electrolyte—such as an aqueous salt solution, an acid, or a base—and electrically connected to a counter-electrode in such a way that its potential (the electrode potential, E, relative to a suitable reference electrode) can be varied. *In situ* electrochemical experiments show that the sample reacts to the potential variation by a reversible macroscopic expansion or contraction, with an amplitude that is comparable to that of commercial actuator materials.

The first experimental demonstration of electrochemical actuation with nanoporous metals used nanoporous Pt samples prepared by consolidating Pt nanoparticles.[1] Strain amplitudes in the order of 0.1% were achieved. More recently, similar experiments using nanoporous metals prepared by dealloying reached reversible strain amplitudes in excess of 1%.[9] The deformation can be greatly amplified by combining massive and nanoporous metal layers into a bilayer foil. Such experiments have demonstrated macroscopic displacements as large as a several millimeters.[10] The experiment of Kramer *et al.*[10] was indeed the first where the deformation of a solid by the action of the capillary forces at its surface has been made visible to the naked eye. We shall now briefly discuss capillary phenomena in nanoporous solids.

7.2.2 Description of Actuation in a Continuum Picture

The microscopic processes behind actuation in conventional actuator materials are typically as follows. The control parameter is the electric field between two metal electrodes applied to opposing surfaces of the material. This field penetrates the material, and it is experienced by any given materials point in the bulk. The local strain at equilibrium, under conditions of no stress, is dependent on the local value of the field. Microscopically, this may be the result of a field-dependent electronic structure within each crystallographic unit cell. For instance, the field may polarize the local charge distribution, thereby leading to a distortion of the unit cells. The macroscopic dimension change reflects these field-induced local strains at zero stress.

In metals, the large density of mobile charge carriers results in an efficient screening of electric fields at the surface. As a consequence, the interior of metals must be free of electric field, so that metals cannot exhibit the afore-mentioned piezoelectric behavior. The local strain in the solid skeleton phase of nanoporous metals can therefore only be understood as the consequence of a change in local stress. As it turns out, this stress is the consequence of a change in the bond forces between the surface atoms: changing the electrode potential leads to a transport of charge to or from the space-charge region within the metal surface that forms one half of the electrochemical double layer. This region can accommodate an excess charge up to a few tenths of an electron per surface metal atom in a layer of practically atomic dimensions. The corresponding value of the local charge density is several orders of magnitude larger than the values in the space-charge layers of semiconductor devices, such as field-effect transistors. *Ab initio* computation shows that the large local space-charge density at the metal-electrolyte interface will significantly affect the bonding between the metal atoms,[11–13] and mechanical equilibrium requires stress in the bulk for compensating the modified bond forces.[1,14]

In a continuum description, the parameter that quantifies the mechanical interaction of the matter at the surface of a solid with the underlying bulk is the surface stress, **s**, the derivative of the surface tension with respect to the tangential strain.[15,16] To illustrate the situation, let us consider the surface-induced stress and strain in a nanowire (Figure 7.1) in a simplified analysis. The wire, which may be viewed as a cartoon of an individual ligament in a nanoporous metal, is taken to have radius r and length l, with $l \gg r$ so that processes near the ends can be ignored. Take the free energy densities of bulk, B, and surface, S, as $\Psi(\varepsilon)$ and $\psi(\varepsilon)$, respectively, with ε the strain tensor, and define the bulk and surface stresses as $\mathbf{S} = \partial\Psi/\partial\varepsilon$ and $\mathbf{s} = \partial\psi/\partial\varepsilon$, respectively. The densities are defined so that the net Helmholtz free energy is $F = \int_{\mathrm{B}} \Psi \, \mathrm{d}V + \int_{\mathrm{S}} \psi \, \mathrm{d}A$. Mechanical equilibrium requires that the variation in F vanishes for any arbitrary small variation $\delta\varepsilon$ in strain:

$$0 = \delta F = \int_{\mathrm{B}} \mathbf{S}\delta\varepsilon \, \mathrm{d}V + \int_{\mathrm{S}} \mathbf{s}\delta\varepsilon \, \mathrm{d}A. \tag{7.1}$$

Eqn (7.1) needs to be satisfied for any arbitrary test strain field $\delta\varepsilon$, and one may therefore consider convenient special cases, specifically uniform $\delta\varepsilon$. To simplify the analysis further, attention is restricted to regions sufficiently far from the fibre ends, where S is also uniform. Furthermore, the surface stress is assumed isotropic in the plane on S (as must be the case for some important low-index crystal surfaces) and uniform in magnitude, $f = \frac{1}{2}$ trace s. Then, $d\psi = f\,de$, where de quantifies the relative change in surface area by elastic strain. Eqn (7.1) here takes on the simpler form

$$0 = SV\delta\varepsilon + fA\delta e. \tag{7.2}$$

If $\delta\varepsilon$ is an *axial* strain of magnitude $\delta\varepsilon$, then eqn (7.2) simplifies to

$$0 = S_A V\delta\varepsilon + fA\delta\varepsilon, \tag{7.3}$$

with S_A the axial component of S. Similarly, for $\delta\varepsilon$, a planar strain, isotropic in the cross-sectional plane of the fibre, and with S_R the radial component of S, one obtains

$$0 = 2S_R V\delta\varepsilon + fA\delta\varepsilon. \tag{7.4}$$

Since $A/V = 2/r$ for the cylindrical fibre, one finds that axial and radial stresses in the fibre at equilibrium are[14]

$$S_A = -2f/r \text{ and } S_R = -f/r. \tag{7.5}$$

These results have first been derived, starting from more general principles, in Weissmüller and Cahn.[14] They form a special case of the generalized capillary equation for solids, which describes the equilibrium between the capillary forces at the surface of solid body of arbitrary shape and size, and the mean stress in the bulk.[14]

The results in eqn (7.5) show that the surface-induced stress in the bulk of a fibre increases as the inverse of the radius. As an estimate, consider $f = 1$ N m^{-1} and $r = 2$ nm. Then, the axial stress is $S_A = -1$ GPa, well beyond the strength of conventional engineering materials. Furthermore, the anisotropy of the stress state implies that there is a significant shear character. The maximum of the shear stress is found in planes inclined by $45°$ to the fibre axis, and its magnitude is $\frac{1}{2}f/r$. In fact, experiment indicates that fast dealloying leads to ligaments sufficiently thin for the surface-induced shear stress in their bulk to induce plastic yielding. This is believed to be the origin of the large irreversible volume contraction during electrochemical dealloying at a large potential, where structure sizes of nanoporous metal reach down to a few nanometers.[9] Computer simulation for nanowires[17] and for nanoporous material with a geometry matched to that of nanoporous gold[18] confirms this suggestion.

Consider the strain that results from the stresses of eqn (7.5). Hooke's law for a solid with isotropic elasticity, characterized by Young's modulus Y and Poisson number v, suggests for the axial and radial strains[19]

$$\varepsilon_A = -\frac{2f}{r}\frac{1-v}{Y}, \quad \varepsilon_R = -\frac{f}{r}\frac{1-3v}{Y}, \tag{7.6}$$

respectively. These results testify to a significantly anisotropic strain. For the example of gold, where $v = 0.44$, the radial strain, ε_R, is even opposite in sign to the axial strain, ε_A. The right-hand side of Figure 7.1 illustrates this.

Atomistic computer simulation[19] as well as experiments using *in situ* lattice parameter data derived from X-ray diffraction[20] have confirmed the important distinction between radial and longitudinal components of the surface-induced strain in elongated nano-objects. These findings underline the important distinction between stress and strain in response to fluid pressure (this gives rise to isotropic stress and strain) and the effects (in general anisotropic) prompted by changes in the surface stress.

In general, averages of the surface-induced stresses and strains in nanoscale bodies depend on the geometry via the specific surface area (area per volume of the solid) and via the orientation distribution function of the surfaces.[14] General solutions are available for the volumetric average of the stress tensor, but closed-form expressions for the strain measures of interest, specifically the macroscopic dimension change, have only been derived for special geometries.[19] It is emphasized that the capillary parameter—surface stress—that determines the surface-induced stress in a solid is fundamentally different from the parameter—surface tension—that determines the pressure in a fluid

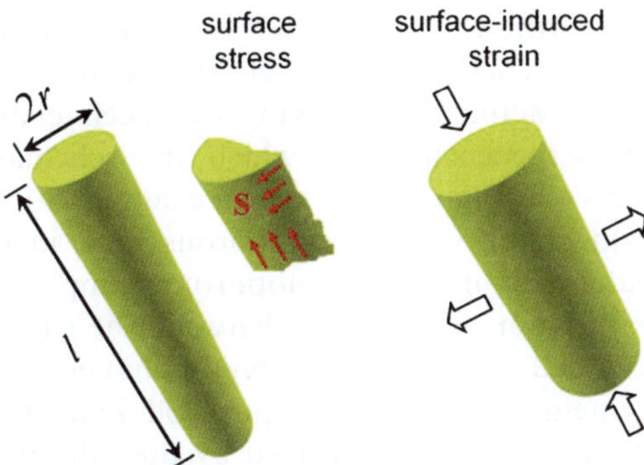

Figure 7.1　Schematic illustrations of the surface stress and surface-induced strain in a circular nanofibre of radius r and length l. Note the longitudinal contraction and radial expansion of a gold nanofibre under the action of a surface-stress tensor, **s**, with positive-valued entries. Reproduced from Jin and Weissmüller.[8]

droplet.[16] Surface stress and surface tension have identical units but a different dimensionality (second rank tensor *vs.* scalar) and generally quite different numerical magnitudes. Among the distinguishing features are the existence of negative surface stress values—as opposed to the invariably positive surface tension—and the linear dependence of surface stress on superficial charge density as opposed to the quadratic charge dependence of the surface tension. Furthermore, the surface-induced stress in a solid depends on the area per volume, which is always positive, as the geometry parameter, whereas the surface-induced (Laplace-) pressure in a fluid depends on the surface curvature, which can be of either sign depending on whether the surface is convex or concave. For a more detailed discussion of the distinction between surface stress and surface tension, and of related common misconceptions in the contemporary literature, the reader is referred to Kramer and Weissmüller.[16]

The considerations in this section have so far considered the stress and strain by which the bulk of the porous solid compensates the absolute value of the surface stress. In fact, reliable *absolute values* for this parameter have rarely if at all been determined from experiments. The known database is essentially derived from *ab initio* simulation.[21,22] However, *changes* in the surface stress in response to variation of the adsorbate coverage or superficial charge density are readily accessible to experiments. It is these changes that drive the change in bulk stress and strain that is exploited by the nanoporous metal actuation experiments.

7.2.3 Surface-Stress-Induced Actuation: Experimental Characterization

The surface-induced strain in the bulk of a crystal can be characterized by measuring the lattice parameter changes using X-ray diffraction,[1,20] or by measuring the sample outer dimension changes using a dilatometer. The former is more straightforward, while the latter is more convenient and could be carried out easily in chemical or electrochemical environments. In this chapter, most of the actuation behavior has been characterized by using *in situ* dilatometry experiments, for which the schematic illustration and photograph are shown in Figure 7.2.[23] As an example, Figure 7.2(c) shows the simultaneous length changes of a nanoporous gold in electrolyte when its electrode potential is changed. A nanoporous sample, here cuboidal with dimensions of $1 \times 1 \times 2 \, mm^3$, is mounted between the support and pushrod in an electrochemical cell. The sample length change is transmitted to an inductive displacement sensor via the pushrod loaded by a constant and small contact pressure. The stress exerted by the dilatometer probe is insignificantly small compared to the yield strength (*e.g.* 10–100 MPa[24,25] for nanoporous gold). An analogous setup, with the electrochemical cell replaced by a simple gas perfusion cell at atmospheric pressure, can be used for probing chemical actuation[26] when the sample is exposed alternately to different gases.

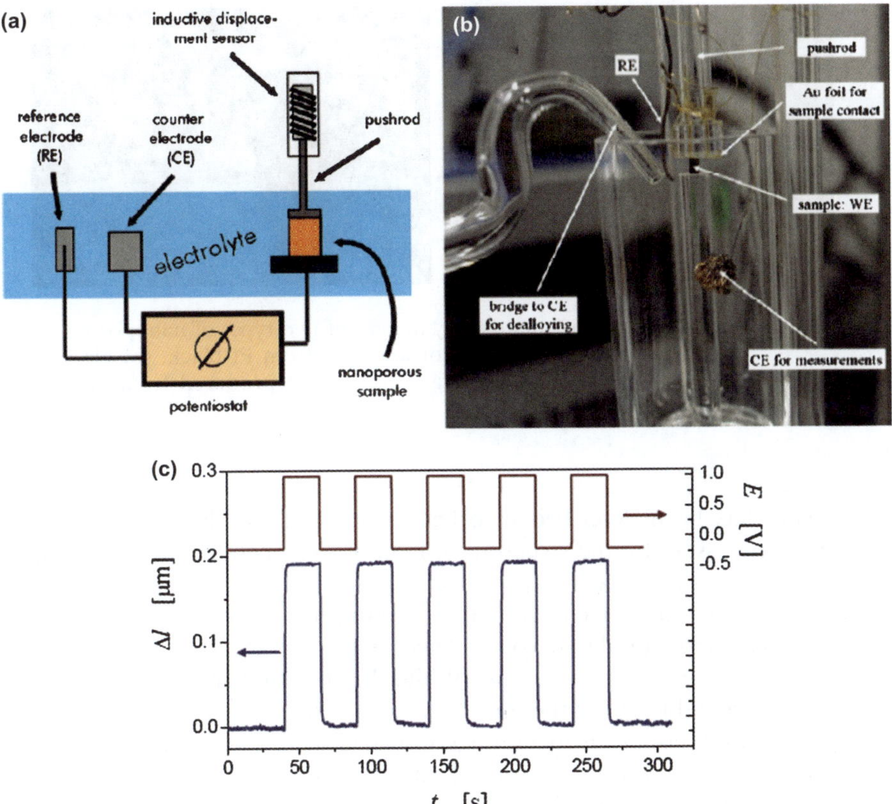

Figure 7.2   (a) Schematic illustration of dilatometry experiments carried out *in situ* in an electrochemical environment. (b) Photograph of an electrochemical cell used for *in situ* dilatometry. (c) Example of reversible length change Δl, measured in nanoporous gold in response to the potential jumping. Reproduced from Kramer *et al.*[10] and Jin *et al.*[23]

7.3   Structure

7.3.1   Nanoporous Metals

Nanoporous metals can be fabricated either by consolidation of metal nano-particles,[1,27] or by an alloy corrosion process known as dealloying.[7,28] Both methods are simple and can yield macroscopic samples with a uniform bi-continuous structure, consisting of an interconnected network of nanometer-sized metals and an interconnected pore space (see Figure 7.3). Both types of samples have also shown similar surface-stress-driven actuation behavior. Compared with consolidated nanoporous materials, nanoporous metals by dealloying are mechanically more stable, more amenable to microstructure controlling and tuning, and thus more suitable for actuator applications. The

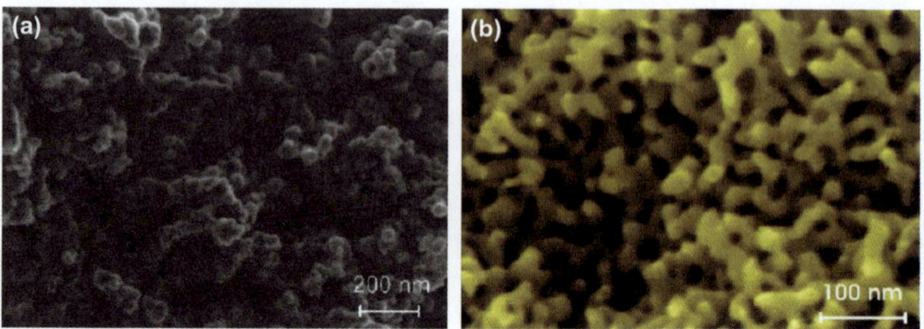

Figure 7.3 Scanning electron microscopy images of two types of nanoporous metals. (a) Nanoporous Pt sample consolidated from Pt black. (b) Nanoporous gold fabricated by dealloying $Ag_{75}Au_{25}$. Reproduced from Weissmüller *et al.*[1] and Kramer *et al.*[10]

dealloying has been described in other sections of this book. Here we will address some issues on the structure of nanoporous metals that are important to actuation performance.

For surface-stress-driven actuation, nanoporous materials with small structure size and thus large volumetric-specific surface area are essential to achieving the large actuation strain. While the structure size of a consolidated nanoporous metal is set by the initial particle size, the ligament and pore sizes of a dealloyed nanoporous metal are controlled by dealloying conditions (*e.g.*, dealloying potential and temperature) and the post-dealloying treatment. Due to the large total excess surface energy, nanoporous metals are inherently unstable against coarsening. Severe coarsening has been observed in nanoporous gold, even at room temperature, for instance, during free corrosion of AuAg alloys in concentrated nitric acid.[29] The coarsening can be accelerated by enhancing the surface diffusivity, alternatively by adding suitable anions such as Cl^-, by varying the electrode potential, or by heating. It is thus possible to prepare nanoporous structures of different length scales, from several to hundreds of nanometers.[30] On the other hand, strategies have also been developed to hinder the surface diffusion and thus stabilize the nanoporous structure. Maintaining the surface oxide formed during dealloying, or deliberately introducing one or two monolayers of oxide on ligament surfaces, can stabilize the nanoporous structure.[23] Adding a small amount of Pt has also been proven effective in maintaining the structure smallness of nanoporous Au.[9,31] The ligament size of nanoporous AuPt can stay small ($\sim 4\,nm$) even after removal of surface oxide by reduction.[9] Under similar circumstances, the ligament diameter of nanoporous Au is considerably larger, in excess of $20\,nm$.[23]

Dealloying is usually accompanied by a macroscopic sample contraction.[32] This effect is detrimental to the mechanical performance, since it introduces cracks at a macroscopic scale. Shrinkage will also need to be avoided if dealloying is to be used as a near net-shape-forming process of nanoporous metal components for devices. Incorporation of second phase into the

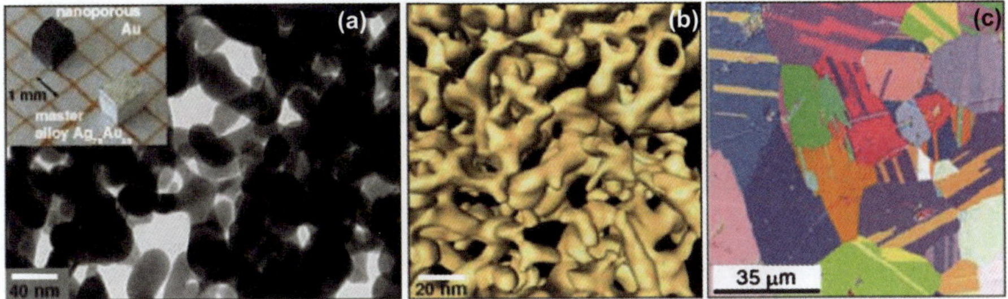

Figure 7.4   Length scales in nanoporous materials. (a) Bright-field transmission electron microscope, image of a nanoporous gold sample. Inset: photograph showing the millimeter-sized master alloy and the bulk dealloyed nanoporous sample containing about 10^{15} ligaments around 10 nm in size. (b) Binarized representation of the three-dimensional reconstruction nanoporous gold obtained by electron tomography in a transmission electron microscope. (c) Electron backscatter diffraction image showing a micrometer-sized grain structure of nanoporous gold. Reproduced from Weissmüller *et al.*[7]

nanoporous metal matrix has been proven to be successful in reducing the crack density and increasing the material strength, which however impairs the actuation performance.[33] Careful choice of the process parameters (dealloying potential and temperature) affords monolithic nanoporous samples in which the macroscopic sample geometry is preserved, and crack formation is prevented.[25,34,35] However, these processes usually involve slow dealloying or elevated temperatures, which in turn give rise to large structure sizes. Monolithic crack-free samples of nanoporous gold have so far only been obtained with ligament diameters of 15 nm or above.

Another distinguishing feature of the dealloyed nanoporous metals is the large grain size (much larger than the ligament size, see Figure 7.4) since the micrometer-scale grain size of the master alloy can be retained after dealloying,[29,36,37] while in the compacted nanoporous metals the grain size is small and is identical to the ligament (particle) size. The grain boundary area is therefore negligible compared to the surface area. It is beneficial to minimize the number of grain boundaries for material stability. Besides surface defects such as step edges, grain boundaries also provide sources for the flux of adatoms that propagates coarsening. Contrary to the surface sources, the action of the grain boundary sources entails the removal of crystal lattice planes and, thereby, macroscopic shrinkage (or sintering) of the material.[32] This irreversible contraction is an undesirable degradation mechanism.

7.3.2   Carbon Nanomaterials

As outlined above, nanoporous materials achieve their functionality as actuators by exploiting the variation of their surface stress for inducing a large compensating stress and strain in the bulk of their solid skeleton. A similar

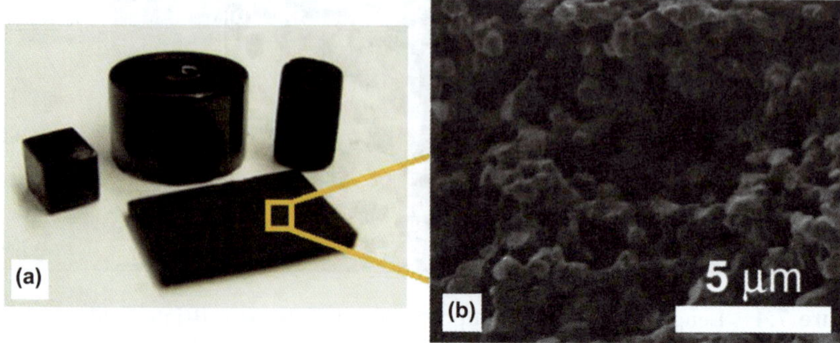

Figure 7.5 Pictures of a carbon aerogel. (a) Macroscopic picture and (b) scanning electron microscopy images of carbon aerogels. Reproduced from Baumann *et al.*[41]

behavior is also exhibited by other large surface area materials and specifically by carbon nanomaterials.[38–40]

Carbon aerogel (CA), a bulk carbon-based material with a nanometer pore structure, can be made by the sol–gel polymerization of resorcinol (R) and formaldehyde (F), followed by supercritical drying, pyrolysis, and activation in a CO_2 atmosphere at controlled temperature and pressure (see, for instance, Baumann *et al.*[41]). Such materials can reach mass-specific surface areas as high as $3200\,\mathrm{m^2\,g^{-1}}$. A picture of macroscopic bulk CAs is shown in Figure 7.5(a), and an SEM image in Figure 7.5(b) shows that the skeletal structure of CAs defining a continuous porous network consists of interconnected carbon ligaments which, themselves, are porous at the nanometer or even sub-nanometer scale.[41] These ligaments appear to be made up of spherical primary particles that have fused together during network formation, which shows that CAs have a similar morphology to nanoporous metal. The bulk density, ρ, of the sample in the image is $0.14\,\mathrm{g\,cm^{-3}}$, corresponding to a solid volume fraction, φ, of around 0.07.

Among the carbon family, carbon nanotubes (CNTs) and graphene are also of current interest in the context of actuation. CNT sheets or their composite have been investigated in this respect.[42–48] They can be formed by vacuum filtration of suspensions as detailed in several publications.[42,48–50] Single-walled carbon nanotubes (SWNTs) in a sheet are interconnected by mechanical entanglement and by van der Waals forces (see Figure 7.6(a)).[42,48] The first demonstration of a CNT actuator[42] (Figure 7.6(b)) consisted of two strips of SWNT sheet that were glued to opposite sides of a piece of double-sided Scotch tape. Applying a potential difference between the two sheets results in a deflection, which can be reversed on reversal of the potential.

Graphene, a single graphite basal plane, can be made in sizes up to $100\,\mu\mathrm{m}$.[51] Recently, graphene films and their composite have also been suggested for electrochemical actuator applications.[52–54] Similar to CNT sheets, graphene films (Figure 7.7(a)) are made by filtration of a suspension of reduced

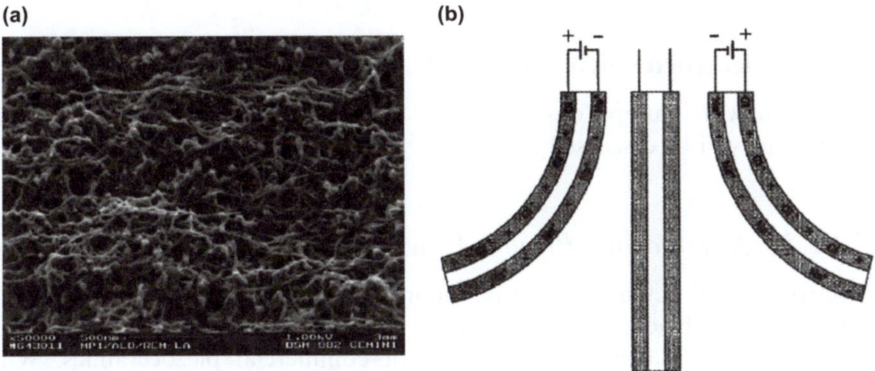

Figure 7.6 (a) Electron micrograph of Buckypape surface showing entangled ropes forming the SWNTs sheet. Reproduced from Fraysse *et al.*[48] (b) Schematic edge view of a cantilever-based actuator, formed from two strips of SWNTs shaded, laminated together with double-sided white Scotch tape, in aqueous NaCl. The Na^+ and Cl^- ions represent ions in the double layer at the nanotube bundle surfaces, which are compensated by the corresponding charges that are injected into the nanotube bundles. Reproduced from Baughman *et al.*[42]

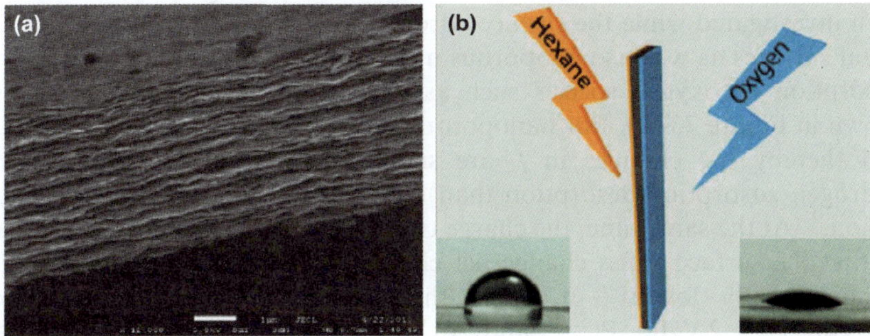

Figure 7.7 (a) Scanning electron microscope image of the cross-section of a graphene film with the scale bar of 1 µm. (b) Plasma treatments of graphene film with hexane and oxygen, and the corresponding surface wettabilities. The water-contact angles for the Hex-P (b) left inset and O_2-P (b) right inset treated surfaces are ~90° and ~15°, respectively. Reproduced from Xie *et al.*[54]

graphene-based species. Thicknesses of several micrometers give free-standing, mechanically flexible graphene film samples. Their opposing surface can be chemically treated in different ways (Figure 7.7(b)), so that one surface is more hydrophobic, thereby suppressing wetting and electrochemical charging. The asymmetric surface properties thus introduce an asymmetric electrochemical response, which produces the driving force responsible for the deflection of the film when the electrode potential is varied.[54]

7.4 Actuation

7.4.1 Electrochemical Actuation with Nanoporous Metals

Actuation with nanoporous metal has been studied for Pt,[1,27] Au,[10,23] and Pt–Au,[9] as will be discussed below.

7.4.1.1 *Nanoporous Pt and Au with Clean Surfaces*

Electrochemical actuation with nanoporous metals was first demonstrated in compacted Pt black, showing sample dimension changes in the order of 0.15%,[1] which is comparable to that of commercial piezoceramics. When nanoporous Pt is brought to contact with electrolyte, it expands upon potential sweeping during capacitive charging in the double-layer potential region (Figure 7.8(a)). Nanoporous gold prepared by dealloying shows a similar behavior (Figure 7.2(c)). This trend of the sample dimension changes agrees well with the trend of surface stress changes as established for metal surfaces, thus confirming a negative-valued surface-stress charge coefficient, $\varsigma = \mathrm{d}f/\mathrm{d}q|_e$ with q the superficial electron charge density and e a tangential strain variable. The magnitude of the surface-stress response to charging is in excellent agreement with density functional theory computation.[11–13]

The actuation behavior and the surface stress–charge response have also been investigated while the surface adsorption/desorption was involved. Clean metal surfaces as well as nanoporous metals as a whole tend to expand during adsorption of oxygen species such as OH^- ($\varsigma < 0$) and hydrogen ($\varsigma > 0$) as shown in Figure 7.8(b). For nanoporous Pt, the macroscopic strain amplitude, and thereby the changes in f, are significantly larger during oxygen and hydrogen adsorption/desorption than in the capacitive double-layer charging region.[27] At the same time, the charge transferred to the electrode is also larger, so that the surface stress change per charge, and thereby the magnitude of ς, is significantly lower ($-0.71\,\mathrm{V}$)[27] than that measured during capacitive charging ($\sim -2\,\mathrm{V}$ for Pt[55,56] and Au[57] in dilute solutions with weak anion adsorptions). An even lower ς value ($-0.16\,\mathrm{V}$)[9] was obtained for nanoporous gold during oxygen adsorption and desorption. For nanoporous gold, the total strain amplitude during oxygen adsorption and desorption is small, only equivalent to that induced by capacitive charging in the double-layer region (see Figure 7.9).

Though the net length change induced by surface stress variation is small in nanoporous metals, it can be amplified by using a bimetallic foil cantilever strategy to achieve millimeter-scale displacement. Gold and AuAg alloy foils can be bound together by cold rolling to form a composite foil. Dealloying this (AuAg side of) AuAg/Au composite material produces a new bimetallic nanoporous gold/Au composite foil strip.[10] Nanoporous gold tends to expand and contract in response to potential changes while the dimensions of the solid gold foil remain constant. This small difference in length changes between two bonded layers leads to foil curling and, hence, to a large transverse

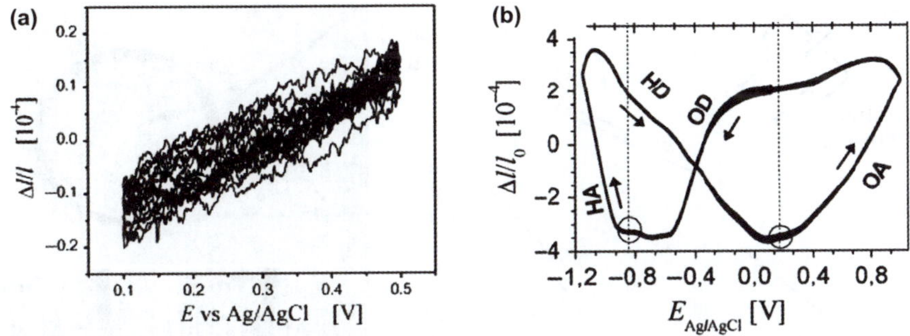

Figure 7.8 Macroscopic dimension changes of nanoporous Pt measured *in situ* with dilatometer in electreochemcial environments. (a) $\Delta l/l_0 \sim E$ measured in capacitive charging double layer region in 0.5 M H_2SO_4 solution. (b) $\Delta l/l_0 \sim E$ measured in 0.7 M NaF while scanning in a wide potential interval which involves both oxygen adsorption/desorption OA/OD, and hydrogen adsorption/desorption HA/HD, processes. Reproduced from Weissmüller *et al.*[1] and Viswanath *et al.*[27]

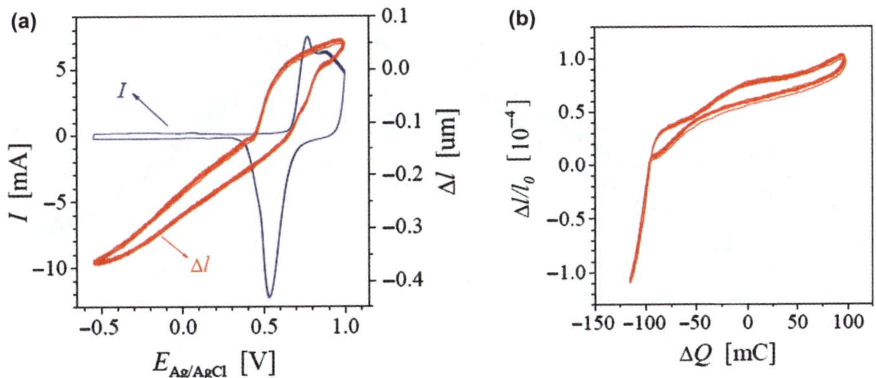

Figure 7.9 *In situ* dilatometry experiments of nanoporous gold in 1 M $HClO_4$. The ligament size is approximately 55 nm. (a) Cyclic voltammogram and length change at a potential sweeping rate of $10 \, mV \, s^{-1}$. (b) $\Delta l/l_0$ versus superficial surface charge derived from (a).

displacement of the foil tip. Figure 7.10 shows the tip positions of a nanoporous gold/Au bimetal foil that was charged in $HClO_4$ solution. When a voltage is applied between two bimetallic foils and is switched back and forth between $+1$ V and -1 V, the tip moves back and forth. The tip displacement reaches 3 mm, which is large enough to be visible to the naked eye.[10]

7.4.1.2 Nanoporous Au and Pt with Oxidized Surfaces

The capacitive charging and adsorption processes discussed above took place on a surface that is initially 'clean'. Pt surfaces can be readily covered with an

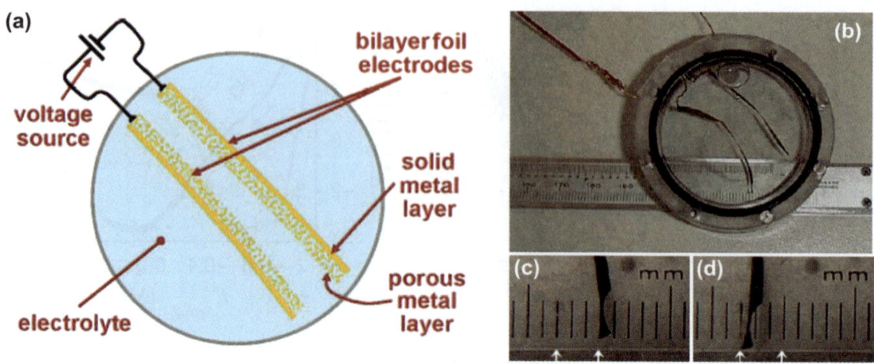

Figure 7.10   Surface-stress-induced cantilever bending of Au/nanoporous gold composite foils. (a) Schematic illustration and (b) photograph of an electrochemical cell with two bimetallic stripe electrodes. The electrolyte is 1 M $HClO_4$. (c, d) Two enlarged views of the cell in (b), showing the tip positions of one foil while the voltage applied between two foils changes between $+1$ V (c) and -1 V (d). Reproduced from Kramer *et al.*[10]

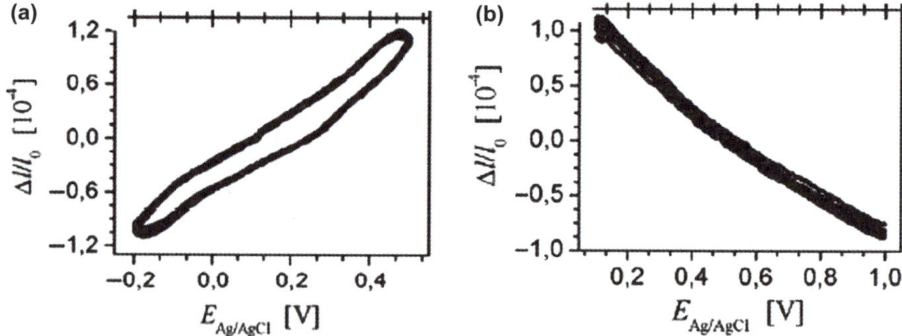

Figure 7.11   *In situ* dilatometry studies $\Delta l/l_0 \sim E$, of (a) oxidized and (b) oxide-free porous platinum immersed in 0.7 M NaF electrolyte. Reproduced from Viswanath *et al.*[27]

oxygen adsorbate monolayer or an oxide layer of atomic thickness that is stable in potential windows of several hundred millivolts. With the adsorbate layer in place, *in situ* dilatometry experiments reveal that the surface stress variation during pseudo-capacitive charging is distinctly different from that of the clean surface (see Figure 7.11). As will be discussed below, this offers an opportunity to develop a high-performance nanoporous metal actuator.

For both Pt and Au, and in double-layer capacitive charging as well as the oxygen adsorption regimes of cyclic voltammograms, the surface stress decreases, and nanoporous metal expands when the surface is charged positively. These trends imply a negative-valued surface stress–charge response parameter, ç. While sweeping the potential in the oxygen adsorption region, it was found that the Pt surface can be (partially) oxidized, and the ç eventually

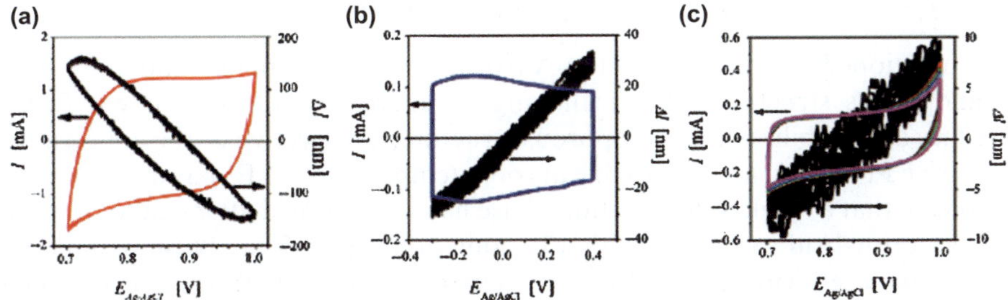

Figure 7.12 Cyclic voltammogram and length change of nanoporous gold measured in 1 M HClO$_4$ at a scan rate of 10 mV s^{-1}: (a) as-dealloyed sample with initial oxidized surface measured in the potential range 0.7–1 V. (b) Reduced sample measured in the double-layer potential region. (c) Reduced sample measusred in the potential range 0.7–1 V. The sample has been pre-treated by by scanning in the same potential range for full oxygen coverage. Reproduced from Jin *et al.*[23]

becomes positive-valued.[27] Interestingly, the negative-valued ς can be restored after removal of oxide by reduction, and the positive-valued ς can be re-gained by repeating the anodic oxidization process.

A similar positive-valued surface stress charge response has been observed in nanoporous gold.[23] This can only be observed in the as-dealloyed state and in samples that were dealloyed at very positive potentials. Figure 7.12(a) shows a nanoporous gold sample with an oxidized surface contracting during positive charging, indicating an 'abnormal' positive ς similar to oxidized nanoporous Pt. The surface state of nanoporous gold can also be converted to a clean state by reduction, which restores the normal negative-valued ς (in the double-layer region) as can be expected (Figure 7.12(b)). Anodic oxidation has also been attempted in nanoporous gold, in order to regain the positive surface stress–charge response as in nanoporous Pt. However, these attempts were unsuccessful. Dilatometry experiments always detect a positive Δf–E relation, and thus a negative ς value, even after the sample has been oxidized by polarizing at very positive potentials for long periods (hours to days).[23]

Although the underlying mechanisms of the abnormal positive ς at oxide-covered metal surfaces are not well understood, and the structure of surface oxide has not been completely identified, the presence of a surface oxide may be advantageous for the actuation performance. The surface atom mobility is lowered when the surface is covered with oxide. This stabilizes the nanoporous structure at very small sizes (a few nanometers), which is beneficial to the actuation performance. Removal of the surface oxide immediately gives rise to the structural coarsening, from ∼4 nm to around 20 nm for nanoporous gold.[23] Accordingly, the maximum strain amplitude is observed to be ∼0.24% for the as-dealloyed nanoporous gold samples (with an oxygen-covered surface). The amplitude was decreased by one order of magnitude (to <0.03%) during reduction, accompanied by the sign inversion of strain-potential response (or surface stress–charge response).[23]

7.4.1.3   *Actuation with Nanoporous AuPt*

As mentioned above, adding Pt to AuAg alloys yields a smaller and more stable nanoporous structure after dealloying.[9,31] Increasing the Pt content further refines and stabilizes the nanoporous structure, irrespective of the concern that massive Pt and Au are immiscible at room temperature.[9] TEM and XRD results indicate that a single solid solution phase has been formed after dealloying in all nanoporous AuPt samples, with a Pt content of up to 50 at.% (in nanoporous structure after dealloying). However, electrochemical experiments reveal 'phase separation' in the surfaces of all these nanoporous AuPt samples. Features of elemental Au and Pt can be easily identified from the cyclic voltammogram (CV) curves. For each electrochemical process—capacitive charging or adsorption— of each element, the characteristics of surface stress–charge response agree qualitatively with that observed previously in pure nanoporous Au and Pt samples.[9]

Similar to nanoporous gold, the initial surfaces of nanoporous AuPt alloys after dealloying are covered with oxygen. Before and after reduction, the surface stress–charge responses are also different in this material, with reversed signs for samples with oxidized and clean surface states. Unlike nanoporous gold, the oxide of which can be removed during the very first cathodic reduction scan, in nanoporous AuPt samples the oxide was removed gradually in first several potential cycles. This leads to a remarkable gradual change in both strain amplitude and sign inversion of potential–strain response (see Figure 7.13(a)). As well as in nanoporous gold, the large electrochemical actuation strain amplitude has been observed in samples with oxidized surface, in NaF solution. The largest linear strain amplitude, up to 1.3%, has been obtained in

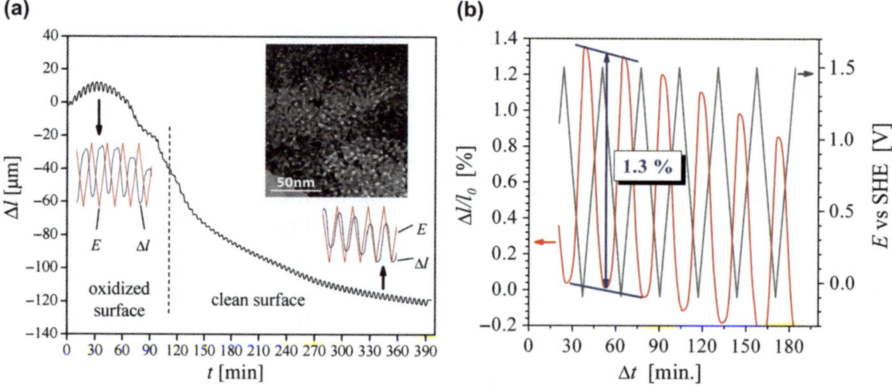

Figure 7.13   (a) Length change, Δl, of a nanoporous $Au_{0.5}Pt_{0.5}$ sample 2 mm long, during repeated cycling of potential, E, between 0.05 and 1.5 V versus SHE in 1 M $HClO_4$ at 10 mV s^{-1}. Note the inversion of sign in $\Delta l - E$ response after cathodic reduction insets. The inset TEM image shows the small structure size of a sample after cathodic reduction. (b) Cyclic length change of nanoporous $Au_{0.8}Pt_{0.2}$ sample with the oxidized surface, along with the potential sweep between -0.1 and 1.5 V in 0.7 M NaF solution at 2 mV s^{-1}, showing the largest strain up to 1.3%. Reproduced from Jin *et al.*[9]

nanoporous $Au_{80}Pt_{20}$ sample as shown in Figure 7.13(b). Further increasing the potential range triggers significant volume shrinkage, indicating that the associated stress may have approached the elastic limit of this material.[9]

As will be discussed in more detail below, a high-performance actuator requires a large and stable actuation strain amplitude and a high material strength so that maximum electrochemical or electrical energy can be converted to mechanical work. The above observations suggest that such an 'ideal' actuator may be built from nanoporous metals with oxidized surfaces. The strain amplitude is usually the largest for nanoporous metals with oxidized surfaces, partly because of the small structure size. The oxidized nanoporous metals are also very stable against structure coarsening, which is the pre-requisite for acquiring the stable actuation strain. Furthermore, nanoporous gold with an oxidized surface is significantly stronger than one with a clean surface,[58] so it can bear a larger mechanical load and convert more energy into mechanical work during actuation.

7.4.2   Chemical Actuation with Nanoporous Metals

The concept of surface-stress induced strain can been generalized to a reversible, chemically driven actuation.[26] Instead of changing the surface charge density in an electrochemical environment as mentioned above, the surface stress here was changed by surface chemistry at the solid–gas interface. The dilatometry experiments detected significant reversible length changes while the nanoporous gold was exposed to ozone (O_3) and carbon monoxide (CO) alternately. The procedure generates and switches back and forth between two distinct surface states: the oxygen-covered surface (after exposing to O_3) and the clean surface (adsorbed oxygen was removed during exposure to CO). The difference of surface stress between two surface states gives rise to the reversible sample-length changes. The largest strain amplitude obtained is 0.5%, as shown in Figure 7.14. The work demonstrates a concept of making metallic

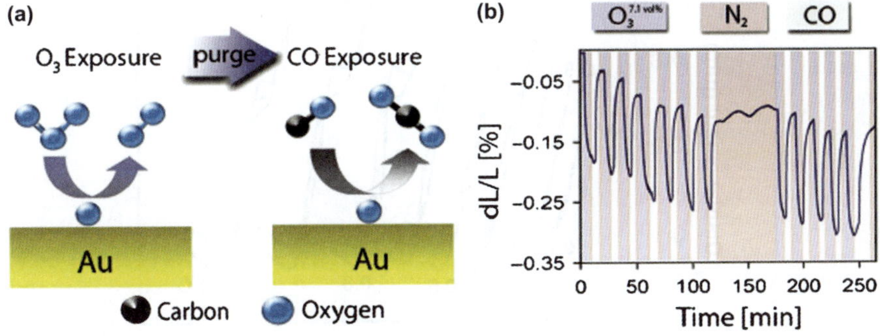

Figure 7.14   Illustration of surface adsorption/desorption processes pompted by gas exchange (a) and of the reversible strain (b) of a dry, nanoporous gold sample while being alternately exposed to a mixture of O_3 and O_2, and pure CO. Reproduced from Biener *et al.*[26]

muscles that could convert the chemical energy directly into mechanical work, analogous to actuation in biological systems.

7.4.3   Electrochemical Actuation with Carbon Nanotubes and Graphene

Sheets of SWNT in an electrochemical environment (see Figure 7.6) have been shown to undergo appreciable strain during potential cycles.[42] Similarly to other electrochemical actuators, the SWNT sheet actuator responds reversibly to the applied potential in an electrolyte; this is shown in Figure 7.15. The behavior is approximately linear, with the potential between 0.1 and 0.4 V (with negligible hysteresis), but it develops a nonlinearity and hysteretic behavior at higher potentials. A pneumatic SWNT sheet actuator has also been presented, which can expand by up to 300% in the thickness direction and contract by up to 3% in the in-plane direction along with the gas evolution during the electrochemical cycling.[59] As an alternative to CNTs, assemblies of multiwalled carbon nanotubes (MWNTs) in various forms have since been investigated for actuation.[45,60,61] MWNTs have been promoted in this context due to their purity, low cost, and the fact that their electrical properties are less dependent on chirality.[60]

As we have mentioned before, among the carbon family, graphene is also of current interest in the context of actuation. Figure 7.16 shows potential-induced bending of the graphene film in Figure 7.7. The film bends to the hydrophobic side at a negative potential and to a hydrophilic side at a positive potential.[54] This phenomenon has been explained in terms of electron injection into a graphene sheet inducing an expansion, while a contraction occurred for hole injection.[54] The in-plane dimension change in graphene sheets has also been studied, and the maximum available strain response was found at 0.85%.[53]

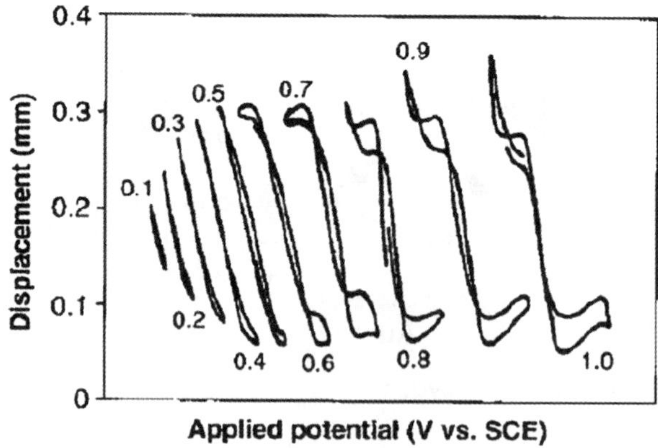

Figure 7.15   Displacement of a carbon-nanotube-based cantilever bimorph during potential cycles. Different potential scan ranges are indicated by labels in the figure. Reproduced from Baughman *et al.*[42]

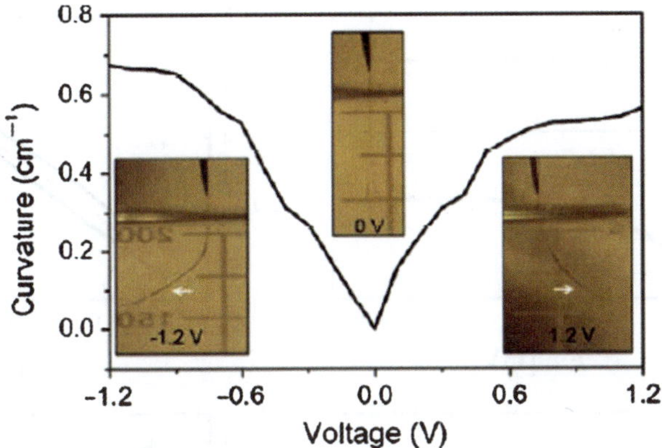

Figure 7.16   Bending of a surface-modified see Figure 7.7, graphene strip during potential change in electrolyte. The graph shows the curvature variation, and the insets show the strip at different potential values. Reproduced from Xie *et al.*[54]

The mechanism behind the potential-controlled expansion/contraction of carbon nanometerials has been investigated in several studies with *ab initio* density functional theory.[62–64] It remains under discussion as to how far (1) electrostatic repulsion due to the space-charge in the double layer and (2) attraction or repulsion due to the population or depletion of bonding/anti-bonding orbitals contributes to the strain. Note that Coulomb repulsion would result in a quadratic variation in surface stress and strain with charge, analogous to what is predicted for fluid electrodes by the Lippmann equation.[16] It has therefore been argued that the linear charge-strain behavior found in experiments with nanoporous metals as well as CA (see below) argues strongly against a significant effect of Coulomb repulsion.[16,38]

We have reviewed the progress in carbon nanotube and graphene-based actuators above, and both materials show a relatively high strain with lower electrode potential. However, the further development is currently hampered by the fact that typical CNT arrays and graphene paper cannot be loaded in compression. It is therefore of interest to search for alternative materials that can have a large strain amplitude, while the stiffness and strength are maintained, and can be loaded compressively.

7.4.4   Electrochemical Actuation with CA

This section explores the potential of CA for actuator application. Obvious benefits of this material are its low mass density and low cost.

Figure 7.17 shows typical *in situ* dilatometry results with a CA. The CV of current *I vs.* electrode potential *E* in Figure 7.17(a) is featureless, consistent with dominantly double-layer charging. In Figure 7.17(b), the graph of charge, *Q, vs. E* is closed, indicating that the charge is conserved. Figure 7.17(c) shows

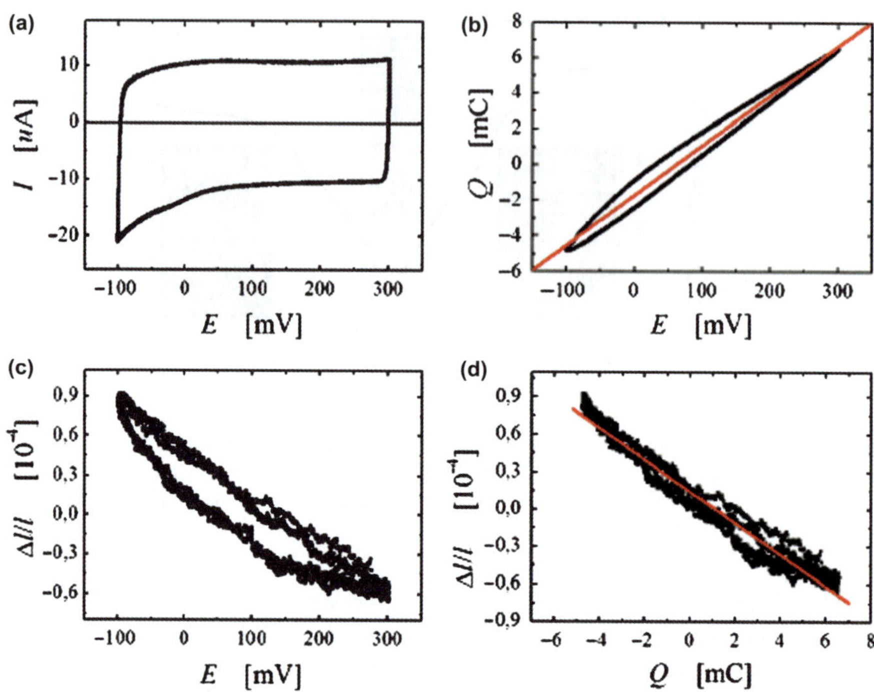

Figure 7.17 Results of *in situ* dilatometry during cyclic scans of the potential, E, in 0.7 M NaF solution and at a scan rate of 0.4 mV s^{-1}. (a) Current I versus E. (b) Total charge Q versus E. (c) strain $\Delta l/l$ versus E. (d) $\Delta l/l$ versus Q. Reproduced from Shao *et al.*[38]

the strain, $\Delta l/l$, as a function of E. The behavior is essentially linear, though with a noticeable hysteresis. Figure 7.17(d) presents the strain–charge correlation, which is also almost linear. The hysteresis is significantly smaller than that of the strain-potential plot shown in Figure 7.17(c). The data thus support the notion that the strain is primarily a function of the charge.

Some experiments with CA show an irreversible contraction superimposed to the reversible length change. This irreversible part might be related to plastic deformation and/or to an electrochemical activation process (*e.g.*, removal of surface contamination) during the cycling. The irreversible process is slow and has little impact on cycles at a higher scan rate. Figure 7.18 exemplifies the stable and reversible expansion/contraction cycles. Note that the deformation is isotropic, so the volume strain emerges as three times the linear strain. The largest linear and volumetric strain amplitudes achieved in this way with CA are 2.2 and 6.6%, respectively.[39] Both characteristics exceed what has been reported with nanoporous metals so far.[1,9] The strain amplitudes are also larger than those of the most common piezoelectric actuation materials.

Figure 7.19 presents the strain amplitudes of CA samples with different specific surface areas (as measured by BET adsorption isotherms), for two different ranges. The figure demonstrates a linear dependence of strain on

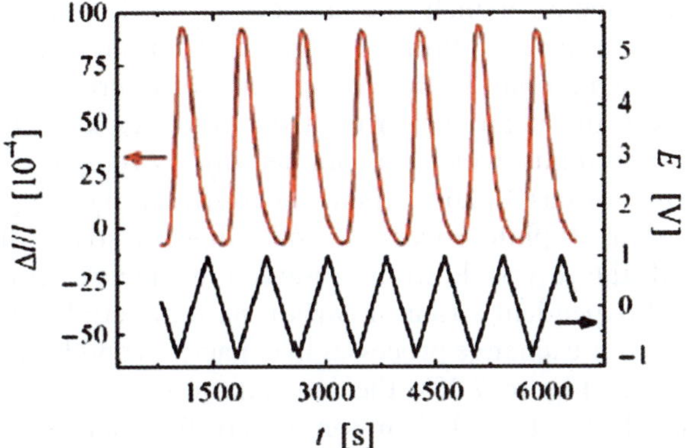

Figure 7.18 *In situ* dilatometry measurement of CA with the potential range from -1.0 to 1.0 V in 0.7 M NaF solution. Strain $\Delta l/l$ in response to the potential E variation at a scan rate of $5\,\text{mV}\,\text{s}^{-1}$. Note that the strain changes reversibly.[39]

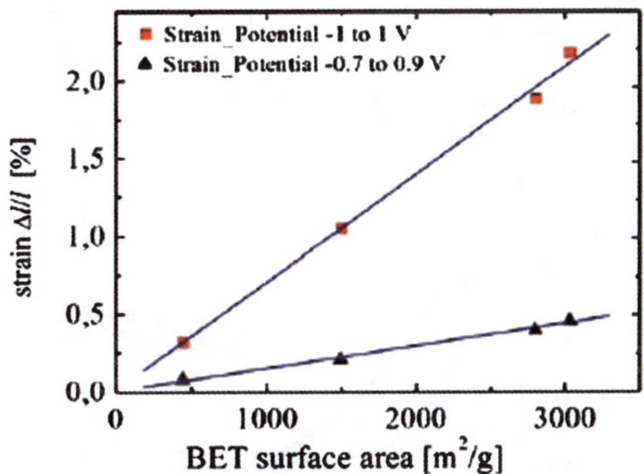

Figure 7.19 *In situ* scans with different samples in 0.7 M NaF at a scan rate of $1\,\text{mV}\,\text{s}^{-1}$ and in a potential range from -0.7 V to 0.9 V (dark triangle) and -1.0 to 1.0 V (red square), respectively.[39]

specific surface area. The significant implication is that the strain is governed by a process at the surface, and not by intercalation (which is a bulk property). The same conclusion is also suggested by the electrochemical signature, which is consistent with processes at the electrode surface. The reversible length changes in the CA samples are therefore driven by processes similar to surface stress change in the nanoporous metals. Yet, in view of the extremely large specific surface area, it may not be appropriate to separate volume- and surface excess properties here. It is also not adequate to distinguish between bulk stress and surface stress. The analogous processes that lead to surface-stress changes in a nanoporous metal, namely the changes in local charge density and the resulting changes in the equilibrium atomic bond length, extend here essentially through the entire material. In this context, it is of interest to note the analogy between charge transfer at carbon electrode surfaces and charge transfer in intercalation compounds, which was pointed out by Baughman *et al.*[42] In fact, the compilation[42] of basal plane expansion/contraction during intercalation of various species in graphite covers a total range in strain of $\sim$2%. This is compatible with the reversible strain observed in the CA actuators.

The potential-dependent strain of carbon nanomaterials has also been used as a signature of ion-exchange processes in capacitors and battery electrodes. In Hahn *et al.*[40] (see Figure 7.20), the intercalation in activated carbon black immersed in 1 M tetraethylammoniumtetrafluoroborate (TEABF$_4$) in

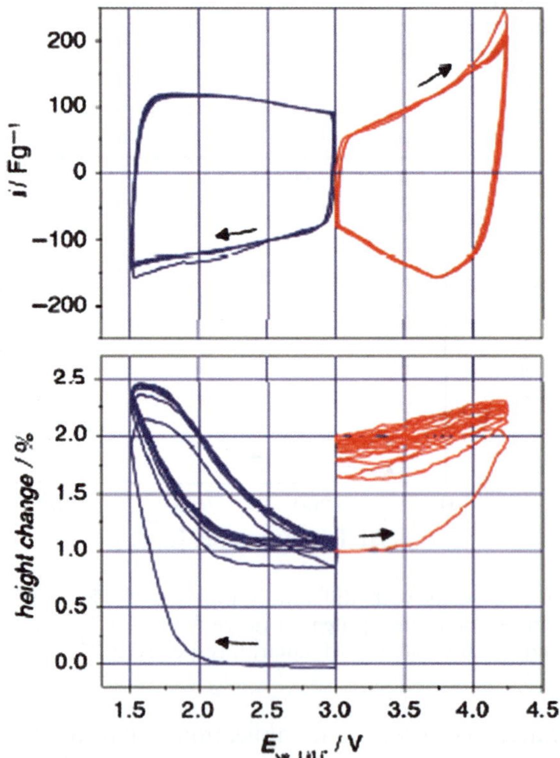

Figure 7.20   Cyclic voltammograms (top) and simultaneous height change (bottom) of the activated carbon black in an organic electrolyte. Two potential ranges are shown in the potential range from 1.5 to 3 V (blue) and 3 to 4.25 V (red) *vs.* Li/Li$^+$. The current is normalized to the weight of the dry carbon. Reversible straining is observed after four subsequent cycles. The amplitude is 1.5% in the potential window 1.5–3 V. Similar behavior is observed in the anodic potential range 3–4.25 V, but with a smaller strain amplitude, less than 0.5%. Reproduced from Hahn *et al.*[40]

acetonitrile (AN) electrolyte results in a strain amplitude in the order of 1%. Hahn *et al.*[40] associate the length change with intercalation/de-intercalation. This phenomenon is common in battery electrodes, and more generally the absorption of species in the bulk is also found in some schemes for hydrogen storage. The process differs fundamentally from the mechanism discussed above for CA samples, namely surface stress change.[38,39] For an experiment where both processes act simultaneously and can be quantitatively separated, see the studies of hydrogen absorption in nanocrystalline Pd.[65,66]

7.5 Two Important Characteristics

7.5.1 Response Time

For actuator applications, the response time is important. Figure 7.21(a) shows the time-dependent length change in a millimeter-sized sample of nanoporous gold during a series of potential jumps from -0.2 V to 1 V and back again in 50 mM sulfuric acid. The halftimes of the jumps in current and strain are similar, 220 and 270 ms, respectively. The strain amplitude (in the form of the percentage of the maximum strain amplitude) at a frequency of 0.3 Hz is almost identical to that during slower switching (Figure 7.21(b)), which is consistent with the response time given above. The bilayer foil samples as shown in Figure 7.10 above react with a considerably faster intrinsic timescale. The time constant of the charging current here is 25 ms,[10] considerably faster than in the thicker dilatometer samples. This agrees qualitatively with the expectation that the transport of ions into or out of the pores will be accelerated as the sample dimension is reduced and the transport path shortened.

In contrast to nanoporous metal, the halftimes of the jumps in current and strain of CAs are typically much longer. For example, with the same sample in Figure 7.18, the values are 145 s and 165 s, respectively. Experiments by the authors support the expectation that the reaction will be faster within solutions of higher electric conductivity.

7.5.2 Work Density

As an important figure of merit for actuation, let us inspect the work density, w, of an actuator, and discuss its scaling with the ligament size in a nanoporous metal. We assume that the actuator is a linear elastic solid with an effective macroscopic Young's modulus Y_{eff}. Since the maximum strain amplitude, ε_{max}, is independent of the load, let the actuator be loaded so that the effective (uniaxial) stress in the material is σ, compensating the load σ_A with A the macroscopic cross-sectional area. When the actuator is driven to its maximum strain, the work density (mechanical work per volume) will be:

$$w_{\text{V},1} = \varepsilon_{\text{max}}\sigma. \tag{7.7}$$

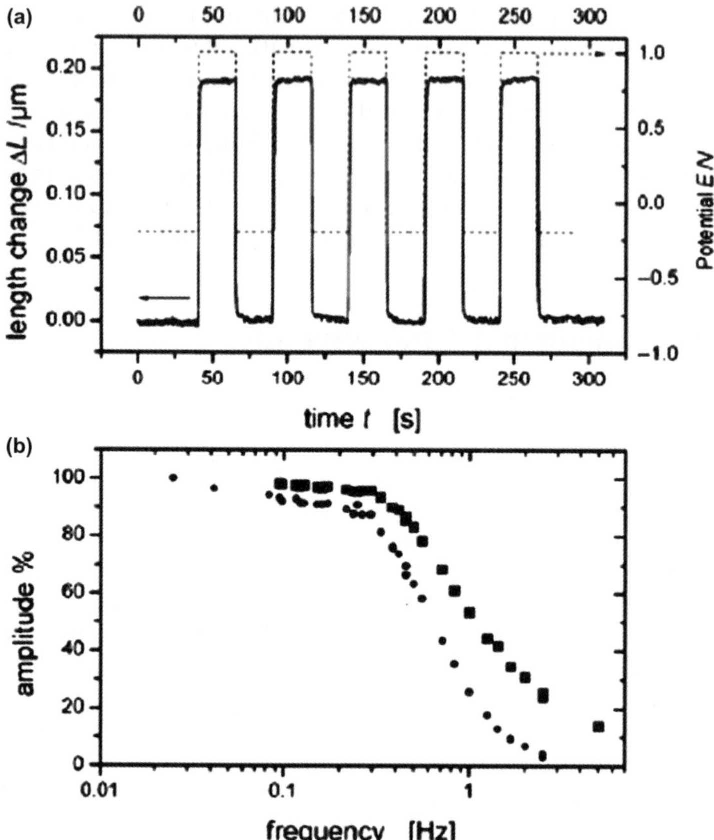

Figure 7.21 (a) Length change ΔL of a millimeter-sized nanoporous gold sample versus time, measured during a series of successive potential jumps (dashed line): potential. (b) Frequency dependence of the amplitude during potential jumps for the sample in (a). Large squares: amplitude of the charge curve. Small circles: amplitude of the length change. The dilatometer's maximum sampling rate of $10\,\mathrm{s}^{-1}$ limits the experimental strain amplitude at a high frequency. Reproduced from Kramer *et al.*[10]

This equation appears counterintuitive, since it may suggest that the work density of linear elastic actuators can be increased simply by increasing the load. The apparent inconsistency arises because eqn (7.7) considers only part of an actuation cycle. Useful estimates of the work density can only be obtained by considering full cycles. If we were to stick with the constant load scenario, then the return stroke would give work density $-\varepsilon_{\mathrm{max}}\,\sigma$, so that the net work in this cycle would be zero. This simply reflects the fact that at the end of the cycle the load has returned to its initial position, so no useful action has occurred. A more interesting cycle considers the following four successive steps: 1, loading to the stress σ; 2, extension to strain $\varepsilon_{\mathrm{max}}$ under load; 3, unloading; and 4, contraction at no load. This cycle does useful work, since the load remains displaced after the cycle is completed, and the actuator is free to lift another load in the next cycle. Here, work is only done in cycle 2. Since the actuator contracts by $-\sigma/Y_{\text{eff}}$ during step 1, the useful stroke in step 2 is reduced from $\varepsilon_{\max}$ at no load to $\varepsilon_{\max}-\sigma/Y_{\text{eff}}$ with the load in place. This implies that the useful work density here is[9]

$$w_{\text{V}} = (\varepsilon_{\max}-\sigma/Y_{\text{eff}})\sigma. \tag{7.8}$$

It can be seen that the work density is now a quadratic function of the load. This function is found to exhibit a maximum at the 'matched load',

$$\sigma_{\text{M}} = \tfrac{1}{2}Y_{\text{eff}}\varepsilon_{\max}. \tag{7.9}$$

At the matched load, the work density is

$$w_{\text{V},\max} = \tfrac{1}{4}Y_{\text{eff}}\varepsilon_{\max}^2. \tag{7.10}$$

This expression scales with the effective Young's modulus, illustrating the benefit of stiffness for achieving a high strain energy density.

To achieve the maximum work density without mechanical failure, the actuator's yield strength, σ_{Y}, must be at least equal to σ_{M}. If the load is limited to a lower value by the insufficient strength, the maximum work density is given by eqn (7.8) with σ_{Y} substituted for σ:

$$w_{\text{V},\max} = (\varepsilon_{\max}-\sigma_{\text{Y}}/Y_{\text{eff}})\sigma_{\text{Y}}. \tag{7.11}$$

For nanoporous metals, σ_{Y} and (through the specific surface area) σ_{M} will vary systematically with the ligament size, L.[7] Even though the details of this variation are the subject of current research,[7] it is of interest to inspect the link. The local yield strength of individual ligaments increases with L according to[7,26,67,68]

$$\sigma_{\text{Y}}^{\text{L}} \propto L^{-\alpha}, \tag{7.12}$$

with α an empirical exponent that takes on values of 0.50–0.75. Furthermore, the macroscopic yield strength satisfies the Gibson–Ashby scaling equation,[69]

$$\sigma_{\text{Y}} \propto 0.3\sigma_{\text{Y}}^{\text{L}}\varphi^{3/2}, \tag{7.13}$$

with φ the solid volume fraction. The scaling equation for Young's modulus is[69]

$$Y_{\text{eff}} \propto Y_0\varphi^2, \tag{7.14}$$

with Y_0 Young's modulus for the massive solid phase. Finally, the maximum strain scales as (see Section 7.2.2)

$$\varepsilon_{max} \propto 1/L. \tag{7.15}$$

Eqns (7.12)–(7.15) imply in particular that the strength can be increased by decreasing L, so that the work density will be compliance-limited, as in eqn (7.10), and not yield-limited as in eqn (7.11). Inserting the above results into eqn (7.10) then results in

$$w_{V,max} \propto \varphi^2/L^2. \tag{7.16}$$

In other words, the maximum work density is achieved with the smallest ligament sizes.

The work density of actuator materials is typically specified in terms of a different stored energy density parameter, which is given by $1/2 Y_{eff} \, \varepsilon^2_{max}$. By comparison to eqn (7.10), it can be seen that this parameter overestimates the maximum work density of linear elastic actuator materials by a factor of 2. In the interest of comparability, the larger value ($1/2 \, Y_{eff} \, \varepsilon^2_{max}$) is here used for specification.

A mass-specific strain energy density, w_M, is related to w_V and to the mass density ρ by $w_M = w_V/\rho$.

The strain energy density triples when one considers the work done by the volume change, since the surface stress-driven macroscopic strain is isotropic in the nanoporous samples. In Jin *et al.*,[9] the energy density parameters w_V and w_M were determined for nanoporous $Au_{80}Pt_{20}$ to be $6.0 \, MJ \, m^{-3}$ and $1200 \, J \, kg^{-1}$ respectively, which is more than 100 times that of skeletal muscle ($\sim 40 \, kJ \, m^{-3}$).[70] The values of w_V and w_M for CA actuators, as specified in Shao *et al.*,[39] are $0.2 \, MJ \, m^{-3}$ and $1455 \, J \, kg^{-1}$, respectively.

A comparison of characteristic figures of merit for different types of actuator materials is presented in Figure 7.22. While the largest strain amplitudes are reached by dielectric elastomers[2] and by skeletal muscles,[70] the highest stiffness is observed for piezoceramics[3] and nanoporous metals.[9] Among the high-stiffness and high-strength materials, nanoporous metals are indeed competitive in regard to the work density. If the oxygen-covered surface state can be used, their work density is even significantly larger than that of the representative piezoceramics. Due to their extremely low density, CAs show an attractive mass-specific strain energy density, especially in view of their low cost. Furthermore, by making comparisons with all alternative materials shown in the chart, nanoporous metals and CAs are distinguished by their low operating voltage. The combination of low voltage with large strain amplitude and high stiffness and strength characterize nanoporous metals as materials for effective conversion of electric energy into mechanical work. The remarkably high mass-specific strain energy density also makes CAs good candidates for actuation, since they are inexpensive materials compared to some noble metals.

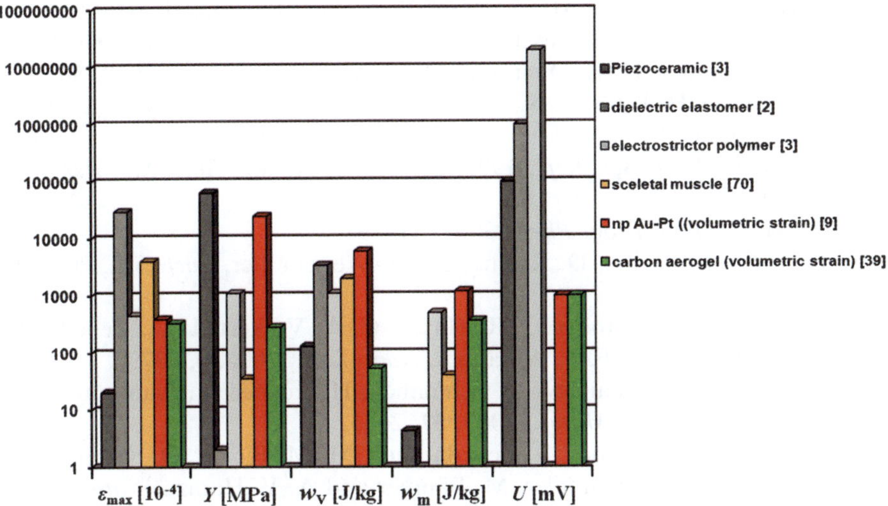

Figure 7.22   Column chart plot of characteristics of various actuator materials. ε_{max}, maximum strain; Y, effective modulus; w_V, volume-specific strain energy density; w_m, mass-specific strain energy density; U, operating voltage for 100 μm actuator size. Adopted from Jin *et al.*[9]

References

1. J. Weissmüller, R. N. Viswanath, D. Kramer, R. Würschum and H. Gleiter, *Science*, 2003, **300**, 312.
2. R. Pelrine, R. Kornbluh, Q. Pei and J. Joseph, *Science*, 2000, **287**, 836.
3. Q. Zhang, V. Bharti and X. Zhao, *Science*, 1998, **280**, 2101.
4. Z. Chen and Q. Zhang, *MRS Bull.*, 2008, **33**, 183.
5. J. Erlebacher and R. Seshadri, *MRS Bull.*, 2009, **34**, 561.
6. Y. Ding and M. W. Chen, *MRS Bull.*, 2009, **34**, 569.
7. J. Weissmuller, R. C. Newman, H. J. Jin, A. M. Hodge and J. W. Kysar, *MRS Bull.*, 2009, **34**, 577.
8. H.-J. Jin and J. Weissmüller, *Adv. Eng. Mater.*, 2010, **12**, 721.
9. H.-J. Jin, X. L. Wang, S. Parida, K. Wang, M. Seo and J. Weissmüller, *Nano Lett.*, 2010, **10**, 187.
10. D. Kramer, R. N. Viswanath and J. Weissmüller, *Nano Lett.*, 2004, **4**, 793.
11. F. Weigend, F. Evers and J. Weissmüller, *Small*, 2006, **2**, 1497.
12. Y. Umeno, C. Elsässer, B. Meyer, P. Gumbsch, M. Nothacker, J. Weissmüller and F. Evers, *Europhys. Lett.*, 2007, **78**, 13001.
13. Y. Umeno, C. Elsässer, B. Meyer, P. Gumbsch and J. Weissmüller, *Europhys. Lett.*, 2008, **84**, 13002.
14. J. Weissmüller and J. W. Cahn, *Acta Mater.*, 1997, **45**, 1899.
15. R. Shuttleworth, *Proc. Phys. Soc. A*, 1950, **63**, 444.
16. D. Kramer and J. Weissmüller, *Surf. Sci.*, 2007, **601**, 3042.

17. J. Diao, K. Gall and M. L. Dunn, *Phys. Rev. B*, 2004, **70**, 075413.
18. D. A. Crowson, D. Farkas and S. G. Corcoran, *Scripta Mater.*, 2009, **61**, 497.
19. J. Weissmüller, H.-L. Duan and D. Farkas, *Acta Mater.*, 2010, **58**, 1.
20. L. H. Shao, H. J. Jin, R. N. Viswanath and J. Weissmüller, *Europhys. Lett.*, 2010, **89**, 66001.
21. H. Ibach, *Surf. Sci. Rep.*, 1997, **29**, 193. H. Ibach, *Surf. Sci. Rep.*, 1999, **35**, 71.
22. W. Haiss, *Rep. Prog. Phys.*, 2001, **64**, 591.
23. H. J. Jin, S. Parida, D. Kramer and J. Weissmüller, *Surf. Sci.*, 2008, **602**, 3588.
24. J. Biener, A. M. Hodge, J. R. Hayes, C. A. Volkert, L. A. Zepeda-Ruiz, A. V. Hamza and F. F. Abraham, *Nano Lett.*, 2006, **6**, 2379.
25. H. J. Jin, L. Kurmanaeva, J. Schmauch, H. Rösner, Y. Ivanisenko and J. Weissmüller, *Acta Mater.*, 2009, **57**, 2665.
26. J. Biener, A. Wittstock, L. Zepeda-Ruiz, M. M. Biener, D. Kramer, R. N. Viswanath, J. Weissmüller, M. Bäumer and A. V. Hamza, *Nature Mater.*, 2009, **8**, 47.
27. R. N. Viswanath, D. Kramer and J. Weissmüller, *Electrochim. Acta*, 2008, **53**, 2757.
28. J. Erlebacher, M. J. Aziz, A. Karma, N. Dimitrov and K. Sieradzki, *Nature*, 2001, **410**, 450.
29. Y. Ding, Y.-J. Kim and J. Erlebacher, *Adv. Mater.*, 2004, **16**, 1897.
30. R. Li and K. Sieradzki, *Phys. Rev. Lett.*, 1992, **68**, 1168.
31. J. Snyder, P. Asanithi, A. B. Dalton and J. Erlebacher, *Adv. Mater.*, 2008, **20**, 1876.
32. S. Parida, D. Kramer, C. A. Volkert, H. Rösner, J. Erlebacher and J. Weissmüller, *Phys. Rev. Lett.*, 2006, **97**, 035504.
33. H. J. Jin, D. Kramer, Y. Ivanisenko and J. Weissmüller, *Adv. Eng. Mater.*, 2007, **9**, 849.
34. Y. Sun, K. P. Kucera, S. A. Burger and T. J. Balk, *Scripta Mater.*, 2008, **58**, 1018.
35. N. A. Senior and R. C. Newman, *Nanotechnology*, 2006, **17**, 2311.
36. A. J. Forty and P. Durkin, *Philos. Mag. A*, 1980, **42**, 295.
37. S. Van Petegem, S. Brandstetter, R. Maass, A. M. Hodge, B. S. El-Dasher, J. Biener, B. Schmitt, C. Borca and H. Van Swygenhoven, *Nano Lett.*, 2009, **9**, 1158.
38. L.-H. Shao, J. Biener, D. Kramer, R. N. Viswanath, T. F. Baumann, A. V. Hamza and J. Weissmüller, *Phys. Chem. Chem. Phys.*, 2010, **12**, 7580.
39. L.-H. Shao, J. Biener, H.-J. Jin, T. F. Baumann and J. Weissmüller. Unpublished results.
40. M. Hahn, O. Barbieri, R. Gallay and R. Kötz, *Carbon*, 2006, **44**, 2523.
41. T. F. Baumann, M. A. Worsley, T. Y. Han and J. H. Satcher, Jr., *J. Non-Cryst. Solids*, 2008, **354**, 3513.
42. R. H. Baughman, C. Cui, A. A. Zakhidov, Z. Iqbal, J. N. Barisci, G. M. Spinks, G. G. Wallace, A. Mazzoldi, D. D. Rossi, A. G. Rinzler, O. Jaschinski, S. Roth and M. Kertesz, *Science*, 1999, **284**, 1340.

43. J. N. Barisci, G. M. Spinks, G. G. Wallace, J. D. Madden and R. H. Baughman, *Smart Mater. Struct.*, 2003, **12**, 549.
44. R. H. Baughman, A. A. Zakhidov and W. A. de Heer, *Science*, 2002, **297**, 787.
45. Y. H. Yun, V. Shanov, Y. Tu, M. J. Schulz, S. Yarmolenko, S. Neralla, J. Sankar and S. Subramaniam, *Nano Lett.*, 2006, **6**, 689.
46. S. Liu, Y. Liu, H. Cebeci, R. Guzmán de Villoria, J.-H. Lin, B. L. Wardle and Q. M. Zhang, *Adv. Funct. Mater.*, 2010, **20**, 3266.
47. C. Li, E. T. Thostenson and T.-W. Chou, *Compos. Sci. Technol.*, 2008, **68**, 1227.
48. J. Fraysse, A. I. Minett, O. Jaschinski, G. S. Duesberg and S. Roth, *Carbon*, 2002, **40**, 1735.
49. J. Liu, A. G. Rinzler, H. Dai, J. H. Hafner, R. K. Bradley, P. J. Boul, A. Lu, T. Iverson, K. Shelimov, C. B. Huffman, F. Rodriguez-Macias, Y.-S. Shon, T. R. Lee, D. T. Colbert and R. E. Smalley, *Science*, 1998, **280**, 1253.
50. A. G. Rinzler, J. Liu, H. Dai, P. Nikolaev, C. B. Huffman, F. J. Rodríguez-Macías, P. J. Boul, A. H. Lu, D. Heymann, D. T. Colbert, R. S. Lee, J. E. Fischer, A. M. Rao, P. C. Eklund and R. E. Smalley, *Appl. Phys. A*, 1998, **67**, 29.
51. A. K. Geim and K. S. Novoselov, *Nature Mater.*, 2007, **6**, 183.
52. J. Liang, Y. Huang, J. Oh, M. Kozlov, D. Sui, S. Fang, Ray H. Baughman, Y. Ma and Y. Chen, *Adv. Funct. Mater.*, 2011, **21**, 3778.
53. X. Xie, H. Bai, G. Shi and L. Qu, *J. Mater. Chem.*, 2011, **21**, 2057.
54. X. Xie, L. Qu, C. Zhou, Y. Li, J. Zhu, H. Bai, G. Shi and L. Dai, *ACS Nano*, 2010, **4**, 6050.
55. V. Tabard-Cossa, M. Godin, I. J. Burgess, T. Monga, R. B. Lennox and P. Grutter, *Anal. Chem.*, 2007, **79**, 8136.
56. R. N. Viswanath, D. Kramer and J. Weissmüller, *Langmuir*, 2005, **21**, 4604.
57. M. Smetanin, R. N. Viswanath, D. Kramer, D. Beckmann, T. Koch, L. A. Kibler, D. M. Kolb and J. Weissmüller, *Langmuir*, 2008, **24**, 8561.
58. H. J. Jin and J. Weissmüller, *Science*, 2011, **332**, 1179.
59. G. M. Spinks, G. G. Wallace, L. S. Fifield, L. R. Dalton, A. Mazzoldi, D. D. Rossi, I. I. Khayrullin and R. H. Baughman, *Adv. Mater.*, 2002, **14**, 1728.
60. M. Hughes and G. M. Spinks, *Adv. Mater.*, 2005, **17**, 443.
61. T. Mirfakhrai, J. Oh, M. Kozlov, E. C. Fok, M. Zhang, S. Fang, R. H. Baughman and J. D. W. Madden, *Smart Mater. Struct.*, 2007, **16**, S243.
62. G. W. Rogers and J. Z. Liu, *J. Am. Chem. Soc.*, 2011, **133**, 10858.
63. G. Sun, Jenö Kürti, M. Kertesz and R. H. Baughman, *J. Am. Chem. Soc.*, 2002, **124**, 15076.
64. M. Verissimo-Alves, B. Koiller, H. Chacham and R. B. Capaz, *Phys. Rev. B*, 2003, **67**, 161401R.
65. J. Weissmüller and C. Lemier, *Phys. Rev. Lett.*, 1999, **82**, 213.

66. C. Lemier and J. Weissmüller, *Acta Mater.*, 2007, **55**, 1241.
67. C. A. Volkert and E. T. Lilleodden, *Phil. Mag.*, 2006, **86**, 5567.
68. A. M. Hodge, J. Biener, J. R. Hayes, P. M. Bythrow, C. A. Volkert and A. V. Hamza, *Acta Mater.*, 2007, **55**, 1343.
69. L. J. Gibson and M. F. Ashby, *Cellular Solids: Structure and Properties*, Cambridge University Press, Cambridge, 1997.
70. J. D. W. Madden, B. Schmid, M. Hechinger, S. R. Lafontaine, P. G. A. Madden, F. S. Hover, R. Kimball and I. W. Hunter, *IEEE J. Oceanic Eng.*, 2004, **29**, 706.

Surface Chemistry and Catalysis

ARNE WITTSTOCK,[*][a,b] JÜRGEN BIENER[a] AND
MARCUS BÄUMER[b]

[a] Nanoscale Synthesis and Characterization Laboratory, Lawrence Livermore
National Laboratory, 7000 East Avenue, Livermore, CA 94500, USA;
[b] Center for Environmental Research and Sustainable Technology and
Institute of Applied and Physical Chemistry, University of Bremen, Leobener
Strasse, 28359 Bremen, Germany
*Email: wittstock1@llnl.gov

8.1   Introduction

The history of chemical catalysis started about 200 years ago in the early 19th
century, along with the evolution of chemical science. One can say that the
concept of catalysis in chemistry is nearly as old as the atomic theory published
by John Dalton in 1806. Scientists made the observation that a chemical
reaction would not proceed unless a material such as a metal was present.
Seemingly, it was only the presence of this material that 'magically' facilitated
the chemical reaction, as it was not consumed, and its physical state was not
changed. In 1836, the Swedish scientist Jöns Jacob Berzelius tried to combine
those observations and dubbed it catalysis. He said that the 'catalytic
power means that substances are able to awake affinities that are asleep at this
temperature by their mere presence'.

Most people nowadays probably know the term catalysis from the automotive
catalyst, also dubbed automotive converter, which became a self-evident part of
the exhaust-gas management in modern cars, for example converting poisonous
constituents such as carbon monoxide or nitrogen oxides in non-poisonous

RSC Nanoscience & Nanotechnology No. 22
Nanoporous Gold: From an Ancient Technology to a High-Tech Material
Edited by Arne Wittstock, Jürgen Biener, Jonah Erlebacher and Marcus Bäumer
© Royal Society of Chemistry 2012
Published by the Royal Society of Chemistry, www.rsc.org

carbon dioxide and nitrogen. Less commonly known is that catalysts in the form of biological molecules such as enzymes are the very bases for life itself or that in the modern chemical industry, the vast majority of products are produced with the aid of catalysts. The reason for this is that a catalyst can provide the very handle and control of a chemical conversion. Life would not be possible with exact control of chemical reactions inside the biological cell.

Today we know that it is the specific interaction, the chemical bonding of the catalyst and the reactants in the course of chemical reaction that opens up a specific kinetic pathway for the reaction. It is noteworthy that a catalyst will not change the thermodynamic restrains of a reaction such as the free enthalpy ΔG and, as a consequence, the equilibrium of the reaction. For example, the heat (i.e. enthalpy) released by the oxidation of carbon monoxide to carbon dioxide using molecular oxygen remains unchanged ($\Delta H° \sim -283$ kJ mol^{-1}) whether it is catalyzed or not. However, it is the temporary chemical bonding of at least one of the reactants with the catalyst that opens a 'faster' pathway for the reaction. In this way, the course of the reaction is altered so that the activation energy of the overall reaction is lowered, and the reaction proceeds faster and at lower temperatures, respectively. Thus, in the light of green chemical technologies, novel energy harvesting, and storage concepts, catalysis plays a central role.

Depending on whether the catalyst and the reactants form a homogeneous compound (one liquid phase) or form a heterogeneous compound consisting of at least two different phases (gas/liquid/solid phase), one distinguishes between homogeneous and heterogeneous catalysis. As nanoporous gold is a solid that inevitably forms heterogeneous compounds with the reactants, we will make reference hereafter only to heterogeneous catalysis (either liquid or gas).

In heterogeneous catalysis, the reactions take place on the very surface region of the catalyst (see Figure 8.1). Reactants will have to impinge on the catalyst surface and be chemically bonded, *i.e.* 'chemisorbed', in order to become activated for a specific chemical reaction. The chemistry of the catalyst surface

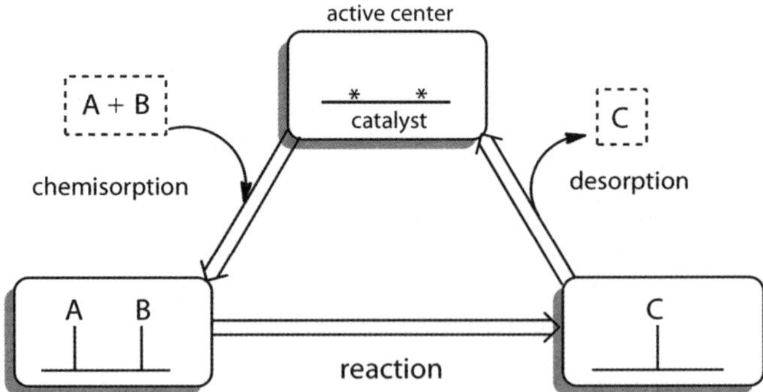

Figure 8.1   Schematic of an ideal catalytic cycle. First, the reactants are chemisorbed and bonded to the catalyst surface. By this means, they are activated for the chemical reaction. The evolving products can desorb from the surface, leading to the initial state of the surface.

thus is of crucial importance for the catalytic performance in terms of activity and selectivity of the catalyst. In order to achieve a high catalytic activity, one has to balance out the strength of the chemical bond formed. This is also dubbed the principle of Sabatier. In order to achieve an activation of the adsorbed molecule, it has to be chemically bonded, and electron density has to be transferred from the catalyst surface into and out of the bonds within the adsorbed reactant ('activation') on the one hand. On the other hand, in order to achieve a sufficiently high turn over, the reactants must not stick strongly to the surface in order to avoid poisoning of the catalyst by blocking the available active sites.

Considering the group of transition metals, one can draw a simplistic curve, the so-called volcano curve, which reflects the ideal catalytic activity as a function of the position of the particular metal in the periodic table of the elements. In this simple picture, with increasing atomic number (*i.e.* 'from left to right'), the strength of chemisorption (formally the heat of adsorption) decreases. The bonding of, for example, oxygen with a metal surface such as Ti is too strong to allow a sufficiently high turnover. Yet, with increasing atomic number (*e.g.* from Ti to Cu), the bonding becomes weaker, allowing a sufficiently high turnover, before the chemisorption becomes too weak to observe any appreciable catalytic activity. As an easy measure, this curve can explain why so many highly catalytically active metals can be found in the middle, of the periodic table of the elements (group 8–10, *e.g.*, Fe, Co, Ni, Pd, and Pt).

Although Au can be found close to these metals, in group 11 (Cu, Ag, and Au) it is distinguished by its nobleness and exceptionally weak interaction with adsorbates.[1] First reports on gold as an active catalyst being even better than other metals were published for the hydrogenation of olefins in the 1970s by Bond *et al.*,[2] later in the mid 1980s for the low-temperature oxidation of CO by Haruta *et al.*,[3] or for the chlorination of olefins by Hutchings.[4] Gold was formerly known as 'catalytically dead' due to its apparent inertness. The interaction of gold surfaces with reactants, such as hydrogen or molecular oxygen, is very weak;[1] as a consequence, gold remains in its metallic form without showing any signs of corrosion, for example, but also was considered to be a poor catalyst, for the reasons explained above. Remarkably, if gold is in the form of very small particles in the range of a few nanometers on a suitable oxidic support, it may indeed be catalytically highly active. The reason for this kind of behavior was controversially discussed throughout the last two decades.[5,6] The chemisorption and activation of molecular oxygen especially are hardly detectable on pure gold surfaces and benefit greatly from the presence of the oxidic support and the nanosized particles.

In 2006, Zielasek *et al.*,[7] and in 2007, Ding *et al.*,[8] independently discovered that unsupported nanoporous gold is a highly active catalyst for the low-temperature oxidation of CO. This came as a surprise, as npAu is not made up of nanoparticles and does not contain an oxidic support material, which was thought to be essential for high catalytic activity. At first, it was speculated whether that abundance of low-coordinated surface atoms within the material is the source of its high catalytic activity revealing the genuine catalytic activity of pure nanostructured but unsupported gold.[8,9] Later reports focused on the remaining traces of less noble metals, such as Ag, that remain in small quantities below 1 at.% in the material after preparation.[10] Since metals such as Ag and Cu, which can also be used for the generation of npAu, are known to bond molecular oxygen chemically, very much in contrast to Au, it is very likely that their presence contributes to the catalytic activity.[11–13] Nevertheless, the overall catalytic characteristics of the material in terms of activity and selectivity were found to be governed by the surface chemistry of Au alone; for example, it was observed that CO oxidation takes place at temperatures well below 0 °C,[7] and the highly selective oxidation of alcohols already proceeds at room temperature.[14]

Among others, it is the weak interaction with adsorbates that enables high activity at low temperatures, making gold a great candidate for a predictable and green catalytic concept.[15] The absence of an oxidic support material reduces the degree of complexity of the surface chemistry and catalysis of npAu. Recent studies showed that insights from well-controllable UHV studies can be directly transferred to npAu catalysts working under ambient pressure conditions.[14,16] The latest studies have reported on the combination of npAu with small amounts of metal oxides such as TiO_2 and a greatly enhanced catalytic activity.[11,17]

In the following sections, we summarize and report on the various catalytic applications of npAu starting with surface chemistry and principles derived from UHV model studies followed by applied studies under ambient pressure conditions in the gas and liquid phases. First, we will discuss the interaction of the gold surface (single crystals as well as npAu) with oxygen, as this is the key for oxidation reactions and the source of reactivity. We will then discuss the reaction of oxygen with carbon monoxide and alcohols, increasing the complexity of reaction by adding the possibility of forming several products. In the latter case, not only the activity of the catalyst but also its selectivity will be discussed and interpreted at a molecular level. Following these sections, we will report on important work in the liquid phase, the oxidation of D-glucose and non-carbon-based compounds, organosilanes to organolianols, and then conclude with examples of tuning and designing the catalytic activity of the npAu by doping with metal oxides such as titania. The latter modification of the material is a very remarkable example of the wide structural and chemical adjustability of npAu.

8.2   Surface Chemistry of Au

The key element of any catalytic cycle is the interaction, namely the bonding and activation, of the particular reactant and the catalyst surface.[18–23] Perhaps due to its early industrial relevance and the lack of suitable experimental techniques, for a long time the use and application of catalysts outweighed the scientific understanding of the underlying molecular processes on the catalyst surface. Experimental techniques as well as theoretical approaches such as density functional theory (DFT) calculations to investigate the key steps of catalysis on the atomic level have been developed only in the last four to five decades, with the aim to close this important knowledge gap. For example, in the early stages of the 20th century, the development of the Haber–Bosch process for the catalytic generation of ammonia from hydrogen and nitrogen, one of the most important catalytic processes to date, was honored with two Nobel Prizes (in 1918 to Fritz Haber and 1931 to Carl Bosch). However, almost one hundred years later, in 2007, a Nobel Prize was awarded to Gerhard Ertl for his contributions to deriving an atomistic understanding (*inter alia*) of this catalytic process.

The biggest hurdles to overcome in the study of adsorption and binding of chemical entities on surfaces are preparing and conserving clean adsorbate-free surfaces over the duration of the experiment (at least several minutes). From simple kinetic gas theory, one can deduce that at 1 mbar of gas pressure (*e.g.* He) approximately 10^6 monolayers (equivalent to one atomic thick layer) of gas molecules impinge on the metal surface per second. Even if only a fraction of those molecules stick to the surface, it will be covered within milliseconds. As a result, pressures in the range of 10^{-10} mbar—a regime called ultrahigh vacuum—are necessary to study clean and well-characterized (single crystalline) surfaces. The required experimental techniques and appliances have been developed since the 1950s, leading to the establishment of a new scientific field, *surface science*.

8.2.1 Interaction of Au with Oxygen

The presence of active oxygen is the first and key step in the catalytic cycle for oxidation reactions over gold surfaces. The presence of oxygen on Au surfaces was found to be mandatory for inducing reactivity.[24–26] Although gold-based catalysts have been deployed as catalysts for a variety of oxidation reactions at ambient pressures, the activation and dissociation of molecular oxygen, which is the natural source for (atomic) oxygen in these experiments, seems unclear. The first experiments in the 1970s revealed some dissociation of molecular oxygen at temperatures above 500 °C on flat single- and polycrystalline gold surfaces,[27–29] yet later experiments showed that the presence of hardly detectable amounts of impurities such as calcium was critical.[30] It is widely agreed that the activation barrier for the splitting of molecular oxygen on Au is very high[31] so that thermal activation leads to desorption from the surface rather than dissociation of the oxygen molecule. The dissociation probability of molecular oxygen on flat gold surfaces was assessed to be below 10^{-6}.[32] As a consequence, molecular oxygen remains mostly physisorbed, for example, on Au(110) surfaces at 28 K, and generally desorbs at temperatures below 60 K under ultra-high vacuum conditions.[33] In order to study atomic oxygen on Au surfaces, different techniques were developed to overcome the activation barrier for splitting molecular oxygen, for example by irradiation of physisorbed layers of molecular oxygen with electrons,[33] O_2 ion bombardment,[34] decomposition of NO_2,[35] oxygen atom impingement,[36] or catalytic decomposition of ozone.[27,37,38] All these techniques result in chemisorbed atomic oxygen, yet, also induce different surface morphologies. For example, when sputtering the surface with oxygen ions, the surface becomes rough, and a pit-and-mount structure evolves due to the impact of the ions on the surface.[39,40] In this way, oxygen atoms can be implanted well below the surface of the metal.[39]

In contrast, the comparatively mild catalytic decomposition of ozone on gold surfaces was shown to lead to about 1 monolayer of atomic oxygen on the surface.[41,42] As revealed in scanning tunneling microscopy studies by Friend *et al.* on Au(111) surfaces, the surface reconstruction is lifted after deposition of oxygen, and small gold oxide clusters in the size of about 2 nm are created (Figure 8.2).[24] With increasing coverage of oxygen (>0.3 monolayer) and temperature (400 K), the clusters aggregate and build two-dimensional arrays of oxide on the surface. It was found that the initially formed small oxygen gold clusters are most reactive.[43] The fact that these clusters are indeed gold–oxygen compounds was verified by reacting the oxygen with CO at low temperatures. The small clusters could still be detected, yet without containing oxygen.[24]

Temperature-programmed desorption studies[44] and photoemission experiments on Au(111) and Au(110) surfaces confirmed at least two different oxygen species on the gold surface generated in this way: one very reactive and more weakly bonded oxygen, dubbed 'chemisorbed oxygen' and corresponding to the cluster-like oxide found by scanning tunneling microscopy, and another slightly more tightly bonded oxygen species called 'surface oxygen' corresponding to the two-dimensional, more ordered oxygen structures. Theoretical studies do not exclude the presence of subsurface oxygen at elevated

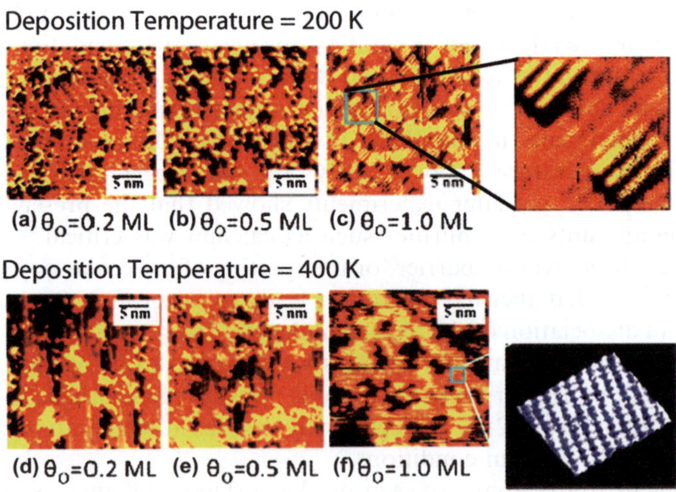

Figure 8.2   Scanning tunneling microscopic images after depostion of various amounts of atomic oxygen on an Au(111) surface (reproduced from Baker *et al.*[24] by permission of the PCCP Owner Societies).

temperatures and oxygen pressures,[24] yet it has not clearly been detected in experiments so far.

The latest studies on npAu surfaces using highly resolved photoemission spectroscopy and ozone decomposition support the fact that these oxygen species can also be formed on npAu surfaces. Due to the low pressure required for these studies, ozone was used as a source of chemisorbed atomic oxygen,[45] yet an interesting difference was noticed: in contrast to flat Au surfaces where, with increasing coverage of oxygen, the ordered two-dimensional 'surface oxide' prevails, the oxygen on npAu remains rather disrupted as 'chemisorbed oxygen', even at high coverages. A possible explanation is that the absence of larger crystallite facets on the strongly curved npAu surface prevents ordering of the oxygen species. This is noteworthy, especially in light of the higher activity of this disrupted chemisorbed oxide, and may also contribute to the high catalytic activity found in ambient pressure studies.

8.2.2 Interaction of Au with CO

In general, CO is molecularly adsorbed on Au surfaces. This means that the scission of the CO bond leading to atomic carbon or oxygen is not detected.[46–48] The adsorption and bonding of CO on Au surfaces were found to depend very much on the local geometry, namely the coordination number (CN) of the surface atoms with the tendency to form a stronger bond with decreasing coordination of the surface atom. For example, Gottfried *et al.* studied adsorption of CO on flat Au(110) surfaces and found weakly bonded physisorbed CO desorbing well below 100 K as well as smaller quantities of more strongly bonded chemisorbed CO desorbing above 100 K.[47] The latter was assigned to CO adsorbed on low-coordinated surface atoms consistent with the observation that the fraction of this chemisorbed CO increased after sputtering and roughening of the surface, respectively. At least two different chemisorbed CO species can be identified by their desorption temperatures at around 120 K and 170 K.[48] Various experimental[31,47–50] as well as theoretical studies[12,49] showed that this chemisorbed CO is bonded to surface atoms with a coordination number of 7 and below, corresponding to atoms at steps (CN 7) and kink/defect sites (CN 6). CO physisorbed on terraces (CN 9) generally desorbs at temperatures well below 100 K.

As mentioned in the introduction, the bonding (chemisorption) of molecules on the catalyst surface is critical for the 'activation' of the molecule and thus the catalytic activity. The presence of step and kink sites can be expected to greatly facilitate the catalytic oxidation of CO, accordingly. Specifically, nano-structured materials such as npAu contain a large fraction of low-coordinated surface atoms due to the presence of curved surfaces and the absence of large crystalline facets. Indeed, when exposing npAu to CO at low temperatures (*e.g.* liquid-nitrogen temperature) one observes large quantities of chemisorbed CO desorbing at temperatures above 120 K, corresponding to CO bonded to step and kink sites.[51] In the presence of coadsorbed oxygen, further desorption at temperatures above 200 K is observed (Figure 8.3). Theoretical calculations

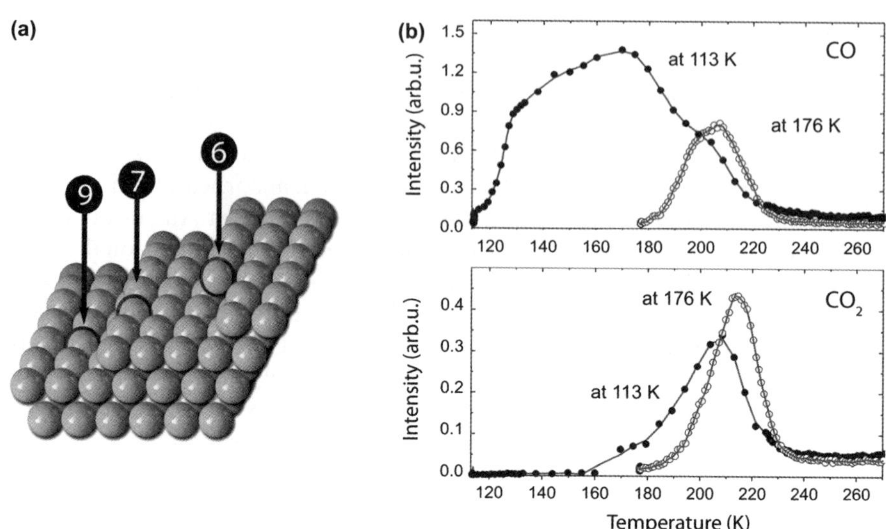

Figure 8.3 (a) Representation of a stepped surface containing for example atoms at terraces (CN 9), steps (CN 7), and kinks (CN 6). (b) TPD of CO ($m/z = 28$, top panel) and CO_2 ($m/z = 44$, bottom panel) after exposing a catalytically active npAu sample disk to 100 L of CO. Spectra were obtained after exposure at 113 K (full circles) and 176 K (open circles), respectively. The detected CO_2 indicates that there is surface oxygen on the npAu sample available for reaction (reproduced from Moskaleva *et al.*[12] by permission of the PCCP Owner Societies).

confirmed that the co-adsorption of atomic oxygen, for example, can increase further the binding energy of CO on the Au surface. As a first result, the bonding and activation of CO on the npAu surface benefit greatly from the large fraction of low-coordinated surface atoms and the co-adsorption of atomic oxygen.

8.2.3 Alcohol Oxidation

The selective oxidation of alcohols plays an important role in the production of a wide range of bulk and commodity chemicals/materials, as alcohols can be synthesized by conventional petrochemistry or are derived from renewable resources such as landfill gas, biomass, or municipal waste.[52] The use of gold as a selective and green catalytic material for the selective oxidation of alcohols has gained considerable interest (a more detailed discussion can be found in Section 8.3.2). In case of alcohols, various bonds such as C–H, C–C, and C–O bonds can be subject to catalytic activation/scission. A metal that activates all bonds equally may be very active for the oxidation, but poorly selective, resulting in full combustion. Examples of such reactive catalytic elements are Pt and Pd. On Au and Ag surfaces, the scission of the C–H bond, however, is greatly hampered,[46] thus making these two materials interesting candidates for the selective oxidation of alcohols. Indeed, Ag is employed as a catalyst in the commercial oxidation of methanol to the corresponding aldehyde, formaldehyde, for example.[52–54]

The corresponding reactivity of gold surfaces for the oxidation of various alcohols has been intensively investigated over the last 25 years.[55] Friend *et al.* used temperature-programmed reaction spectroscopy to investigate the molecular transformation of alcohols on Au surfaces.[53] In these experiments, the reactants are adsorbed on the metal surface at a low temperature (~ 100 K), and the reaction products desorbing from the metal surface are monitored by mass spectrometry while linearly increasing the sample temperature. The desorption temperature of the particular products is correlated to the specific activation barrier of formation (this technique is limited to reactions where desorption of the products is not the rate-limiting step). In combination with other techniques, such as infrared spectroscopy and isotope labeling, a mechanistic picture of the reactions on the surface can be derived.[53]

In case of Au surfaces, no reactivity can be observed in the absence of atomic oxygen, pre-adsorbed on the surface. As soon as atomic oxygen is present, though, a very rich and selective surface chemistry evolves (Figure 8.4). First, in a Brønstedt-type acid–base reaction, the alcoholic proton of the alcohol reacts with atomic oxygen on the surface and forms an adsorbed alcoxy group

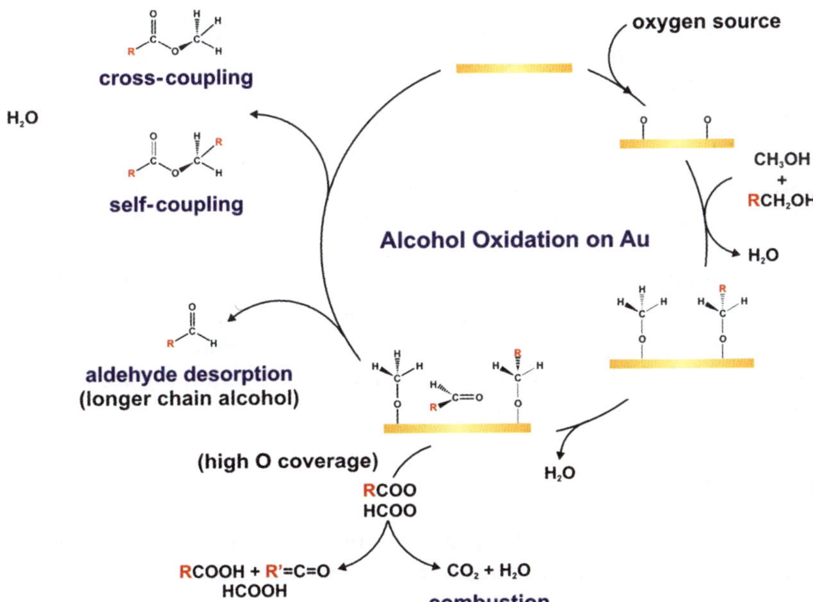

Figure 8.4 Schematic of self- and cross-coupling of alcohols on Au surfaces.[25] The reaction is initiated by a Brønstedt-type acid–base reaction of surface oxygen with the alcoholic proton (image courtesy of K. M. Kosuda, Harvard University).

(RO-Au). In a subsequent step, this surface-bonded alcoxy can be subjected to further deprotonation by scission of the alcoholic C–H bond (β-elimination), by reaction with either surface oxide or an adjacent alcoxy. The resulting surface-bonded aldehyde can (a) react with an adjacent alcoxy forming the corresponding coupling product, the ester (*e.g.* methyl formate), (b) desorb from the surface resulting in 'free' aldehyde (*e.g.* formaldehyde), or (c) be further oxidized to the carboxylate (*e.g.* formate), eventually leading to the full combustion product, CO_2. The tendency to form the aldehyde by β-hydrogen elimination increases with increasing alcohol carbon chain length. As this is considered to be the rate-limiting step, it determines the overall selectivity of the reaction.[25] In the case of methanol, which contains only one C atom, the rate for β-elimination is the lowest, resulting in low concentrations of formaldehyde that will predominantly react with methoxy, to form the ester, methyl formate. This is very much in contrast to Ag where the formaldehyde is the favored product. With increasing chain length (*e.g.* ethanol, propanol, and *n*-butanol) the formation of the aldehyde is favored, and the product distribution will change, in favor of the formation of aldehyde.

8.3 Gas-Phase Catalysis over Nanoporous Gold

In order to study the catalytic activity of catalysts under 'working conditions', higher, *i.e.* ambient, pressures are typically deployed. The catalyst is placed in a reactor with environmental and temperature control, and either the reactants are fed continuously into (and out of) the reaction chamber ('continuous flow reactor'), or the reactor containing the reactants is sealed for a specific reaction time ('batch-type reactor'). In any event, the reactants will continuously be able to react with the catalyst surface, and a large number of catalytic cycles can be investigated. Later, gas-phase catalytic experiments will be reported that were mainly obtained in a continuous flow reactor (as described in Wittstock *et al.*,[14] for example). This type of steady-state fixed-bed turbular flow reactor is especially suitable for determining the activity and kinetics of catalytic reactions.[56] In the studies reported below, the monolithic npAu samples were in the form of free-standing membranes (*e.g.* disks with a diameter of 5 mm and a thickness of 200–300 μm) so that the reactant gases could diffuse into and out of the inner sections of the material from all sites.

8.3.1 CO Oxidation

The catalytic oxidation of CO with molecular oxygen certainly represents one of the most investigated reactions in heterogeneous catalysis that is compelling from an applied as well as from a fundamental point of view. The oxidation of harmful CO to non-poisonous CO_2 plays an important role in exhaust gas cleaning (*e.g.* in the automotive converter). The superior activity of gold for low-temperature oxidation of CO and hydrocarbons compared to typical metals used in car catalysts (Pt, Pd, Rh) may solve the issue of the very low efficiency of the converter during the cold start-up period. Consequently, intense research efforts have been initiated in this respect.[36,57,58] The first commercial Au-based automotive converters have recently become available (NS Gold™ from Nanostella Inc.). Additionally, the ability of gold to selectively oxidize CO in the presence of H_2 renders applications in hydrogen-fuel-cell applications promising.[59] Very small quantities of CO present in the hydrogen feed gas poison the Pt electrode catalyst over time; efficient removal by oxidation with an Au catalyst thus represents a viable solution to this problem. Besides this technological potential, the oxidation of CO was widely used for investigations of the surface chemistry. A more detailed description of this topic and the interaction of CO with Au surfaces can be found in Section 8.2.2. Due to the comparatively simple structure of this diatomic molecule, the oxidation resulting in only one product, it is particularly suitable as a test reaction probing the catalytic activity of a material for oxidation reactions.

Initial reports on the catalytic activity of npAu consequently emphasized the oxidation of CO with molecular oxygen. In 2006 and 2007, two groups independently reported on the remarkable activity of npAu for this reaction at temperatures as low as $-30\,°C$.[7,8] Depending on the preparation conditions and the storage of the sample, the material needed to be 'activated' before conversion of CO was detected. Pre-treatment and activation, respectively, of the catalyst are not uncommon in ambient pressure catalysis and refer to a potential cleaning and structural/chemical evolution of the catalyst surface so that the actual catalytically active surface evolves, and the 'active sites' on the catalysts surface become available. It seems, however, that immediately after preparation of npAu samples, no activation period is necessary, and samples can be active from the very first moment on.[8] In other investigations, the sample was exposed to a stream of CO and oxygen at temperatures above $60\,°C$ for a period of several hours before activity was detected.[10] The fact that no activation could be observed when no CO was present in the feed indicates a reaction, possibly a reduction, of the surface that leads to the catalytically active state. On the other hand, when samples were stored for a longer period of time (several weeks), longer activation periods were necessary, an observation that implies a concurrent cleaning procedure.

However, after activation of the npAu catalyst, a stable conversion of CO is shown over several hours to days. A typical curve for the conversion of CO adding different quantities of oxygen to the feed gas is shown in Figure 8.5. The conversion of CO increases with increasing supply of reactants until saturation is reached. In these experiments, no poisoning or decline of conversion at or above a certain concentration of either CO or oxygen was detected, in contrast to other transition metals, such as platinum or palladium at these temperatures.

As already mentioned in Section 8.2.1, the bonding of molecular oxygen on Au surfaces is very weak, and the probability of dissociation, leading to reactive atomic oxygen on the surface, is very low. Although different theoretical studies predict that, in case of low-coordinated Au atoms, the bonding of molecular oxygen becomes stronger,[60–62] this is not a straightforward explanation for the high catalytic activity of npAu. Different than in case of Au particles that were

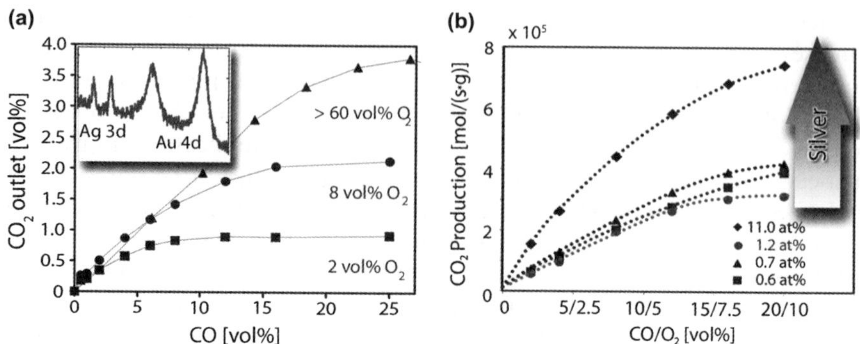

Figure 8.5 Catalytic activity of npAu: oxidation of CO with molecular oxygen at 40 °C. (a) Production of CO_2 increasing with the supply of CO and O_2 but is eventually limited by oxygen supply. Inset: XP spectrum of the Au 4d and Ag 3d region proving the presence of Ag on the surface of the catalyst. (b) Activity of samples containing different amounts of residual silver (Ag contents were determined by AAS). For elevated Ag contents (*e.g.* 11.4 at.%) the catalytic activity is increased by almost a factor of two, compared to samples containing around 1 at.% residual Ag (reproduced from Wittstock *et al.*[11] by permission of the PCCP Owner Societies).

found to be active only for gas-phase CO oxidation when the particle diameter was smaller than 5 nm, the ligaments in npAu can be 10 times larger, 40–50 nm in diameter and still highly catalytically active. The catalytic activity of npAu rather scales with the surface area of the material.[17] Accordingly, other possible methods of efficient activation of molecular oxygen have been discussed.[10,63] After dealloying of the material, small fractions of the less noble constituent in the starting alloy remain in the material. For example, when dealloying Au–Ag alloys, the concentration of residual Ag in the bulk is virtually below 1 at.%. This is possibly below the detection limit of energy dispersive X-ray spectroscopy (EDX), for example, often applied in combination with scanning electron microscopy for morphological and compositional characterization. In some studies, this low concentration of the less noble constituent obviously was an obstacle when identifying possible contributors to the catalytic activity of the material. However, surface-sensitive X-ray photoelectron spectroscopy at the outer surface of the npAu material revealed that the Ag is present and can be enriched at the surface of the material to several percent (see Figure 8.5).[10,45] From thermodynamic considerations, it is clear that during the corrosion, the activity of the Ag ions in the solution is not zero and—although becoming very small—some Ag has to remain in the Au material. By choosing the preparation conditions carefully (*e.g.* adjusting the time of dealloying or the potential for corrosion), one can adjust the amount of residual Ag within the material. In this way, samples containing amounts of up to about 15 at.% of Ag were prepared. The upper limit of Ag remaining in the material is given by the formation of a homogeneous three-dimensional porous structure. The samples used in this study contained different amounts of Ag, yet all showed a similar homogeneous size of ligaments and pores around 30 nm throughout the cross-section of the particular npAu disk. Experimental results testing those samples for the CO oxidation are shown in Figure 8.5. The general trend was that with increasing Ag content, the activity for CO oxidation increased. This finding is in line with results from bimetallic Au–Ag particles, where the combination of Au and Ag was found to be more active than either alone. Depending on the preparation conditions and the type of support, a fraction of Ag between 10 at.% and about 40 at.% was found to be optimal in case of CO oxidation.[64,65]

Besides the overall activity of the catalyst, the kinetics of the reaction in terms of the reaction constant and the reaction order are important means of characterizing a catalytic system.[66] To determine the kinetic parameters of the reaction, double logarithmic plots of the activity as a function of the partial pressure were used.[10,16] As inferred from Figure 8.6, the apparent reaction order of O_2 is approximately 0.5 for all investigated samples, while the reaction order of CO is around 1, and in some studies, lower values of approximately 0.7 have been reported.[10,66] Considering mass-transport phenomena,[66] these results correspond to an actual reaction order of 0 for O_2 and between ~ 0 and 1 for CO. A reaction order of zero implies fast saturation of adsorption sites for oxygen, owing to either strong interaction (large adsorption enthalpy) or a very limited number of active sites. This is compatible with a model where O_2 is primarily adsorbed on Ag sites that constitute the minority of sites on the catalyst surface. On the other hand, a reaction order of approximately 1, as measured for CO, implies that the adsorption, *i.e.* the supply of CO from the gas phase, is rate-limiting, either because of a weak interaction (low adsorption enthalpy) or because of a large number of available sites for adsorption. This is compatible with CO adsorption primarily taking place on Au sites that constitute the majority of the npAu surface. However, with respect to different

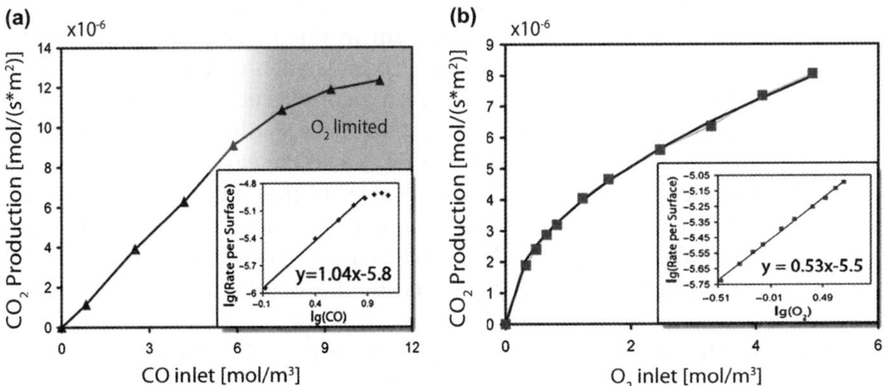

Figure 8.6 Kinetics of CO oxidation with npAu. The partial pressure of the particular reactant was varied under excess co-reactant. A double logarithmic plot reveals the exponent, *i.e.* reaction order of the particular component (reproduced from ref. 10, Copyright 2009 American Chemical Society).

amounts of Ag within the material, no substantial change in the reaction order of O_2 was observed. This in turn implies that even though the overall activity is altered for increasing amounts of residual Ag, the reaction mechanism is not noticeably changed.

One experiment further highlighting the surface chemistry of npAu concerns the influence of additives such as water on the catalytic conversion. Water adsorbs (physisorbs) only weakly on metal surfaces, especially on Au surfaces. Only if oxygen is present can the water molecule be activated to transiently form OH species.[67] This effect is discussed to have an impact on the activity of surface bonded oxygen, *e.g.* by disrupting the oxygen layer which leads to higher activity.[43] In case of supported gold catalysts, water can amplify the reaction rate by orders of magnitude.[68] Here, it is however mostly the interaction with the oxidic support (such as TiO_2) which is thought to cause this effect.[69,70] In case of npAu and a stoichiometric supply of reactants, the conversion of CO is increased by about 100% just by adding 0.01 vol% water to the gas stream.[16] Neither the apparent reaction orders nor the activation energy were altered when increasing the water content in the gas feed. When no oxygen was present, the reaction immediately ceased; accordingly, water is not an additional source for oxygen. This finding implies that water works as a co-catalyst; it is not consumed in the process, as it is not contained in the products or educts. Nevertheless, it amplifies the total activity of the catalyst. These experimental results are compatible with the observed reactivity of Au surfaces[67,71] and exemplify that support-free metal catalysts working under ambient conditions are also influenced by moisture.

8.3.2 Oxidation of Alcohols

As already pointed out in Section 8.2.3, the oxidation of alcohols represents a key catalytic conversion in the industrial production of commodity and bulk chemicals in the range of tens of millions of tons per year in volume.[52] Materials generated from alcohols are important in our daily lives in the forms of plastics and paints, carpets, and so on. Products such as PET bottles, DVDs, and car dashboards are manufactured from alcohols such as methanol. Methyl alcohol (methanol or wood alcohol) is the smallest alcohol, containing just one carbon atom, and has the formula CH_3OH. Since the early stages of industrial production of alcohols, methanol was the alcohol that received most attention and also provided a starting-point for the generation of longer-chain alcohols such as ethanol and butanol. In the early stages of industrial generation of this alcohol from about 1830, it was generated by dry distillation of wood and was dubbed wood alcohol, accordingly. With the onset of petrochemistry in the beginning of the 20th century, the source of methanol shifted to natural gas. Interestingly, the development of the related process used iron as a catalyst and was strongly correlated to the industrial production of ammonia using the Haber–Bosch process (see introduction to Section 8.2) at BASF in Germany. Until now, most methanol (and many other industrial alcohols, accordingly) was generated from syngas (synthesis gas), a mixture of CO and H_2 generated by cracking of natural gas.[52] In light of the ever-increasing costs of these resources and their forecast shortage, it is very likely that we will see another shift in production of this important alcohol towards renewable and green resources. It is of particular importance that today considerable amounts of methanol are already generated from such sustainable resources as landfill gas, biomass degradation or simply waste.[72] Methanol constitutes a viable green resource for chemicals and fuels, accordingly.

The oxidation of methanol to formaldehyde and formic acid via the hydrolysis of methyl formate represents one of the major processing routes for methanol. Both chemicals are generated from methanol in the range of several thousand tons (formic acid) to millions of tons per year (formaldehyde).[73,74] Consequently, both processes are critical in view of selectivity, waste generation, and energy consumption. Catalytic research has focused on *green* and *sustainable* catalytic concepts using renewable, ubiquitous, and non-hazardous resources, consuming minimal energy, such as heating the reactor. Two catalytic materials are of considerable interest regarding the selective oxidation (in contrast to total oxidation): Ag and Au, as they primarily break the O–H bond of the alcohol (see Section 8.2.3). Indeed, crystalline Ag or Ag gauze is employed as an industrial catalyst for the generation of formaldehyde by selective oxidation of methanol using molecular oxygen with a conversion of nearly 100% at temperatures of about 700 °C.[73] As can be deduced from model experiments, Au is supposed to be a very selective catalyst for the generation of methyl formate and formic acid, accordingly. The current industrial process for the generation of methyl formate is based on the carbonization of methanol using metallic sodium as a catalyst. Several publications during the last decade have focused on gold-based catalysts as part of a 'greener' catalytic process using the selective oxidation of methanol with molecular oxygen, the ideal resource for reactive atomic oxygen.[75–82] Most of this work emphasized the liquid phase oxidation of alcohols using gold particles with a size of around a few nanometers on oxidic supports. The activity and selectivity of the reported processes are very promising, reaching turnover frequencies (TOF) in the order of $0.1–0.2\,s^{-1}$.[78,80,82,83] Still, issues that constitute problems for nanometer-sized gold particles on oxidic supports include: sintering of particles, contamination, and an unclear role of the support material. A further obstacle is the low activity of the catalysts in the absence of a base as co-catalyst.

NpAu as an unsupported gold catalyst constitutes a very interesting catalyst for this type of reaction. The first experimental work in 2010 focused on the gas-phase oxidation of methanol using molecular oxygen.[14] The npAu was found to be a highly active catalyst for this type of reaction already at temperatures as low as 20 °C (Figure 8.7). Besides small amounts of CO_2 stemming from the total oxidation of methanol, the only detected product was methyl formate. As anticipated from UHV experiments (Section 8.2.3), the ester is formed by coupling of the surface-bonded methoxy and formaldehyde on the npAu surface. As the tendency to undergo β-hydrogen elimination is less pronounced for methanol on Au surfaces, the selectivity is governed by the fast reaction of methoxy and aldehyde. The selectivity for the formation of

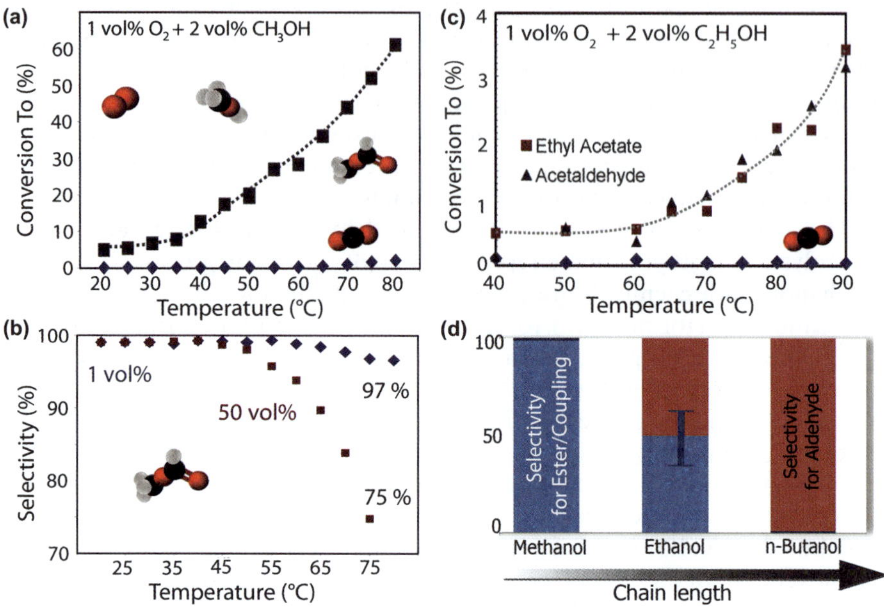

Figure 8.7   Gas-phase oxidation of alcohols over npAu. (a) Oxidation of methanol: conversion of methanol into methyl formate as a function of the reaction temperature. The reaction proceeds at temperatures as low a 20 °C (*i.e.* room temperature). The conversion to CO_2 accounts only for small traces. (b) Selectivity of methanol oxidation over npAu. At low oxygen concentrations, almost all methanol is converted into methyl formate (selectivity close to 100%). However, when drastically increasing the oxygen concentration at elevated temperatures, increasing total oxidation can be detected, and still the production of methyl formate prevails (image courtesy of AAAS/Science). (c) Oxidation of ethanol: coupling to ethyl acetate as well as the oxidation to the aldehyde can be observed in about the same proportion. The total oxidation is still hardly detectable. Overall, the activity is reduced. (d) The selectivity for either the coupling of the alcohol (formation of the ester) or the oxidation (formation of the aldehyde) is a function of the chain length of the alcohol and the tendency to undergo β-hydrogen elimination, accordingly.

the coupling product was close to 100% under the experimental conditions (temperatures below 100 °C, oxygen concentrations between 1 vol% and 50 vol%). The activity in terms of TOF for this reaction was around $0.2\,s^{-1}$ and very comparable to that of highly active supported catalysts, yet without the need of additional base as a co-catalyst. Additionally, by using a continuous flow reactor and deployment of methanol in the gas phase a recovery of the catalyst, as for example necessary in a batch-type approach, became obsolete. This finding gained considerable interest, as the experimental conditions applied do reflect an exemplary green catalytic process.[15] Importantly, the surface chemistry during a reaction can be anticipated and predicted based on the mechanistic understanding derived from UHV model experiments.[84]

However, this pattern of reactivity is not restricted to methanol but can be extended to the gas-phase oxidation of longer-chain alcohols such as ethanol or *n*-butanol containing two and four carbon atoms, respectively.[85] By using npAu as a catalyst and molecular oxygen, both alcohols can be oxidized already at temperatures well below 100 °C. Yet, the product distribution is considerably altered. In the case of ethanol, the aldehyde (acetaldehyde) as well as the coupling product (ethyl acetate) are observed in almost similar proportions (Figure 8.7). In the case of oxidation of *n*-butanol, no coupling but exclusively the formation of the aldehyde (*n*-butanal) was detected in the entire temperature range (up to 180 °C). In general, the activity in terms of TOF was found to decrease with increasing chain length, too. While the TOF in case of methanol oxidation was around $0.2\,\mathrm{s}^{-1}$, it decreased to $0.01\,\mathrm{s}^{-1}$ in case of ethanol oxidation and $0.008\,\mathrm{s}^{-1}$ for the oxidation of *n*-butanol (all numbers determined at 80 °C). Although, for the latter two cases, the activity of the catalyst was not tested for a maximum conversion (*e.g.* by supplying higher amounts of reactants possibly reaching a maximum of activity), the trend underlines the fact that the activity decreases with increasing chain length. Ding *et al.* reported TOFs in the range of $0.1\,\mathrm{s}^{-1}$ in the case of aerobic gas-phase oxidation of benzyl alcohol to benzaldehyde over npAu. Yet, temperatures above 200 °C were necessary to achieve TOFs in this range.[86] This observation can be similarly made using supported Au catalysts.[78] However, due to different supports and experimental conditions, a straightforward correlation between the observed activities as a function of the chain length of the alcohol is very difficult in these cases. With respect to unsupported npAu, a possible explanation for the decreasing activity for the long-chain alcohols is the more pronounced tendency to form a carboxylate species with increasing chain length.[25] As the experiments are performed at low temperatures and very dry conditions (5 ppm of water in the feed gas), preventing a reaction of water forming the free acid, this carboxylate species may block active sites on the surface.

The concept of oxygen-mediated coupling reactions on the npAu surface can be further explored by cross-coupling of two different alcohols or even two different species such as an alcohol and an aldehyde. For example, when co-dosing similar proportions of methanol and acetaldehyde, the cross-coupling product of both is observed (exclusively), the methyl acetate.[14] This can be understood in terms of the reactivity of the aldehyde on the surface, rapidly reacting with the surface-bonded methoxy. The competing β-hydrogen elimination of methoxy, on the other hand, which would lead to the formaldehyde and the self-coupling, is largely suppressed due to the comparatively higher activation barrier of this step.[25] However, two different alcohols can also be co-dosed, and the cross-coupling products are formed. When methanol and ethanol are co-dosed, only the methyl acetate is observed, stemming from the cross-coupling of methoxy and acetaldehyde on the surface.[85] The formation of either ethyl formate or methyl formate is not detected; both reactions are suppressed due to the higher activation barrier for β-hydrogen elimination of methoxy.

After all, the oxidation and coupling of alcohols over npAu shed light on a rich surface chemistry of Au towards selective and predictable oxidation reactions. Basic principles of reactivity can be understood and anticipated based on model experiments under UHV conditions. The absence of an oxidic support material and the extended gold type surface of npAu open the door for this correlation.

8.4   Liquid-Phase Catalysis

In many cases, the vapor pressure of one of the reactants is low, and it is more convenient and practical to investigate the catalytic activity in the liquid phase. Regarding the interpretation of the surface chemistry and the underlying mechanism, the situation is mostly distinguished from the gas-phase catalysis due to the inherent higher concentration of spectator species on the surface. For example, catalysts such as gold nanoparticles on a carbon support, which do not show any catalytic activity for CO oxidation in the gas phase, can become catalytically active in the liquid phase.[87–90] In these cases, the abundance of hydroxyl groups in the liquid (water) phase is discussed to bridge the impediment of activation of molecular oxygen.[90] The liquid-phase catalysis by npAu is still an evolving field; parameters such as the residual Ag or Cu concentration may have a very different, perhaps even no, impact anymore. In the following, we will discuss two recent and well-studied examples of the catalytic aerobic oxidation in liquid phase, namely the oxidation of organosilanes and D-glucose. Both reactions are important from an academic and industrial point of view. They are suited to demonstrate the broad potential of npAu as a catalyst also in liquid phase.

8.4.1   Aerobic Oxidation of D-Glucose

The oxidation of D-glucose is the reaction exclusively used to yield gluconic acid. This acid is a natural compound in the human body as part of the carbohydrate metabolism. It is also used for example in the food industry as an acidity regulator (E574) or in cleaning products as a mild acid. The annual production of gluconic acid amounts to about 60 000 tons.[91]

$$\text{glucose} \xrightarrow[\text{+ O}_2]{\text{cat. npAu}} \text{gluconic acid} \tag{8.1}$$

Besides biotechnological production (*i.e.* fermentation) using enzymes such as *Aspergillus niger* and *Gluconobacter suboxydans*, the catalytic oxidation of glucose using molecular oxygen and metal catalysts is a viable alternative.[91,92] Finely dispersed Pt group metals on oxidic supports, such as alumina, or activated charcoal (carbon black) have been employed for this reaction since the 1970s.[82,93] The fast decay of catalyst activity and the insufficient selectivity of these catalysts fueled ongoing research on an alternative catalyst design. The lack of reactivity of Au surfaces with respect to the unfavorable activation of C–H and C–C bonds has already been discussed in the context of oxidation of alcohols, such as methanol and ethanol, in Sections 8.2.3 and 8.3.2. Accordingly, Au as a promising highly selective catalyst was investigated for this type of reaction. In 2002, Biella *et al.* reported finely dispersed Au on carbon supports being highly selective and active catalysts for this type of reaction already at temperatures in the range of 330 K using molecular oxygen and water as solvent. Carbon appeared to be the support of choice regarding liquid-phase oxidation reactions; this is noteworthy, as, in the case of gas-phase oxidation reactions, carbon is known as an inert support leading to poor catalytic activity. Besides a different mechanism of activation of molecular oxygen in the liquid phase,[90] it seems that an inert support material such as carbon prevents strong sticking and thus poisoning of the catalyst by either the reactants or intermediates, and thus leads to higher activity. Rossi *et al.* demonstrated that the absence of a support is even more favorable.[94] Unsupported colloidal gold particles of about 3.6 nm in size were remarkably active for glucose oxidation.[94] The detected turnover frequency (TOF) was in the range of $14\ s^{-1}$, thus about 14 times higher than the maximal TOF measured for supported gold particles.[95] This high activity was achieved using glucose not only as a reactant but also as a stabilizing agent so that the gold colloids were stable but 'naked' particles. Unfortunately, the high activity of particles led to deactivation within minutes due to sintering of particles from 3.6 nm to over 10 nm in size.

In 2008, Ding *et al.* demonstrated that unsupported nanoporous gold is an active and stable catalyst for this type of reaction, too.[96] The experiments were performed in a glass reactor containing a solution of 0.1 M D-glucose and 20 mg of crushed npAu. The pH of the solution was controlled and adjusted, respectively, using an automated potentiometric titrator. The oxygen was bubbled through the reaction solution from underneath. Every hour, a small volume of the mixture was sampled and the conversion determined. The resulting conversion over time is depicted in Figure 8.8. The selectivity of the oxidation reaction was found to be close to 100%, thus meaning that almost no isomerization into fructose or other biproducts was detected.

The reaction proceeded already at very mild temperatures slightly above room temperature (303 K). As expected for a chemical reaction, the conversion of reactants proceeded faster with rising temperature. By increasing the temperature to 323 K, the conversion was doubled and resulted in appreciable yields. The pH value of the reaction media was found to be critical for catalytic conversion. As also reported for supported Au particles, alkaline conditions were favorable for the reaction (pH > 7).[97] In the case of npAu, a pH of 9 was found to be optimal, as the activity was the highest, and the conversion of glucose also increased over the entire duration of the experiment. Besides the catalyst's activity and selectivity, it is the stability of the self-contained

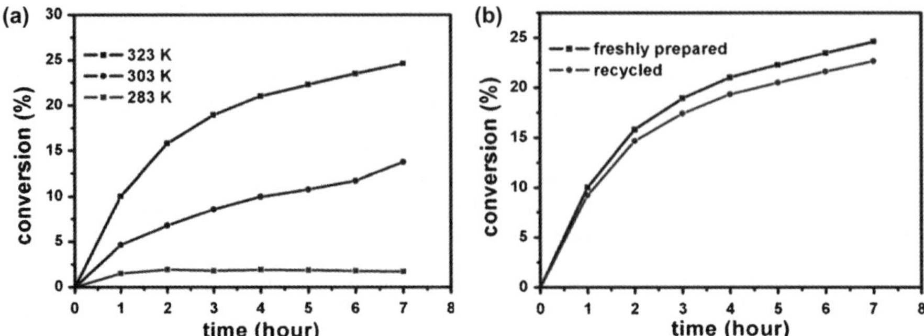

Figure 8.8 Aerobic oxidation of D-glucose to gluconic acid using npAu. (a) Conversion of glucose over time using 20 mg of npAu at pH 9. (b) Conversion of glucose using 20 mg of npAu in two subsequent runs, proving its good recyclability (reproduced from Yin *et al.*,[96] Copyright 2008 American Chemical Society).

nanostructure of npAu that renders this material so interesting. The catalytic conversion using the same sample of npAu in two subsequent experiments (7 h duration each) was found to be virtually the same (Figure 8.8). This finding is in line with scanning electron micrographs of the sample before and after the reaction showing no increase in pore or ligament size (30 nm), which would indicate ripening. Noteworthy, npAu samples containing smaller (*e.g.* 6 nm wide) ligaments were also prepared. Those samples initially showed a higher catalytic activity owing to the increased specific surface area yet lesser catalyst stability. For example, the ligaments in the range of 6 nm increased to about 15 nm after 7 h of catalytic conversion with accordingly reduced activity. Although the activity of npAu in this study was somewhat lower than that of (supported) gold catalysts—likely owing to the higher specific surface area of latter catalysts—it shows a very promising catalyst stability. The use of molecular oxygen as the oxidizing agent and water as the solvent also makes it an environmentally benign process.

8.4.2 Oxidation of Silanes

Hitherto, we reported exclusively on the aerobic oxidation of carbon-based entities, such as methanol, ethanol, or glucose, a transformation that is of utmost importance in academic as well as industrial organic chemistry. Interestingly, in 2010, Asao *et al.* reported on the oxidation and hydrolysis, respectively, of silicon based compounds such as organosilanes to organosilanoles using npAu (eqn (8.2) and (8.3)). In the periodic table, silicon (Si) is positioned in the same group (14) as carbon and contains the same number of valence electrons. As a consequence, some aspects of its chemistry, such as the tendency to form tetrahedral compounds with a valence number of 4, are similar to those of carbon. Organosilanes are derived from silanes (*e.g.* SiH_4), yet one or more hydrogen atoms are replaced by an organic entity. Silanols, or silicon-based alcohols, can be derived by oxidation/hydrolysis of silanes according to the following equation:

$$R_{(4-n)}SiH_n + nH_2O \xrightarrow[\text{Catalyst}]{} R_{(4-n)}Si(OH)_n + nH_2. \qquad (8.2)$$

Traditionally, silanols are used as building blocks in synthetic chemistry to generate metallasiloxanes, for example.[98,99] Very similar to the acidity of alcohols discussed in Section 8.2.3, the proton of the OH group can react in a Brønstedt-type acid–base reaction forming an Si–O–metal bond. Those compounds are called metallasiloxanes. Several attempts aimed at the synthesis of ceramics using for example metallasiloxanes as precursors.[100,101] Silanoles tend to polymerize forming longer-chain Si–O–Si entities (such as in ceramics) by eliminating water. When using the appropriate silanol containing a specific organic group, one can generate various silicon-based polymers.[102] A recent review of the chemistry of organosilanols can be found in Chandrasekhar *et al.*[102]

As an oxidizing agent for the oxidation of organosilanes, chemicals, such as ozone, $AgNO_2$ or $KMnO_4$, have been used, to name just a few.[102] However, molecular oxygen or water as oxidizing reactants would be preferable (according to eqn (8.2)), since they are of course non-poisonous, non-hazardous, and inexpensive, and the resulting by-products, hydrogen or water, are non-polluting, too. In 2009, Kaneda *et al.* reported that gold nanoparticles are a suitable and very active catalyst for this type of reaction.[103] Recently, Asao *et al.* used 40 µm thin nanoporous gold foil in liquid phase to oxidize various organosilanes.[104] As an example, by using finely dispersed 1 mol% of npAu as a catalyst, dimethylphenylsilanol was derived within 1 h at room temperature with nearly 100% yield (eqn (8.3)). The turnover frequency for this reaction was calculated to be about $3\,s^{-1}$. Importantly, the npAu catalyst could be recycled and reused several (*i.e.* more than five) times:

$$PhMe_2Si\text{–}H + H_2O \xrightarrow[\text{acetone, RT, 1h}]{\text{npAu foil (1 mol\%)}} PhMe_2Si\text{–}OH + H_2. \qquad (8.3)$$

When using a gold foil with the same dimensions instead of the npAu foil, the activity dropped drastically, implying that the surface area and the nanostructure of the material play a decisive role. However, the exact mechanism of the catalytic conversion has not yet been clarified.

8.5 Surface Modification of Nanoporous Gold by Metal Oxides

The self-contained nanoporous structure of npAu opens further options for the preparation of high-performance materials by modifying it with certain additives. Of particular advantage is the high electrical and thermal conductivity of the bicontinuous monolithic structure of npAu leading to a large range of interesting applications. For example, Qian *et al.* demonstrated that by coating npAu with alumina, the optical properties of the materials can be tuned.[105,106] Ding *et al.* demonstrated that by applying titania particles on the npAu outer surface, a very interesting photocatalytic electrode material can be generated.[107]

The combination of npAu with metal oxides for catalytic applications is a very promising approach, for several reasons. First, pure gold lacks the ability of efficiently activating molecular oxygen, which is the natural source for reactive atomic oxygen. In the case of npAu, this obstacle can be overcome by traces of a less noble metal (Ag, Cu) in the material. As shown for gold nanoparticles by combining gold with metal oxides such as titania or iron oxide, highly active oxidation catalysts can be generated, too.[108] Accordingly, metal oxides are interesting additives, possibly leading to even more active oxidation npAu catalysts. Second, nanostructured metals are prone to temperature-activated ripping and coarsening of structures, limiting the applicability at high temperatures. Especially in the case of gold, the inherent instability of nanosized structures leads to fast deactivation at elevated temperatures. Biener *et al.* recently demonstrated that by coating npAu with Al_2O_3, for example, the nanostructure of npAu can be conserved, even at temperatures of 1000 °C, while ligaments in pure npAu start coarsening already at temperatures of about 200 °C. This opens the door for applications such as in automotive converters.

In general, two different ways of depositing metal oxides in the nanoporous structure of npAu are possible: (a) deposition from the gas phase, *e.g.* by chemical or physical vapor-deposition techniques (ALD, CVD, PVD) or (b) deposition from liquid phase, meaning by immersion of npAu samples into a solution containing either the metal oxide or a viable precursor. In the following, we will report on two examples showing how modification of npAu with metal oxides can lead to highly active and stable catalysts. Obvious choices for additives are reducible metal oxides such as TiO_2, CeO_2 or FeO_x, as these materials are known to lead to high activity in the case of gold nanoparticle based systems.

8.5.1 Gas-Phase Deposition: ALD-Modified Nanoporous Gold

Atomic layer deposition (ALD) has been shown to be an especially dedicated gas-phase deposition technique for the coating of high-aspect ratio materials.[109,110] In the case of free-standing films of npAu with a thickness of 200–300 μm and a pore size of about 30 nm, the aspect ratio—the length of the pore in relation to its diameter—amounts to several thousand. The slow diffusion of molecules into inner sections of the material is a major obstacle when aiming at a conformal coating of the inner surface and avoiding clogging of pores close to the outer surface. ALD consists of self-limited surface reactions (*i.e.* chemisorption) of precursors (see Figure 8.9). By separately dosing reactants, the maximal amount of material deposited during one cycle of ALD is ideally only 1 monolayer. This limitation makes ALD an ideally suited technique for conformal coating of high-aspect-ratio materials.

In the first study of this kind, Biener *et al.* deposited TiO_2 inside the npAu to generate a highly catalytically active material. Titania was chosen as an

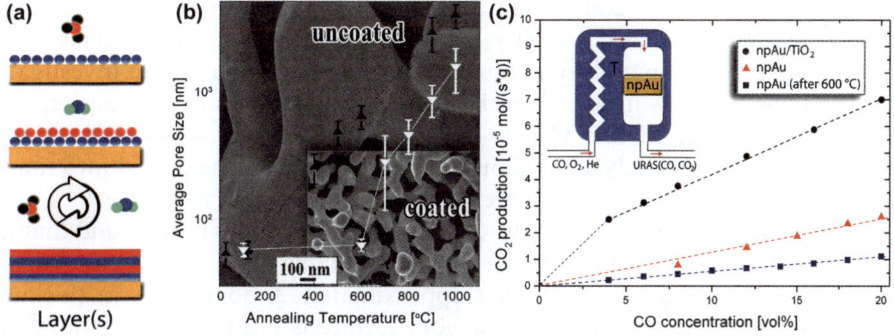

Figure 8.9 TiO$_2$-modified npAu. (a) Schematic of atomic layer deposition (ALD). This deposition technique relies on sequential self-limiting surface reactions of reactants. The resulting deposition is very homogeneous and particularly suited for high-aspect-ratio materials such as npAu. (b) Thermal stability of uncoated (black) and titania-coated (white, 30 cycles of ALD) npAu, displayed is the average ligament size after annealing for 3 h at the particular temperature under He atmosphere. The SEM images show the morphology of both types of npAu sample after annealing at 600 °C. (c) Catalytic activity of coated and uncoated npAu for CO oxidation at 60 °C using a continuous flow reactor (schematic as an inset, 30 vol% of O$_2$, He as carrier gas, total flow of 50 mL min^{-1}). Coated samples (10 cycles of ALD) have to be annealed to 600 °C before catalytic activity is detectable. This is in line with the observation that the initially continuous TiO$_2$ film breaks up so that particles are formed at this temperature (reproduced from Biener *et al.*,[17] Copyright 2011 American Chemical Society).

exemplary reducible oxide leading to highly active gold based catalytic materials. For deposition, a well-established process using titanium tetrachloride (TiCl$_4$) and water (H$_2$O) ALD was deployed. TiCl$_4$ exhibits a sufficiently high vapor pressure so that reasonable partial pressures in the gas phase can be achieved at temperatures slightly above 100 °C. The molecule readily reacts (hydrolyzes) with water and hydroxyl groups on the surfaces forming a 'Cl$_3$Ti–O–Au' layer. As there are no hydroxyl groups left on the surface after reaction, residual TiCl$_4$ not chemisorbed to the surface can be removed by pumping and purging of the system, respectively. By subsequently dosing water, the residual chloride groups are removed by hydrolysis, and a first '(HO)$_2$/OTi–O–Au' layer is formed. This corresponds to one cycle of ALD. The impeded mass transport in and out of the pores can be accounted for by prolonged purge and pump times (*i.e.* 90 s). The average growth rate for the TiO$_2$ deposition was about 0.07 nm per ALD cycle. Considering the lattice spacing (O–Ti–O unit) of the Ti anatase phase of 0.35 nm for the (101) plane,[111] this would correspond to roughly 1/5 of a monolayer as an equivalent of a closed layer of titania on the surface.

The impact of the titania coating on the temperature stability of the material was investigated by annealing (Figure 8.9). While pure npAu samples start coarsening at around 200 °C, the titania-coated samples (30 cycles of titania ALD) showed enhanced stability up to temperatures of 600 °C. This lower temperature as compared to alumina-coated samples, which were investigated in the same work, is in line with the comparatively lower crystallization temperature of titania of 370 °C[112] as compared to alumina ~950 °C.[113] At this temperature, the initially amorphous layer of titania is transformed into titania crystallites (anatase) leading to the break-up of the initial film.

After this initial break-up of the continuous titania film, it was found that the samples were catalytically active. Note that titania itself is not catalytically active, meaning that a closed layer approximately 2 monolayers in thickness (*e.g.* after 10 cycles of ALD) on the gold surface is not expected to show any detectable activity.[114] Yet, after the break-up of the film and the formation of titania particles on the npAu surface, catalytic activity evolves. The activity for CO oxidation (measured at 60 °C) of the titania-modified sample was noticeably amplified by a factor of about 5, as compared to the pure npAu (Figure 8.9). The enhanced catalytic activity of the TiO_2/npAu system is likely due to an increased activation of molecular oxygen, as this is considered to be the reaction step with the highest activation energy and thus most hampered on pure gold surfaces (see considerations in Section 8.2.1).

8.5.2 Liquid-Phase Deposition

Aside from deposition of metal oxides through the gas phase as reported in the previous chapter, another viable route is the deposition of either the metal oxide or its precursor from solution. The advantage of these bench-top techniques is that they are connected with a significantly reduced experimental and instrumental demand. Ding *et al.* first used the deposition of TiO_2 particles suspended in ethanol in 2009.[107] Although the majority of particles were larger than the pores and accordingly were only deposited on the very outer surface of the npAu material, the resulting electrode material showed very promising photocatalytic performance. Wittstock *et al.* deployed a comparable technique to generate TiO_2/npAu composite materials for catalytic gas-phase CO oxidation.[11] Titania particles in the form of a commercially available powder (*e.g.* TiO_2 nanopowder from Sigma Aldrich, particle sizes below 100 nm) were suspended in ethanol (here $2.5\,g\,mL^{-1}$). Droplets of this suspension were applied to the outer surface of a free-standing npAu membrane (200–300 μm thick). The membrane was subsequently rinsed to remove excess of particles from the outer surface. Scanning electron micrographs showed that the majority of particles were clearly deposited on the outer surface of the material. Yet, the catalytic activity for CO oxidation was enhanced by about a factor of 5 as compared to an unmodified npAu sample. Transmission electron micrographs showed that particles in the size of 10 nm are also contained in the titania powder. It is reasonable that particles smaller than the ligament size can be sucked into the pores and thus deposited, even below the outer surface while being suspended in the ethanol due to capillary action.

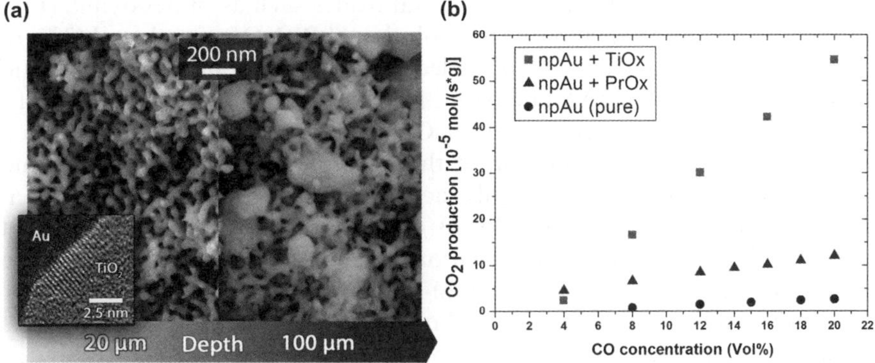

Figure 8.10 Oxide-modified npAu. (a) Cross-sectional scanning electron micrographs of a titania-coated npAu membrane after calcination at 500 °C in air. Structures in the outer layers, *i.e.* 20–25 μm from the outer surface, are apparently stabilized and not coarsened (average ligament diameter ~40 nm); yet the inner sections of the sample (*e.g.* 100 μm from outer surface) show a bimodal ligament size distribution indicating a lower and non-homogeneous distribution of metal oxide. The inset shows a high-resolution TEM of a TiO_2 particle on the npAu surface. (b) Catalytic conversion of CO using pure npAu and oxide modified npAu at 60 °C (30 vol% O_2 and total flow rate of 50 mL min^{-1}).

Besides direct application of titania particles, the application of a precursor is another viable means to generate TiO_2/npAu composites. Wichmann *et al.* demonstrated that by using titanium isopropoxide (TTIP), which is a liquid at room temperature, conformal coating of TiO_2 in npAu can be achieved.[115] After immersion of a free-standing membrane (disk with a diameter of 5 mm and a thickness of about 250 μm) in a TTIP solution and subsequent calcination at 400 °C the npAu (inner-) surface was abundantly covered with TiO_2 particles in the size of about 5 nm. Quantification by EDX showed a content of about 3.5 wt% of titania inside the npAu structure (*e.g.* 30 μm below the outer surface). The catalytic activity of these TiO_2/npAu samples was greatly enhanced as compared to the pure npAu (Figure 8.10). Besides being highly active, the nanostructure of the oxide modified npAu system could be preserved at temperatures as high as 400 °C (Figure 8.10). Even in sections of the membrane being 100 μm far from the outer surface of the membrane, a stabilization of the ligaments could be observed (Figure 8.10). However, in those deep sections, a bimodal ligament size distribution indicated a lower coverage of the surface and thus reduced stabilization. This is in line with a reduced amount of titania detected by EDX in deeper sections of the membrane (*e.g.* 1 wt% as compared to 3.5 wt% close to the outer surface). As a result of the stabilization of the nanostructure, the TiO_2/npAu samples showed considerable catalytic durability at high temperatures; for example, at 250 °C, the catalytic activity for CO oxidation did not decrease considerably over a period of several days. In a similar attempt, Wichman *et al.* showed that this approach can be expanded to other metal oxides such as praseodymia (PrO_x). Immersion of a npAu membrane in a solution of $Pr(NO_3)_3$ in ethanol (20 g L^{-1}) and subsequent calcination to 500 °C led to deposition of praseodymia particles inside the pores of the npAu material. In this case, the catalytic activity of the composite material was enhanced for CO oxidation, too (Figure 8.10).

Summing up, the combination of the npAu material with a suitable metal oxide indeed can result in highly active catalytic materials with superior stability and durability, especially at elevated temperatures. This opens the door for high-temperature applications such as in automotive converters, which has been an obstacle for gold-based catalysts so far. However, it is noteworthy that by coating the npAu surface with metal oxides, a reversed situation as opposed to an oxide-supported metal nanoparticle catalyst, for example, is realized. A detailed comparison of the systems may also contribute to a mechanistic understanding of both systems.

8.6 Summary and Remarks

In this chapter, we have reported on the various aspects of the surface chemistry and catalysis of npAu. We first discussed the interaction of oxygen with a pure gold surface, as this is the starting-point for the catalytic activity of gold surfaces. Yet, the dissociation of molecular oxygen, which is the natural source for oxygen as it is found abundantly in the surrounding atmosphere (about 20 vol%), has a very low probability and is hardly detectable on pure gold surfaces. It is observed that in case of npAu, this hurdle can be overcome by small traces of the residual less noble component (Cu or Ag), which is contained in the material after preparation. The oxygen species present on the npAu surfaces (prepared by decomposition of ozone) are the same, as one can see on single crystalline surfaces.

We then discussed two classes of gas-phase catalytic reactions of molecular oxygen with increasing complexity: first, the reaction of CO and second, various alcohols with increasing chain lengths (methanol, ethanol, etc.). In both cases, we discussed the reactivity in terms of activity and selectivity in strong correlation with results obtained from UHV studies. A pattern of reactivity can be found and explained at the molecular level. It is the extended gold surface and the absence of any support material that makes this material an ideal candidate for transferring knowledge gained from UHV studies to a catalyst working under 'real world' ambient pressure conditions. We subsequently discussed two well-studied and important catalytic reactions performed in the liquid phase, the oxidation of glucose and organosilanes over npAu. Both reactions exemplify the potential of npAu as a versatile catalyst combining activity with durability and good recyclability.

We discussed the latest research on the chemical functionalization of the npAu surface with metal oxides. The bicontinuous and monolithic structure of npAu makes it particularly attractive for a material design involving the npAu as an active and tunable scaffold. We reported on the latest work using liquid- and gas-phase deposition techniques decorating the npAu surface with metal oxides such as titania or praseodymia. In both cases, the catalytic activity for CO oxidation can be drastically amplified. Additionally, by putting metal oxides on the gold surface, its mesoscopic structure can be conserved at temperatures of several hundred degrees Celsius. This opens the door for catalytic applications at elevated temperatures, as in the automotive converter, for example.

The catalysis with npAu is only roughly five years old, starting with the first publication of its activity for the low-temperature CO oxidation in 2006. Yet, in those last five years, several fascinating discoveries were made, resulting in more than 40 papers to date,[116] covering the catalytic conversion of various compounds. Most importantly, in several cases the reactivity of the npAu could be interpreted and thus anticipated at the molecular level based on models derived from UHV and DFT model studies. This is the very fundament for the future development of materials for catalytic applications. Future work should be aimed in three directions: the discovery of new catalytic reactions (including hydrogenation reactions), ongoing efforts to correlate and understand the surface chemistry based on models of the molecular transformations on its surface, and last but not least a material design aimed at an optimized performance of the material in terms of activity, selectivity but also durability and recyclability, as all of these determine the economic viability of the entire catalytic process. Given the fact that the efforts of the last five years will continue in a comparable manner, one can expect further considerable progress in all three fields in years to come.

Acknowledgments

A portion of this work (AW and JB) was performed under the auspices of US Department of Energy by LLNL under Contract DE-AC52-07NA27344.

References

1. B. Hammer and J. K. Norskov, *Nature*, 1995, **376**, 238–240.
2. G. C. Bond, P. A. Sermon, G. Webb, D. A. Buchanan and P. B. Wells, *J. Chem. Soc. Chem. Commun.*, 1973, 444–445.
3. M. Haruta, T. Kobayashi, H. Sano and N. Yamada, *Chem. Lett.*, 1987, **16**, 405–408.
4. G. J. Hutchings, *J. Catal.*, 1985, **96**, 292–295.
5. M. Haruta, *Catal. Today*, 1997, **36**, 153–166.
6. G. C. Bond and D. T. Thompson, *Catal. Rev. Sci. Eng.*, 1999, **41**, 319–388.
7. V. Zielasek, B. Jürgens, C. Schulz, J. Biener, M. M. Biener, A. V. Hamza and M. Bäumer, *Angew. Chem. Int. Edit.*, 2006, **45**, 8241–8244.
8. C. X. Xu, J. X. Su, X. H. Xu, P. P. Liu, H. J. Zhao, F. Tian and Y. Ding, *J. Am. Chem. Soc.*, 2007, **129**, 42–43.
9. C. Xu, X. Xu, J. Su and Y. Ding, *J. Catal.*, 2007, **252**, 243–248.

10. A. Wittstock, B. Neumann, A. Schaefer, K. Dumbuya, C. Kübel, M. M. Biener, V. Zielasek, H.-P. Steinrück, J. M. Gottfried, J. Biener, A. Hamza and M. Bäumer, *J. Phys. Chem. C*, 2009, **113**, 5593–5600.
11. A. Wittstock, A. Wichmann, J. Biener and M. Bäumer, *Faraday Disc*, 2011.
12. L. V. Moskaleva, S. Rohe, A. Wittstock, V. Zielasek, T. Kluner, K. M. Neyman and M. Baumer, *Phys. Chem. Chem. Phys.*, 2011, **13**, 4529–4539.
13. J. L. C. Fajin, M. Cordeiro and J. R. B. Gomes, *Chem. Commun.*, 2011, **47**, 8403–8405.
14. A. Wittstock, V. Zielasek, J. Biener, C. M. Friend and M. Bäumer, *Science*, 2010, **327**, 319–322.
15. C. H. Christensen and J. K. Norskov, *Science*, 2010, **327**, 278–279.
16. A. Wittstock, J. Biener and M. Baumer, *Phys. Chem. Chem. Phys.*, 2010, **12**, 12919–12930.
17. M. M. Biener, J. Biener, A. Wichmann, A. Wittstock, T. F. Baumann, M. Bäumer and A. V. Hamza, *Nano Lett.*, 2011, **11**, 3085–3090.
18. J. K. Norskov, T. Bligaard, B. Hvolbaek, F. Abild-Pedersen, I. Chorkendorff and C. H. Christensen, *Chem. Soc. Rev.*, 2008, **37**, 2163–2171.
19. C. H. Christensen and J. K. Norskov, *J. Chem. Phys.*, 2008, **128**, 8.
20. J. Greeley, J. K. Norskov and M. Mavrikakis, *Annu. Rev. Phys. Chem.*, 2002, **53**, 319–348.
21. J. K. Norskov, T. Bligaard, A. Logadottir, S. Bahn, L. B. Hansen, M. Bollinger, H. Bengaard, B. Hammer, Z. Sljivancanin, M. Mavrikakis, Y. Xu, S. Dahl and C. J. H. Jacobsen, *J. Catal.*, 2002, **209**, 275–278.
22. P. Liu and J. K. Norskov, *Phys. Chem. Chem. Phys.*, 2001, **3**, 3814–3818.
23. B. Hammer and J. K. Norskov, in *Advances in Catalysis, Volume 45*, Academic Press, San Diego, CA, 2000, pp. 71–129.
24. T. A. Baker, X. Liu and C. M. Friend, *Phys. Chem. Chem. Phys.*, 2011, **13**, 34–46.
25. B. Xu, J. Haubrich, C. G. Freyschlag, R. J. Madix and C. M. Friend, *Chem. Sci.*, 2010, **1**, 310–314.
26. R. G. Quiller, X. Y. Liu and C. M. Friend, *Chem. Asian J.*, 2010, **5**, 78–86.
27. N. Saliba, D. H. Parker and B. E. Koel, *Surf. Sci.*, 1998, **410**, 270–282.
28. M. A. Chesters and G. A. Somorjai, *Surf. Sci.*, 1975, **52**, 21–28.
29. D. D. Eley and P. B. Moore, *Surf. Sci.*, 1978, **76**, L599–L602.
30. J. J. Pireaux, M. Chtaib, J. P. Delrue, P. A. Thiry, M. Liehr and R. Caudano, *Surf. Sci.*, 1984, **141**, 211–220.
31. J. M. Gottfried, PhD dissertation, Freie Universität, 2003.
32. X. Y. Deng, B. K. Min, A. Guloy and C. M. Friend, *J. Am. Chem. Soc.*, 2005, **127**, 9267–9270.
33. J. M. Gottfried, K. J. Schmidt, S. L. M. Schroeder and K. Christmann, *Surf. Sci.*, 2002, **511**, 65–82.
34. J. M. Gottfried, N. Elghobashi, S. L. M. Schroeder and K. Christmann, *Surf. Sci.*, 2003, **523**, 89–102.
35. J. Wang, M. R. Voss, H. Busse and B. E. Koel, *J. Phys. Chem. B*, 1998, **102**, 4693–4696.
36. J. L. Gong and C. B. Mullins, *Accounts Chem. Res.*, 2009, **42**, 1063–1073.

37. B. K. Min, A. R. Alemozafar, M. M. Biener, J. Biener and C. M. Friend, *Top. Catal.*, 2005, **36**, 77–90.

38. B. K. Min, X. Deng, D. Pinnaduwage, R. Schalek and C. M. Friend, *Phys. Rev. B*, 2005, **72**.

39. V. Zielasek, B. J. Xu, X. Y. Liu, M. Baumer and C. M. Friend, *J. Phys. Chem. C*, 2009, **113**, 8924–8929.

40. J. Biener, A. Wittstock, M. M. Biener, T. Nowitzki, A. V. Hamza and M. Baeumer, *Langmuir*, 2010, **26**, 13736–13740.

41. J. Kirn, E. Samano and B. E. Koel, *Surf. Sci.*, 2006, **600**, 4622–4632.

42. T. A. Baker, C. M. Friend and E. Kaxiras, *J. Phys. Chem. C*, 2009, **113**, 3232–3238.

43. B. K. Min, A. R. Alemozafar, D. Pinnaduwage, X. Deng and C. M. Friend, *J. Phys. Chem. B*, 2006, **110**, 19833–19838.

44. J. M. Gottfried, K. J. Schmidt, S. L. M. Schroeder and K. Christmann, *Surf. Sci.*, 2003, **525**, 184–196.

45. A. Schaefer, D. Ragazzon, A. Wittstock, L. E. Walle, A. Borg, M. Bäumer and A. Sandell, *J. Phys. Chem. C*, 2012, accepted (01/12).

46. C. G. Freyschlag and R. J. Madix, *Mater. Today*, 2011, **14**, 134–142.

47. J. M. Gottfried, K. J. Schmidt, S. L. M. Schroeder and K. Christmann, *Surf. Sci.*, 2003, **536**, 206–224.

48. C. J. Weststrate, E. Lundgren, J. N. Andersen, E. D. L. Rienks, A. C. Gluhoi, J. W. Bakker, I. M. N. Groot and B. E. Nieuwenhuys, *Surf. Sci.*, 2009, **603**, 2152–2157.

49. W. L. Yim, T. Nowitzki, M. Necke, H. Schnars, P. Nickut, J. Biener, M. M. Biener, V. Zielasek, K. Al-Shamery, T. Kluner and M. Baumer, *J. Phys. Chem. C*, 2007, **111**, 445–451.

50. M. J. Gottfried and K. Christmann, *Surf. Sci.*, 2004, **566–568**, Part 2, 1112–1117.

51. L. V. Moskaleva, S. Rohe, A. Wittstock, V. Zielasek, T. Kluner, K. M. Neyman and M. Baumer, *Phys. Chem. Chem. Phys.*, 2011, **13**, 4529–4539.

52. E. Fiedler, G. Grossmann, D. B. Kersebohm, G. Weiss and C. Witte, in *Ullmann's Encyclopedia of Industrial Chemistry*, Wiley-VCH Verlag GmbH & Co. KGaA, Weinheim, 2000.

53. X. Y. Liu, R. J. Madix and C. M. Friend, *Chem. Soc. Rev.*, 2008, **37**, 2243–2261.

54. I. E. Wachs and R. J. Madix, *Surf. Sci.*, 1978, **76**, 531–558.

55. D. A. Outka and R. J. Madix, *J. Am. Chem. Soc.*, 1987, **109**, 1708–1714.

56. H. Knozinger and K. Kochloefl, *Ullmann's Encyclopedia of Industrial Chemistry*, Wiley-VCH Verlag GmbH & Co. KGaA, Weinheim, 2005.

57. C. L. Bracey, P. R. Ellis and G. J. Hutchings, *Chem. Soc. Rev.*, 2009, **38**, 2231–2243.

58. M. Bowker, *Chem. Soc. Rev.*, 2008, **37**, 2204–2211.

59. N. Bion, F. Epron, M. Moreno, F. Marino and D. Duprez, *Top. Catal.*, 2008, **51**, 76–88.

60. J. Wintterlin, T. Zambelli, J. Trost, J. Greeley and M. Mavrikakis, *Angew. Chem. Int. Edit.*, 2003, **42**, 2850–2853.

61. H. Falsig, B. Hvolbaek, I. S. Kristensen, T. Jiang, T. Bligaard, C. H. Christensen and J. K. Norskov, *Angew. Chem. Int. Edit.*, 2008, **47**, 4835–4839.

62. B. Hvolbaek, T. V. W. Janssens, B. S. Clausen, H. Falsig, C. H. Christensen and J. K. Norskov, *Nano Today*, 2007, **2**, 14–18.

63. M. Haruta, *ChemPhysChem.*, 2007, **8**, 1911–1913.

64. B. Jurgens, C. Kubel, C. Schulz, T. Nowitzki, V. Zielasek, J. Biener, M. M. Biener, A. V. Hamza and M. Baumer, *Gold Bull.*, 2007, **40**, 142–149.

65. A. Q. Wang, C. M. Chang and C. Y. Mou, *J. Phys. Chem. B*, 2005, **109**, 18860–18867.

66. J. M. Thomas and W. J. Thomas, *Principle and Practice of Heterogeneous Catalysis*, VCH Weinheim, New York, 1996.

67. R. G. Quiller, T. A. Baker, X. Deng, M. E. Colling, B. K. Min and C. M. Friend, *J. Chem. Phys.*, 2008, **129**, 9.

68. M. Date and M. Haruta, *J. Catal.*, 2001, **201**, 221–224.

69. L. M. Liu, B. McAllister, H. Q. Ye and P. Hu, *J. Am. Chem. Soc.*, 2006, **128**, 4017–4022.

70. Z. L. Wu, S. H. Zhou, H. G. Zhu, S. Dai and S. H. Overbury, *J. Phys. Chem. C*, 2009, **113**, 3726–3734.

71. H. Y. Su, M. M. Yang, X. H. Bao and W. X. Li, *J. Phys. Chem. C*, 2008, **112**, 17303–17310.

72. M. A. report, Responsible Care and Social Responsibility Report, http://www.methanex.com/ourcompany/documents/Methanex_RCSR_Report_final.pdf, 2010.

73. G. Reuss, W. Disteldorf, A. O. Gamer and A. Hilt, in *Ullmann's Encyclopedia of Industrial Chemistry*, Wiley-VCH Verlag GmbH & Co. KGaA, Weinheim, 2000.

74. W. Reutemann and H. Kieczka, in *Ullmann's Encyclopedia of Industrial Chemistry*, Wiley-VCH Verlag GmbH & Co. KGaA, Weinheim, 2000.

75. B. Jorgensen, S. E. Christiansen, M. L. D. Thomsen and C. H. Christensen, *J. Catal.*, 2007, **251**, 332–337.

76. I. S. Nielsen, E. Taarning, K. Egeblad, R. Madsen and C. H. Christensen, *Catal. Lett.*, 2007, **116**, 35–40.

77. D. I. Enache, J. K. Edwards, P. Landon, B. Solsona-Espriu, A. F. Carley, A. A. Herzing, M. Watanabe, C. J. Kiely, D. W. Knight and G. J. Hutchings, *Science*, 2006, **311**, 362–365.

78. A. Abad, P. Concepcion, A. Corma and H. Garcia, *Angew. Chem. Int. Edit.*, 2005, **44**, 4066–4069.

79. M. D. Hughes, Y. J. Xu, P. Jenkins, P. McMorn, P. Landon, D. I. Enache, A. F. Carley, G. A. Attard, G. J. Hutchings, F. King, E. H. Stitt, P. Johnston, K. Griffin and C. J. Kiely, *Nature*, 2005, **437**, 1132–1135.

80. D. I. Enache, D. W. Knight and G. J. Hutchings, *Catal. Lett.*, 2005, **103**, 43–52.

81. P. G. N. Mertens, M. Bulut, L. E. M. Gevers, I. F. J. Vankelecom, P. A. Jacobs and D. E. De Vos, *Catal. Lett.*, 2005, **102**, 57–61.

82. T. Mallat and A. Baiker, *Chem. Rev.*, 2004, **104**, 3037–3058.

83. S. Biella and M. Rossi, *Chem. Commun.*, 2003, 378–379.

84. R. J. Madix, C. M. Friend and X. Y. Liu, *J. Catal.*, 2008, **258**, 410–413.

85. K. M. Kosuda, A. Wittstock, C. M. Friend and M. Bäumer, *Angew. Chem. Int. Edit.*, 2012, DOI: 10.1002/anie.201107178.

86. D. Q. Han, T. T. Xu, J. X. Su, X. H. Xu and Y. Ding, *ChemCatChem.*, 2010, **2**, 383–386.

87. M. A. Sanchez-Castillo, C. Couto, W. B. Kim and J. A. Dumesic, *Angew. Chem. Int. Edit.*, 2004, **43**, 1140–1142.

88. W. B. Kim, T. Voitl, G. J. Rodriguez-Rivera and J. A. Dumesic, *Science*, 2004, **305**, 1280–1283.

89. W. C. Ketchie, Y. L. Fang, M. S. Wong, M. Murayama and R. J. Davis, *J. Catal.*, 2007, **250**, 94–101.

90. W. Ketchie, M. Murayama and R. Davis, *Top. Catal.*, 2007, **44**, 307–317.

91. H. Hustede, H.-J. Haberstroh and E. Schinzig, *Ullmans' Encyclopedia of Industrial Chemistry*, 2000.

92. M. Comotti, C. Della Pina, E. Falletta and M. Rossi, *J. Catal.*, 2006, **244**, 122–125.

93. Johnson, Matthey and Co. Ltd, *Pat., GB 1208101*, 1970.

94. M. Comotti, C. Della Pina, R. Matarrese and M. Rossi, *Angew. Chem. Int. Edit.*, 2004, **43**, 5812–5815.

95. S. Biella, L. Prati and M. Rossi, *J. Catal.*, 2002, **206**, 242–247.

96. H. M. Yin, C. Q. Zhou, C. X. Xu, P. P. Liu, X. H. Xu and Y. Ding, *J. Phys. Chem. C*, 2008, **112**, 9673–9678.

97. Y. Onal, S. Schimpf and P. Claus, *J. Catal.*, 2004, **223**, 122–133.

98. R. Murugavel, M. Bhattacharjee and H. W. Roesky, *Appl. Organomet. Chem.*, 1999, **13**, 227–243.

99. R. Murugavel, A. Voigt, M. G. Walawalkar and H. W. Roesky, *Chem. Rev.*, 1996, **96**, 2205–2236.

100. K. L. Fujdala and T. D. Tilley, *Chem. Mat.*, 2002, **14**, 1376–1384.

101. C. G. Lugmair, K. L. Fujdala and T. D. Tilley, *Chem. Mat.*, 2002, **14**, 888–898.

102. V. Chandrasekhar, R. Boomishankar and S. Nagendran, *Chem. Rev.*, 2004, **104**, 5847–5910.

103. T. Mitsudome, A. Noujima, T. Mizugaki, K. Jitsukawa and K. Kaneda, *Chem. Commun.*, 2009, 5302–5304.

104. N. Asao, Y. Ishikawa, N. Hatakeyama, Menggenbateer, Y. Yamamoto, M. W. Chen, W. Zhang and A. Inoue, *Angew. Chem. Int. Edit.*, 2010, **49**, 10093–10095.

105. L. H. Qian, W. Shen, B. Shen, G. W. W. Qin and B. Das, *Nanotechnology*, 2010, **21**.

106. L. Qian, W. Shen, B. Das, B. Shen and G. W. Qin, *Chem. Phys. Lett.*, 2009, **479**, 259–263.

107. C. C. Jia, H. M. Yin, H. Y. Ma, R. Y. Wang, X. B. Ge, A. Q. Zhou, X. H. Xu and Y. Ding, *J. Phys. Chem. C*, 2009, **113**, 16138–16143.

108. A. S. K. Hashmi and G. J. Hutchings, *Angew. Chem. Int. Edit.*, 2006, **45**, 7896–7936.
109. J. S. King, A. Wittstock, J. Biener, S. O. Kucheyev, Y. M. Wang, T. F. Baumann, S. K. Giri, A. V. Hamza, M. Baeumer and S. F. Bent, *Nano Lett.*, 2008, **8**, 2405–2409.
110. S. O. Kucheyev, J. Biener, T. F. Baumann, Y. M. Wang, A. V. Hamza, Z. Li, D. K. Lee and R. G. Gordon, *Langmuir*, 2008, **24**, 943–948.
111. Q. Feng, P. H. Wen, Z. Q. Tao, Y. Ishikawa and H. Itoh, *Appl. Phys. Lett.*, 2010, **97**, 131906-1-3.
112. S. Yin, Y. Inoue, S. Uchida, Y. Fujishiro and T. Sato, *J. Mater. Res.*, 1998, **13**, 844–847.
113. S. Jakschik, U. Schroeder, T. Hecht, M. Gutsche, H. Seidl and J. W. Bartha, *Thin Solid Films*, 2003, **425**, 216–220.
114. U. Diebold, *Surf. Sci. Rep.*, 2003, **48**, 53–229.
115. A. Wichmann, A. Wittstock, C. Frank, M. Biener, B. Neumann, L. Mädler, J. Biener, A. Rosenauer and M. Bäumer, submitted to *Science*, 2012.
116. ISI Web of Knowledge as of June 2011, keywords 'nanoporous gold' and catalysis. Also including work on electrocatalytic applications.

CHAPTER 9
Electrocatalytical Properties of Nanoporous Gold

HOUYI MA* AND YI DING*

School of Chemistry and Chemical Engineering, Shandong University,
Jinan 250100, PR China
*Email: yding@sdu.edu.cn; hyma@sdu.edu.cn

9.1 Introduction

Gold has been prized throughout history for its beauty, permanence, and rarity, and is traditionally used for wealth store, jewellery, or decorative purposes. In recent years, rapid developments in nanoscience and nanotechnology have stimulated the application of gold in different disciplines, leading to an exponentially increasing number of scientific publications associated with the processing and applications of various new materials based on gold. For example, gold catalysis has become one of the hottest research topics in the catalysis field in the past decade, and in many applications nanostructured gold demonstrates superior activities not achievable using conventional catalysts. Gold nanostructures such as nanoparticles (AuNPs) also possess novel physical properties, such as high thermal and electrical conductivity and superior chemical stability, which enable them to act as ideal materials for electrocatalysis. In recent years, the number of studies concerning electrochemical applications of AuNPs in the fields of bioassays, biosensors, chemical sensors, and fuel cells has grown increasingly. However, AuNPs have a strong tendency toward agglomeration over time under operating conditions, which reduces their active surface areas to a certain degree, thereby shortening the lifetime of the devices based on AuNPs. In addition, the processing of AuNPs usually involves the

RSC Nanoscience & Nanotechnology No. 22
Nanoporous Gold: From an Ancient Technology to a High-Tech Material
Edited by Arne Wittstock, Jürgen Biener, Jonah Erlebacher and Marcus Bäumer

Published by the Royal Society of Chemistry, www.rsc.org

relatively complicated preparation and assembly procedures, which are detrimental to their actual applications.

In sharp contrast, dealloyed nanoporous gold (NPG) represents an interesting type of metallic nanostructure, which contains bicontinuous void channels with average diameters tunable from a few to hundreds of nanometers.[1,2] With a well-controlled three-dimensional structure, NPG has intriguing properties, such as a high specific surface area, excellent electrical conductivity, selective filtration performance, and good permeability for gas or liquid molecules, which make it well suited to a wide variety of applications. For example, NPG may serve as catalysts,[3,4] sensors,[5] actuators,[6] and electrocatalysts in fuel cells.[7] Naturally, the preparation and characterization of NPG have attracted considerable attention from both fundamental and practical points of view.

In the last two decades, the field of dealloying has undergone significant changes in research efforts, from an originally *pure* electrochemistry topic to true nanotechnology used for some of the most advanced applications such as energy-saving technologies and green chemistry. Accordingly, various gold-based alloys have been investigated to fabricate NPG or NPG-related materials,[2] including single-phase solid solutions such as AuAg and AuCu, intermetallic compounds such as the Au–Al system, and even multicomponent alloys or bulk metallic glasses. The dealloying process may be carried out in corrosive media either under applied anodic potentials (*i.e.* electrochemical dealloying) or at an open-circuit potential (chemical dealloying or free corrosion). While dealloying is a dynamic process, it is noted that the microstructure and morphology (*e.g.* pore size and ligament thickness) of an NPG sample could be tailored well by controlling the reaction temperature, corrosion medium, dealloying time, and applied potential.

Gold has a rich surface chemistry that has been well executed in various disciplines related to surface sciences. Characterized by their excellent stability and biocompatibility, bulk gold electrodes, such as gold-disk electrodes or gold-wire microelectrodes, have long been used in the electrochemistry field as standards. With a novel sponge morphology at the nanometer scale, NPG could act as either an active electrode itself or a novel high-surface-area electrode support for some new applications. Electrochemical energy conversion and electrochemical detection (sensing) are two major areas related to electrocatalysis.[8] In the former case, NPG provides a novel platform for fabricating high-performance fuel-cell catalysts with ultralow precious metal loading; in the latter case, the good conductivity and high specific surface area make NPG a fascinating electrochemical sensor material. In this chapter, we review recent progress in investigations on the electrocatalytical properties of NPG and NPG-based catalysts, and their applications in fuel cells and electrochemical sensors, and put forward prospective aspects for future studies.

9.2 Applications of NPG in Electrocatalysis

In view of global high-energy demands, fossil-fuel depletions, and environmental pollution throughout the world, proton-exchange membrane fuel cells

(PEMFCs) as new energy-converting devices with a high efficiency and low/zero emissions have been attracting worldwide attention in recent decades. Among the low-temperature fuel cells, H_2-PEMFCs, direct methanol fuel cells (DMFCs), and direct formic acid fuel cells (DFAFCs) have been investigated most intensively.[9] Currently, platinum (Pt) remains the most investigated noble-metal catalyst, and also has almost the best performance of all three types of fuel cells. However, not only is Pt expensive, but also it easily suffers from poisoning caused by strong adsorption of chemisorbed carbonaceous intermediates (*e.g.* adsorbed CO species). How to enhance the Pt utilization efficiency and to increase the poison resistance of the present Pt catalysts is thus of critical importance for bringing fuel cells closer to reality. Considering that gold is an oxophilic element and that NPG has several attractive structural properties quite different from those of bulk gold, NPG is proposed to play a unique role in designing and fabricating novel electrocatalysts with higher activities and a stronger poison resistance.

9.2.1 Hydrogen Fuel Cells

The use of hydrogen as a fuel has many advantages. Hydrogen can be produced from water-splitting or from the reformation of some liquid fuels, including methane and methanol. Moreover, when pure H_2 is used as the fuel, the only by-product other than electricity will be water. Therefore, H_2-PEMFCs are considered to be ideal power sources to drive our sustainable society in an environmentally friendly manner.

Among anode catalysts commonly employed, Pt displays the highest catalytic activity toward the hydrogen oxidation reaction (HOR). In order to achieve high Pt utilization, the current strategy is to disperse pre-formed Pt nanoparticles onto high-surface-area support materials such as activated carbon black. Research in this area has focused on the optimization of the interfacial structure of Pt nanoparticles with various kinds of high-surface-area carbon support materials including carbon fibers, carbon nanotubes, or, more recently, graphene. However, the size mismatch and the lack of chemical binding between Pt and the carbon materials suggest that the interfacial structure of Pt/C-like nanocomposites is intrinsically unstable. Under operating conditions, the support materials may be subject to corrosion, while Pt nanoparticles may agglomerate to form larger particles and lose their surface area and suffer from a reduced life time. Moreover, even though Pt nanoparticles may be mass-produced at low cost, for those larger than 3 nm, significant portions of Pt atoms are still buried within the structure and not involved in the surface reactions. While it is well documented that Pt can mix with many metal elements to form stable alloys, at first glance, it would seem unrealistic to replace a carbon support with much more expensive metal materials such as gold. In 2003, Erlebacher and co-workers discovered that commercially available white gold leaves could be easily dealloyed to form low-cost NPG thin films in large scale (Figure 9.1). And by tailoring the processing conditions, more complicated nanostructures such as those with multimodal porosity could

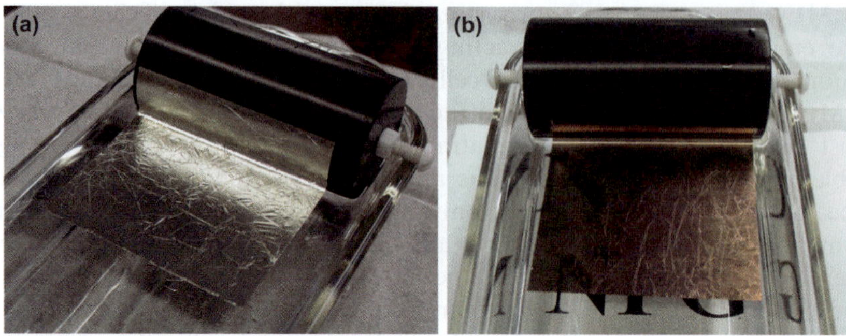

Figure 9.1 Digital pictures of white-gold leaf before (a) and after (b) dealloying in nitric acid.

be fabricated.[10] Being only about 100 nm in thickness, NPG leaves contain very tiny amount of precious metal (<0.1 mg Au cm^{-2}), which is well acceptable for fuel-cell applications.[11] More importantly, Erlebacher's group reported an electroless plating technique that allowed controlled deposition of a tiny amount of Pt onto the NPG leaf substrate.[12] They also demonstrated the feasibility to use Pt-plated NPG leaves as electrode materials for H$_2$-PEMFCs.[7,12]

As discussed above, the major advantage for using porous metal supports such as NPG lies in their ability to tether catalytically active materials such as Pt in a highly dispersed manner so that the overall Pt utilization may be improved. It was therefore a great challenge to decorate the entire three-dimensional network structure of NPG with a pore size of the order of 10 nm. In Erlebacher's work, they invented a novel interfacial electroless plating technique by floating the NPG thin membranes on the solution of metal ions, and flowing the reducing agent such as hydrazine as a vapor on top (Figure 9.2(a)).[10,12] With this method, the plating is in principle confined to occur only within the porous structure and not preferentially on either side. This method was proved to be highly effective to decorate many different materials including Pt, Pd, Ag, Sn, and even MnO$_2$, with a high degree of control, and the resulting hybrid nanostructures were found to exhibit interesting interfacial properties useful for electrocatalysis,[13] supercapacitor,[14] electrochemical sensor,[15] and surface-enhanced Raman scattering.[16]

Figure 9.2(b) and (c) compare the microstructures of NPG samples before and after plating using electron microscopy.[12] Compared to bare NPG, it seems that there is a uniform coating of Pt nanoparticles on the surface upon plating. However, electron diffraction shows that the single-crystal nature of the NPG substrate is well retained (inset in Figure 9.2(c)), indicating a crystallographic relationship between these Pt nanoparticles and the NPG substrate.[12] This assumption is proved by the high-resolution TEM observation that Pt actually grows epitaxially on NPG substrates and forms hemispherical islands with a diameter of around 3 nm and a height of just a few atomic layers (Figure 9.2(d)).[12] These Pt–NPG nanocomposites were also quite stable, and no

(a)

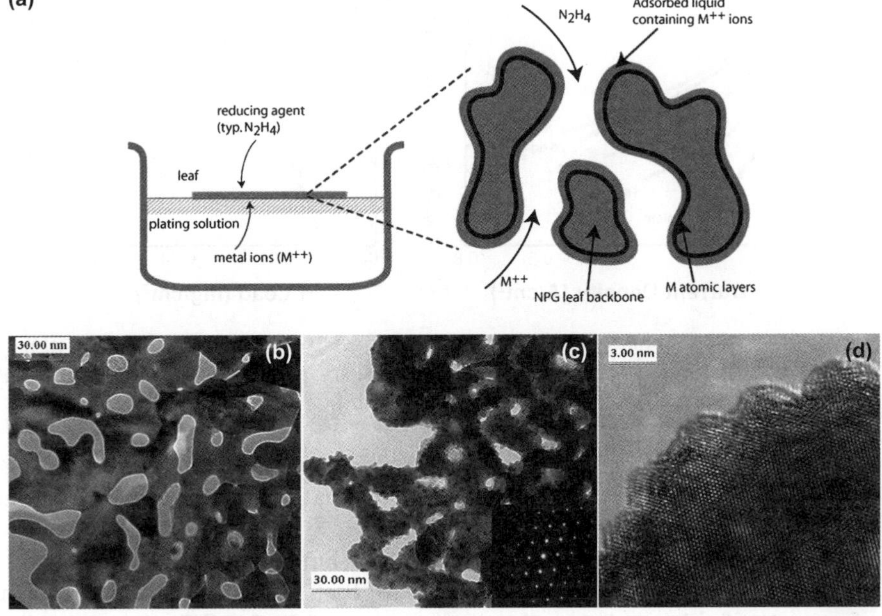

Figure 9.2 (a) Schematic illustration of the interfacial electroless plating process. (b, c) TEM images of NPG leaf before and after Pt plating; (d) HRTEM image of Pt plated NPG. The inset electron diffraction pattern in (c) shows the single-crystal nature of the Pt–NPG nanocomposites.

coarsening of the mesoporous structure was found either after standing at ambient temperature for over 2 months or after annealing at 300 °C for periods of about 1 h.[12]

Pt–NPG leaves typically contain $\sim 0.1\,\text{mg cm}^{-2}$ of Au and less than $50\,\mu\text{g cm}^{-2}$ of Pt. These nanocomposites can be readily integrated into a membrane electrode assembly (MEA), and act as both current collector and catalyst, which simplifies the MEA fabrication.[7,12] The catalytic performance of Pt–NPG leaves in H_2/O_2 fuel cells has been evaluated by Erlebacher's group. The single cell typically generates an open circuit potential around 0.9 V, a short-circuit current density larger than $1\,\text{A cm}^{-2}$, and a maximum power density larger than $250\,\text{mW cm}^{-2}$ at Pt loads of $\sim 0.03\,\text{mg cm}^{-2}$.[12] Upon optimization, the model fuel cell fabricated based on Pt–NPG nanocatalysts could generate a specific power as high as $4.5\,\text{kW g}^{-1}$ of Pt (Figure 9.3), which compares favorably to the best fuel cells reported so far and is also close to the US Department of Energy target of 2015 for Pt utilization.[7]

The oxygen reduction reaction (ORR) is of great importance in fuel cells, because oxygen reduction is inherently more difficult than hydrogen oxidation on Pt. According to Erlebacher's report,[17] NPG could act as the catalyst for ORR in both gas-phase (in fuel cells) and aqueous environments (using a rotating disk electrode), although the overall efficiency was found to be much lower than that of Pt. NPG is found to reduce oxygen via an effective

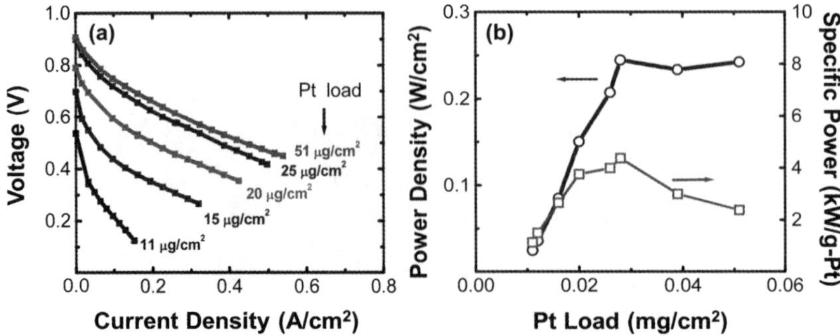

Figure 9.3   (a) Voltage–current polarization of Pt–NPG electrocatalysts with different Pt loads. (b) Maximum power density and specific power of Pt–NPG samples *vs.* Pt load (reproduced with permission from Ding and Chen[2]).

four-electron route comprising a first reduction of oxygen to hydrogen peroxide, and then an unusually active further catalytic reduction of hydrogen peroxide to water.

To date, there have been a very number of limited studies dealing with the processing of NPG-based electrocatalysts for fuel-cell tests. There will be plenty of scope to further enhance the performance of Pt–NPG leaves and related materials and to explore their new opportunities in other fuel-cell technologies.

9.2.2   Electrochemical Oxidation of Methanol

Although the H_2-PEMFCs have a relatively high efficiency and power performance, their ultimate applications rely on the availability of high-quality yet cheap H_2 fuel, and efficient and safe methods for storage and transport of high-density H_2. In contrast, DMFC using renewable methanol as the fuel is considered to be a favorable short-term option in terms of fuel usage and feed strategy. The technological key to improve the DMFC performance is to find an active low-Pt catalyst that is tolerant to CO poisoning, especially on the anode.

9.2.2.1   NPG

It was first reported by Zhang *et al.*[18] that the unsupported NPG made by selective dissolution of Ag component from Au–Ag alloys has a very good catalytic activity toward methanol electro-oxidation in alkaline solutions.

As indicated in Figure 9.4, the CV profile of NPG shows a dramatically different behavior in an alkaline solution upon addition of methanol, characterized by dominant anode waves in both forward and backward potential sweeps. The methanol oxidation on the Au electrode proceeds independently in two potential regions, with different mechanisms. At lower potentials, methanol is oxidized to formates *via* an overall four-electron transfer reaction:[19,20]

$$CH_3OH + 5OH^- = HCOO^- + 4H_2O + 4e^-. \tag{9.1}$$

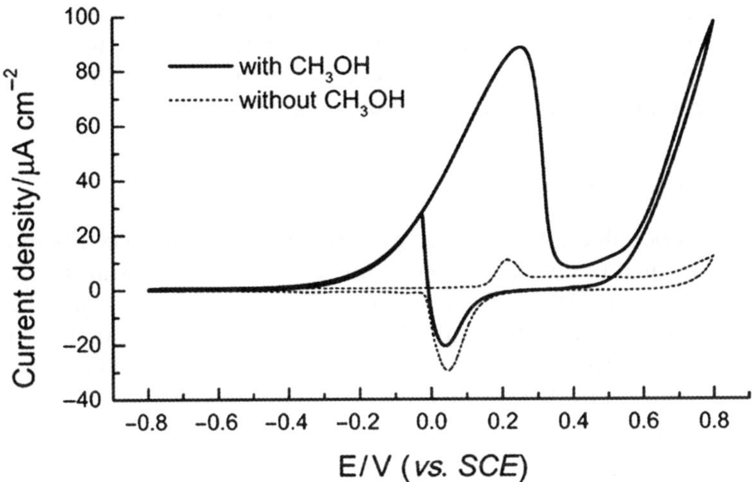

Figure 9.4 Cyclic voltammograms (CVs) of NPG electrodes in 0.5 M KOH solutions with and without 1.0 M CH₃OH (reproduced with permission from Zhang *et al.*[18]).

whereas at higher potentials, the methanol molecules are oxidized to carbonates with the exchange of six electrons:

$$CH_3OH + 8OH^- = CO_3^{2-} + 6H_2O + 6e^-. \tag{9.2}$$

The electro-oxidation of methanol is related to the chemisorbed OH^- anions but depends more strongly on the pre-oxidation species,[19,21] such as $Au–OH_{ads}^{(1-\lambda)-}$ (λ is the charge-transfer coefficient, $0 < \lambda < 1$). The gold surface oxidation needs to consume sufficient $Au–OH_{ads}^{(1-\lambda)-}$ species to form relatively dense and ordered gold oxides. As a result, the gradual exhaustion of $Au–OH_{ads}^{(1-\lambda)-}$ species slows down and eventually stops the methanol-oxidation reaction. NPG does not show any obvious catalytic activity towards the methanol oxidation at low potentials, but its catalytic activity toward the methanol oxidation starts to take place at higher potentials. The four-electron-transfer reaction described by eqn (9.1) cannot proceed on oxidized NPG surfaces, whereas the oxide-induced reaction involves a six-electron charge-transfer process, and the reaction products are carbonates instead of formates (eqn (9.2)). This is an important characteristic of Au as the electrocatalyst, which provides more reaction channels for methanol oxidation, and allows reaction intermediates produced at low potentials to be further oxidized, thereby eliminating any possible catalyst poisoning.[18]

Unfortunately, upon continuous potential cycling, the high density of low coordinated atoms on bare NPG leads to severe coarsening of the porous structure by surface diffusion. This phenomenon can be observed even at room temperature. The study concerning the size effect of NPG on the methanol reaction was also carried out by Zhang *et al.*,[18] and the results showed that smaller porosity generally exhibited a higher catalytic efficiency, but its stability decreased obviously, possibly due to its metastable surface state at a nanoscale. When the pore size or ligament thickness was relatively large, the stability of the porous structure could be improved significantly.[22]

9.2.2.2 Pt–NPG

In view of a possible catalytic synergy between Au and Pt, recently much effort has been devoted to developing bimetallic Pt–NPG electrocatalysts. A series of unique structural and catalytic properties were observed from these novel nanocomposites. According to the existing literature, so far there are three kinds of representative methods for fabricating Pt–NPG nanocatalysts: (1) immersion–electrodeposition (IE);[18,23] (2) underpotential deposition (UPD) coupled with *in situ* redox replacement reaction;[24–27] and (3) interfacial electroless plating.[10,12,13]

IE is actually a modified electrodeposition method, in which the platinum ions remaining in the porous structures are reduced at a certain applied cathodic potential. Zhang *et al.*[18,23] used the IE method to deposit a very tiny amount of Pt atoms onto the whole surface of a NPG sample. The amount of Pt deposited on the NPG substrate can be controlled by changing the numbers of IE cycles. The deposited Pt was found to affect the electrocatalytic activity of NPG toward the methanol oxidation in alkaline solution in the following two aspects: (1) onset of the methanol electro-oxidation occurred at $\sim -0.7\,\mathrm{V}$ (*vs.* SCE), 200 mV more negative than that for the NPG electrode; (2) the current density due to methanol oxidation was significantly enhanced. However, in the negative-going direction, two anodic peaks gradually decreased with the successive potential scan. This may be due to the surface alloying between Au and Pt upon continuous potential cycling.[18]

NPG is a poor electrocatalyst for methanol oxidation in acidic solutions. However, upon surface modification with a small amount of platinum, the bimetallic Pt–NPG catalysts immediately exhibit significantly enhanced electrocatalytic activities in acidic solutions. Moreover, Pt–NPG was found to show a stronger poison resistance than commercial Pt–Ru catalysts, which should be attributed to the synergistic effect of Pt and Au in Pt–NPG.[23]

On the basis of the bi-functional mechanism, the oxidation of the surface CO species at the bimetallic Pt–NPG catalysts proceeds through one of the following three paths:[23]

$$Pt\text{–}CO_{ad} + Au\text{–}OH_{ad} \rightarrow CO_2 + H^+ + Pt + Au + e^- \tag{9.3}$$

$$Au\text{–}CO_{ad} + Au\text{–}OH_{ad} \rightarrow CO_2 + H^+ + Au + e^- \tag{9.4}$$

$$Pt\text{–}CO_{ad} + Pt\text{–}OH_{ad} \rightarrow CO_2 + H^+ + Pt + e^-. \tag{9.5}$$

In order to have a sense of how Au and Pt surface atoms may interplay to show the bi-functional effect, Zhang *et al.*[23] used density functional theory calculations to compare the adsorption of CO and OH species onto pure trimers, Au_3 and Pt_3, and the binary Au–Pt trimer, Au_2Pt. The simulation indicates that the surface reaction of CO with OH on Au–Pt surface alloys is energetically more favorable than that on pure metals. Alloying Pt with Au may greatly increase Pt's reactivity due to an upshift of its surface *d*-states.[28,29] It is therefore inferred that Pt surface clusters on Pt–NPG catalysts may act as active centers during the methanol oxidation, and the surface reaction of CO with OH probably takes place in the following form:[23]

$$CO + 2OH \rightarrow CO_2 + H_2O. \tag{9.6}$$

Based on the simulation results, it is not surprising that Pt–NPG catalysts have obvious advantages in catalytic activity and poison tolerance over the traditional catalysts. These findings are not only of fundamental importance to understanding the synergistic effect of gold and platinum on their combined catalytic ability but also helpful for the development of better bimetallic Au–Pt catalysts for DMFCs.

UPD is a very important method to construct a monolayer or sub-monolayer of noble metal atoms on a foreign metal surface. Liu *et al.*[24] successfully deposited ultrathin Pt films onto NPG by utilizing UPD of Cu onto Au or Pt surfaces, followed by *in situ* redox replacement reaction (RRR) of UPD Cu by Pt (Figure 9.5(a)). The thickness of Pt layers can be controlled precisely by repeating the Cu–UPD–RRR cycles, and at least seven atomic layers of Pt could be decorated in such a way according to this literature (Figure 9.5(b)). Yoo *et al.*[25,26] compared the electrocatalytic activity of Pt decorated NPG nanorods and nanotubes for methanol oxidation. The catalytic activity of NPG nanotubes was found to be better than that of nanorods. Besides Pt, Pd is another commonly used catalyst, and it can also be decorated on NPG by this UPD method. Kiani *et al.*[27] fabricated a low Pd-coated NPG film electrode via UPD and spontaneous galvanic replacement. The synergistic effect between the NPG film electrode and Pd monolayer greatly enhanced the electrocatalytic activity toward the ORR. The success in the fabrication of NPG-based materials provides a new path to prepare electrocatalysts with ultralow precious metal loading and high activity.

Electroless deposition is among the most straightforward and efficient methods used for large-scale fabrication of Pt and Pt-based nanomaterials.[11,13] This method involves the reduction of Pt and possibly other co-catalysts from a metal salt solution onto an electrode surface. A variety of reducing agents have been employed to synthesize Pt-based bimetallic nanomaterials. Ge *et al.*[13] deposited Pt onto NPG by reducing H_2PtCl_6 with hydrazine hydrate and found that the longer the plating time, the more Pt atomic layers uniformly covering the NPG substrates. It was found from Figure 9.6 that the peak current reaches $1060\,mA\,mg^{-1}$ Pt for the sample plated for $0.5\,min$, which is more than twice that for the commercial Pt/C catalyst. This result further proved that the highest Pt utilization and a better performance could be achieved on monolayer (or sub-monolayer)-type structures on NPG.

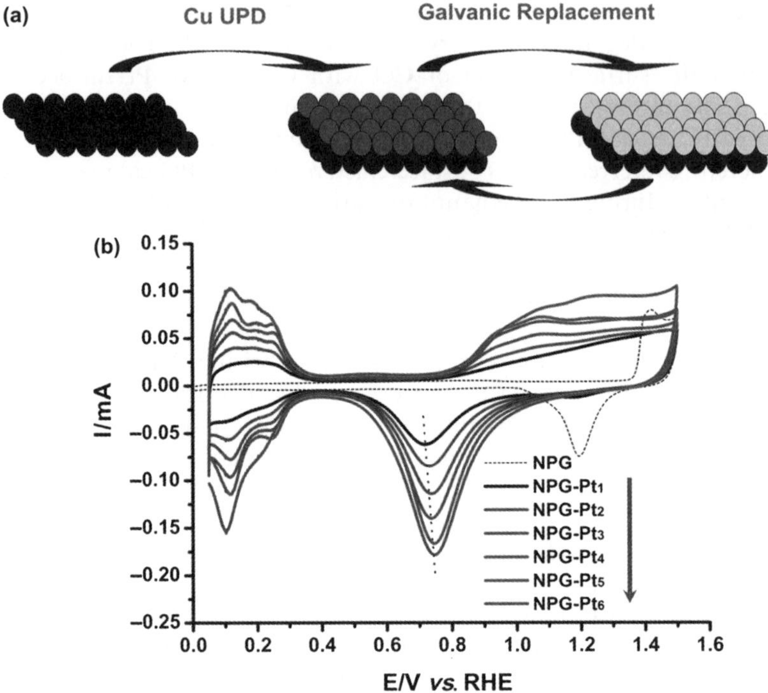

Figure 9.5 (a) Schematic illustration of the deposition process mediated by Cu UPD. (b) Cyclic voltammograms of NPG and NPG-Pt$_n$ in 0.5 M H$_2$SO$_4$, where n (1 ~ 7) denotes the number of cycles of the deposition process.

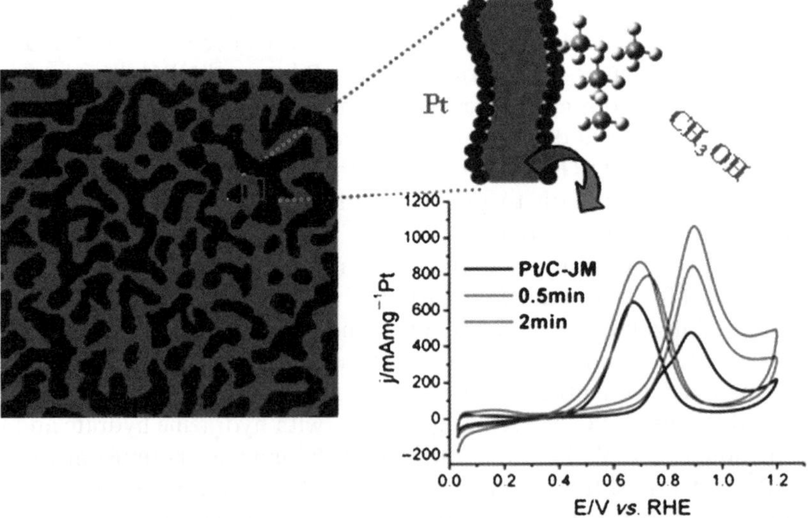

Figure 9.6 Electroless Pt plating onto NPG generating a new type core-shell structured Pt–NPG electrocatalysts that show a high specific activity in methanol electro-oxidation, especially for those with thinner Pt layers.

9.2.3 Electrochemical Oxidation of Formic Acid

Formic acid is much less hazardous to store and transport, and exhibits a smaller crossover flux through Nafion than methanol,[30] which are beneficial to the use of highly concentration fuel solutions and thinner membranes in DFAFCs. Based on the Gibbs free-energy calculation, DFAFCs have a higher electromotive force than either hydrogen fuel cells or DMFCs.

The mechanism for formic acid electro-oxidation on Pt surfaces usually involves two major pathways: dehydrogenation and dehydration.[31]

$$HCOOH \rightarrow CO_2 + 2H^+ + 2e^- \quad \text{(dehydrogenation path)}$$

$$HCOOH \rightarrow CO_{ads} + H_2O \rightarrow CO_2 + 2H^+ + 2e^- \quad \text{(dehydration path)}.$$

The dehydration pathway produces adsorbed carbon monoxide, which requires higher electrode potentials in order to be further oxidized to CO_2. In comparison, an effective bi-functional catalyst would allow the reaction to proceed only via the desired dehydrogenation path for direct formic acid electro-oxidation.

Pt–NPG was found to exhibit remarkable surface structure-dependent catalytic properties toward formic acid oxidation, as shown in Figure 9.7.[32] The most striking observation in Figure 9.7 is that the least plated Pt–NPG samples (0.5 min) exhibit an unusual activity toward formic acid oxidation, and only one broad peak appears at a low potential (–0.05 V *vs.* Mercury/Mercurous Sulfate Reference Electrode (MSE)), instead of two peaks for the other three samples. The activity of the Pt–NPG 0.5 sample recorded at –0.05 V is 5.4 mA cm^{-2}, more than twice as high as that of the Pt–NPG 128 sample, and if

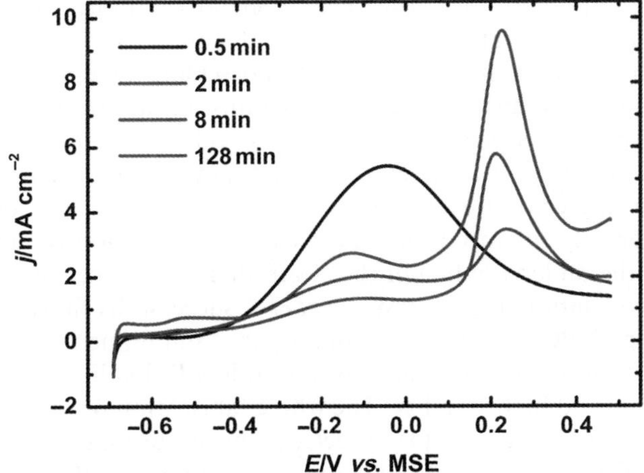

Figure 9.7 Cyclic voltammogram curves for different Pt–NPG samples in 0.1 M HClO$_4$ + 0.1 M HCOOH solution (reproduced with permission from Ge *et al.*[32]).

the Pt mass is counted, the normalized Pt activity would see an impressive 70-fold improvement as compared to the commercial Pt/C catalyst. The markedly enhanced activity at lower potential and almost complete suppression of the formation of adsorbed CO (the second peak) on the 0.5 min sample suggest that this sample is particularly effective in formic acid oxidation by choosing the direct dehydrogenation path.[32] Indeed, studies on this system showed a very clear trend that heavier Pt loading generally resulted in a poorer performance for this reaction. This assumption was further supported by a follow-up study by Ge and co-workers,[33] who correlated the evolution of the structure and property of Pt-decorated NPG samples by thermal annealing. During the annealing process, epitaxial Pt nanoislands were smoothed out and alloyed with the Au substrate, forming a thin alloy layer coating on the NPG surface. As the annealing temperature or time increases, more Au atoms will be exposed on the surface, leading to an Au-rich surface alloy, which indeed exhibited a higher activity for formic acid electro-oxidation.

The low Pt utilization and poor catalyst durability are two main obstacles to the commercial application of DFAFCs. Therefore, different approaches have been carried out to improve the catalyst dispersion and support design. By templating chemically synthesized silver nanowires, Gu *et al.*[34] fabricated interesting gold-based porous nanotubes with a continuous 1D structure, hollow interiors, and high porosity. These porous nanotubes themselves are excellent catalysts. Meanwhile, the gold nanostructures offer a good platform for Pt decoration, and the strong ensemble effect in Pt/Au may entitle them to act as superior low Pt-loading electrocatalysts in DFAFCs. Xu *et al.*[35] prepared high-surface-area nanoporous Au/Pt alloys (NPAs) with uniform pores and ligaments less than 10 nm in size by selective etching of Cu from Au/Pt/Cu alloy. As the composition varies, these Au/Pt NPAs also demonstrated quite similar surface structure sensitive electrocatalytical properties to those observed for Pt–NPG, especially for formic acid electro-oxidation.[35] These novel alloy nanomaterials have particular structural advantages in electrocatalytic reactions, which can be summarized as: (1) possessing extremely clean surfaces readily available for surface electrode reactions; (2) allowing infinite pathways for the transport of electrons and target molecules, thereby greatly facilitating the reaction kinetics; (3) completely eliminating the support corrosion problem and particle aggregation/sintering problem; (4) exhibiting strong synergistic effects with greatly enhanced catalytic activities under appropriate conditions.[35]

Based on the understanding of the electrode reaction mechanisms for formic acid electro-oxidation, recently, Wang *et al.*[36] successfully fabricated interesting sandwich-type NPG-based electrocatalysts that simultaneously fulfill three key requirements for a good electrocatalyst: ultralow Pt loading, great tolerance to CO poisoning, and high stability. The whole fabrication procedure is illustrated schematically in Figure 9.8(a). Electrochemical testing showed that upon deposition of even a few Au atoms onto the monolayer Pt-covered NPG, the as-prepared NPG–Pt$_1$–Au$_x$ demonstrated not only a higher catalytic activity but also a much improved poison resistance. This trend seems to be even more pronounced as more Au atoms are deposited (Figure 9.8(b)). The mass-specific

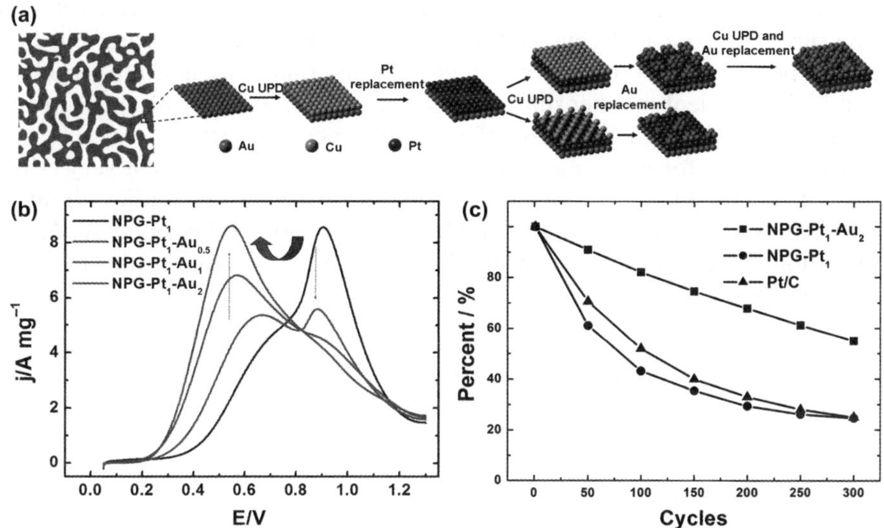

Figure 9.8   (a) Schematic illustration of the fabrication procedure of NPG–Pt_1–Au_x catalysts. (b) Mass-specific forward cyclic voltammogram segments of NPG–Pt_1–Au_x catalysts in 0.1 M $HClO_4$ + 0.05 M HCOOH solution. (c) EAS loss of NPG–Pt_1–Au_2, NPG–Pt_1 and Pt/C after different cycles of cyclic voltammogram excursion in 0.1 M $HClO_4$ between 0.05 and 1.5 V.

current density at 0.5 V on NPG–Pt_1–Au_2 is about 140 times higher than that of Pt/C at the same potential.[36] As discussed above, the direct pathway (*i.e.* dehydrogenation) of formic acid electro-oxidation needs only isolated Pt atoms, but the indirect way (*i.e.* dehydration) needs an ensemble of Pt neighboring atoms. Therefore, Wang *et al.* suggested that the much-enhanced catalytic activity at lower potential on as-prepared bimetallic nanostructures was caused by switching the reaction path from an indirect path on NPG–Pt_1 to a direct path on NPG–Pt_1–Au_x,[36] which was further proved by an *in situ* spectroscopic observation with time-resolved surface-enhanced infrared absorption spectroscopy. More interestingly, the NPG–Pt_1–Au_2 sample showed markedly improved stability as compared to both Pt/C and NPG–Pt_1 samples (Figure 9.8(c)), possibly due to the stabilization effect of Au from both the outer layer and the substrate in a mechanism of raising the Pt oxidation potential. Considering that it is possible to tailor the respective structures and compositions within each structure unit, this research provides a rational design strategy to functional nanocatalysts which is applicable to DFAFCs and other energy saving technologies.

9.2.4   Electrochemical Oxidation of Glucose

Compared with other small organic molecules, glucose can theoretically be completely oxidized to CO_2 and H_2O, releasing 24 electrons per glucose molecule,[37] although the transfer of 24 electrons has not yet been achieved. The direct glucose fuel cells (DGFCs) should thus be very promising high-energy power sources. Furthermore, considering that glucose can be produced in a highly efficient manner during photosynthesis in nature, the entire process is carbon-neutral, which clearly offers environmental benefits.

The catalysts for glucose oxidation and oxygen reduction are a very critical part of the DGFCs. Here, we first focus on reviewing the studies about the gold-based nanocatalysts used for glucose oxidation.[38] The glucose electro-oxidation on gold electrode in alkaline solution proceeds via the following reaction pathway:[38]

$$Au + OH^- \rightarrow Au-OH_{ads}{}^{(1-\lambda)-} + \lambda e^- \tag{9.7}$$

$$\tag{9.8}$$

$$\tag{9.9}$$

The formation of catalytically active sites $Au-OH_{ads}$ is the most important step for glucose oxidation. On the other hand, The generation of an oxide precursor in the potential region prior to gold oxide formation is an important property of Au electrodes,[18] which plays a key role in enhancing the catalytic activity of Au in alkaline solutions. Due to the rough surface morphology of NPG and its availability for OH^- adsorption, NPG exhibits a much better electrocatalytical activity toward glucose oxidation in neutral and alkaline solutions as compared to bulk gold.[38]

Recently, Liu *et al.*[38] found that a tiny amount of Ag components in NPG nanostructures can significantly enhance the rate for electrocatalytic oxidation of glucose. The main reason proposed was that the existence of Ag promotes the chemisorption of OH^- ions on the gold surface, thereby forming more active sites ($Au-OH_{ads}$) for glucose oxidation. This study provides some insight into the reaction mechanism of glucose oxidation and is also helpful for designing and developing high-performance electrocatalysts applicable to DGFCs.

Surface modification of Pt was also found to dramatically improve NPG's activity toward glucose oxidation, especially in neutral and alkaline solutions. Yan *et al.*[15] investigated NPG samples decorated with different amounts of Pt, and found that NPG–Pt 64 (min) demonstrated the highest activity and stability. They also constructed and ran the DGFCs using NPG–Pt 64 as the anode, and commercial Pt/C as the cathode. In an alkaline solution of 2 M $NaOH + 0.5$ M glucose, the maximum power density can achieve ~ 4.4 mW cm^{-2} at 60 °C, which is more than 20 times higher than that in neutral solution

$(0.18\,\mathrm{mW\,cm^{-2}})$, indicating a quicker reaction rate for NPG–Pt in alkaline solution.[15]

Glucose is a promising fuel for miniature fuel cells, since it is abundant in nature, nontoxic, nonvolatile, renewable, and environmentally friendly. Moreover, glucose is also ubiquitously available in body fluids, which makes glucose the most favorable fuel option for implantable fuel-cell systems. It is expected that micro-DGFCs will power autonomous, implanted, medical sensor/transmitter devices. NPG and related nanostructures as promising anode catalysts for this application deserve more attention.

9.3 Applications of NPG in Electrochemical Sensors

NPG and related materials are macroscopic in appearance, yet microscopic in feature dimensions. Their excellent structural integrity and high surface area provide sound basis for the construction of various sensors. NPG-based sensors generally offer higher sensitivity, lower detection limit, and faster electron-transfer kinetics than traditional gold electrodes. All these properties make NPG extremely attractive for a wide range of electrochemical sensors ranging from amperometric enzyme electrodes to DNA hybridization biosensors.

9.3.1 Non-enzymatic Sensors

Recent studies have confirmed that NPG could demonstrate high electro-chemical sensitivity for some important biomolecules, such as glucose,[39] dopamine (DA),[40,41] ascorbic acid (AA),[40,41] H_2O_2,[42] and NADH.[42] Dopamine (DA) is one of the most significant catecholamines belonging to the family of excitatory chemical neurotransmitters. The amperometric detection of DA at conventional solid electrodes is usually interfered with by the co-existing AA. However, NPG exhibits a substantial enhancement in electrochemical sensitivity for DA and AA in a way that their anodic peaks locate at two different potential regions.[40] Further studies showed that the oxidation of AA on NPG was a diffusion-controlled process, while for DA it was an adsorption-controlled process. It is thus possible to use NPG to selectively determine DA in the presence of AA, as can be seen in Figure 9.9. In addition, the oxidation of DA on NPG appears at a lower potential (0.29 V, Figure 9.9(b)) than that on the polished gold-sheet electrode owing to the high electrocatalytic activity of NPG. A detection limit of 17 nM for DA was achieved in the presence of millimolar levels of AA.[40]

Glucose is the primary form of sugar stored in the human body for energy. Its detection is of practical importance in food and fermentation analysis, environmental monitoring, and medical diagnosis. Over the last decades, much effort has been devoted to alleviating the drawbacks of enzymatic glucose sensors. The most common and serious problem is insufficient stability originating from the nature of enzymes. Thus, the development of non-enzymatic catalysts for electro-oxidation of glucose is of critical importance. Seeing that

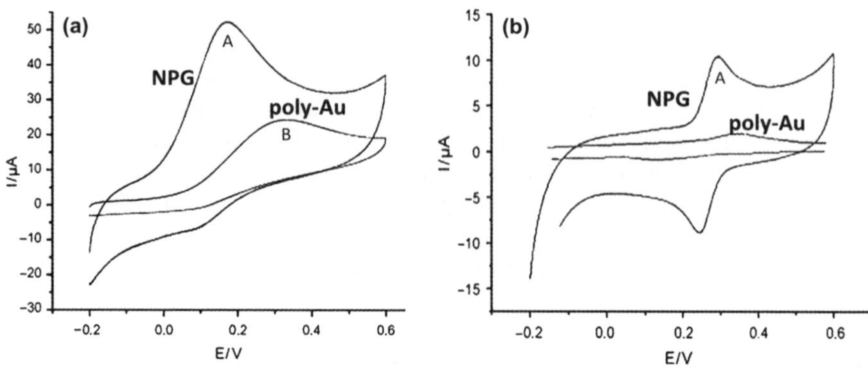

Figure 9.9 Cyclic voltammogram curves of (a) 1.5×10^{-4} M AA and (b) 1.0×10^{-5} M DA on NPG and polished gold sheet (poly-Au) electrodes in 0.1 M phosphate buffer (pH 7.0) (modified with permission from Qiu *et al.*[40]).

NPG shows a high electrocatalytic activity toward glucose electro-oxidation,[39] it is reasonable to explore the feasibility in the fabrication of bimetallic sensors based on NPG. Adding Pt could further enhance the catalytic performance of NPG. When only a small amount of Pt was decorated on NPG, the catalytic activity of NPG in neutral medium could be greatly improved, resulting in the enhanced current response and negatively shifted oxidation peak.[39] In addition, at physiological levels, the major interfering species, AA, 4-acetamidophenol, and uric acid, were found to have negligible influences on glucose sensing, due to their different electrochemical oxidation mechanisms.

Non-enzymatic sensors using NPG-based electrocatalysts offer two apparent advantages over enzymatic glucose sensors: high stability and ease of fabrication. The three-dimensional porous channels of NPG enable direct detection of biomolecules, with increased sensitivity and faster electron-transfer kinetics. The fabrication of binary or multifunctional NPG-based nanostructures may provide further opportunities in improving the detection selectivity and sensitivity.

9.3.2 Enzyme-Based Sensors

Previous examples dealing with NPG-based electrochemical sensors mainly focus on some biomolecules that are electro-active and amenable to direct electrochemical detection. However, not all biomolecules fulfill this requirement, and so enzyme-based sensors are still pursued in many applications where enzymes show very high specificity and affinity for specific biomolecules.

NPG has been used as a novel support material for enzyme sensors. Direct electron transfer from an enzyme to NPG electrode is advantageous because of the potentially fast time response and the lack of intermediate reactions that cause interference. Qiu *et al.*[43,44] immobilized laccase on NPG and studied the direct electron transfer between enzymes and electrode (Figure 9.10).

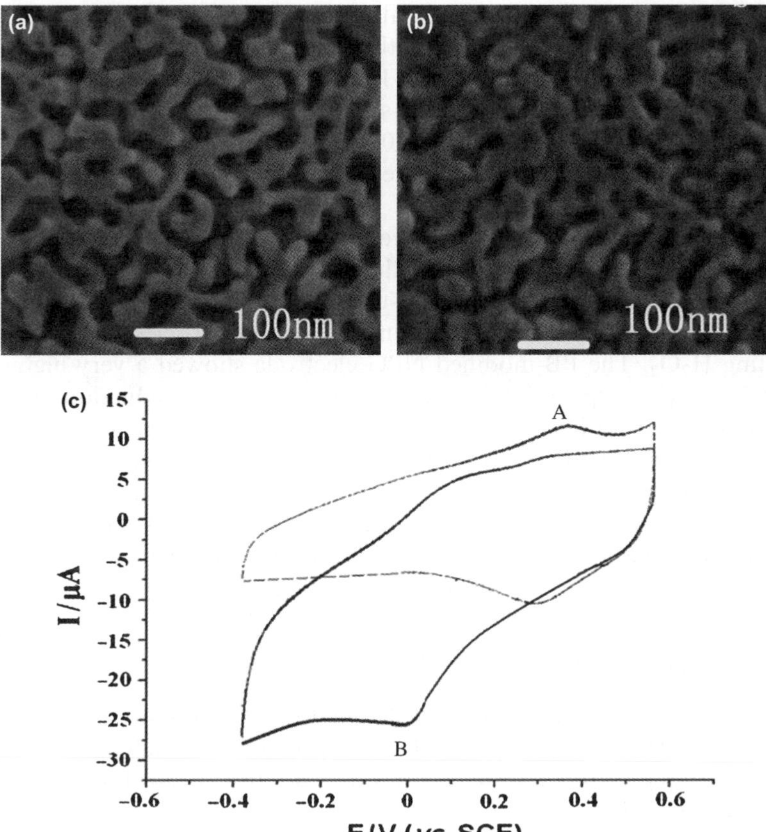

Figure 9.10 SEM images of NPG samples before (a) and after (b) laccase immobilization. (c) Cyclic voltammogram curves of laccase-loaded NPG/GC electrode in (A) deaerated and (B) air-saturated 0.1 M phosphate–citric acid buffer solution (pH 4.4) (modified with permission from Qiu *et al.*[43]).

Their results indicated that the physical adsorption strategy was the best and easiest for immobilizing laccase on NPG. By taking advantage of the interaction of nanoscale gold with NH_2 end groups on laccase, this process is remarkably simple and low cost. The electron transfer between laccase and the electrode surface is very fast, due to the outstanding physicochemical characteristics and the nanoscale pore channels of NPG.[44] The laccase-loaded NPG/GC electrode also exhibited a strong electrocatalytic activity toward O_2 reduction as shown in Figure 9.10(c). In air-saturated solution, the cathodic peak of the laccase-loaded NPG electrode occurred at about 0.02 V, which corresponds to the reduction of oxygen.[43]

The enzyme-modified NPG for electrochemical detection was also investigated by Qiu *et al.*[42] When coupled with alcohol dehydrogenase (ADH) and glucose oxidase (GOD), NPG can be used for sensitive detection of ethanol and glucose based on the high electrocatalytic activity of NPG. It showed a good linear correlation from 1 to 8 mM with a detection limit of 120 μM (signal-to-noise ratio = 3) for ethanol and 1 to 18 mM with a detection limit of 196 μM (signal-to-noise ratio = 3) for glucose sensing. The enzyme-modified NPG also displayed a high stability, with only 5.0% and 4.2% loss of current response for ADH- and GOD-based biosensors respectively after one month's storage at 4 °C.[42]

Prussian Blue (PB) is usually considered an 'artificial peroxidase' due to its high activity and selectivity toward the reduction of hydrogen peroxide, and it has been extensively used in constructing electrochemical biosensors.[45,46] Jia *et al.*[47] investigated the deposition of PB nanoparticles on NPG film for detecting H_2O_2. The PB-modified NPG electrode showed a very high electro-catalytic activity, repeatability, and stability toward the reduction of H_2O_2.

Characterized with unique physical and electrochemical properties, NPG and NPG-related nanostructures form favorable substrates for the immobilization of various enzymes. There is one particular advantage for NPG as it has excellent biocompatibility with various biomolecules and species. It is then expected that NPG will be used more widely in constructing various electrochemical biosensors, either enzymatic or non-enzymatic.

9.3.3 Immunosensors

The sensitivity of immunosensors is dictated by the amount of receptor molecules immobilized on the sensor surface. The use of nanomaterials as a medium of signal amplification affords many opportunities to advance biomolecular and gene detection.

It has been demonstrated that NPG could play a dual role in recognition and transduction events. Hu *et al.*[48] fabricated a DNA biosensor by immobilizing capture probe DNA on the NPG electrode and hybridization with target DNA, which further hybridized with the reporter DNA loaded on the gold nanoparticles (AuNPs). The AuNPs contained two kinds of bio barcode DNA: one was complementary to the target DNA, while the other was not (the scheme of multilayer construction is shown in Figure 9.11(a)). By taking advantage of the dual-amplification effects of the NPG electrode and multifunctional encoded AuNP, their biosensor allowed very high sensitivity for the selective detection of DNA. The chronocoulometry curves in Figure 9.11(b) show an increase in the charges with the increase in target DNA concentration, implying that one could use this DNA sensor to perform quantitive target DNA detection. By using $[Ru(NH_3)_6]^{3+}$ as an indicator, this NPG-based electrochemical DNA biosensor exhibited excellent sensitivity and selectivity, with the detection limit for target DNA down to 28 aM.[48] This research group also reported a 'sandwich-type' hybridization strategy for the detection of DNA sequence connected to PbS nanoparticles (PbS–NPs) tags and NPG electrode.[49] PbS–NPs could exhibit sharp and well-resolved stripping voltammetry signals due to the well-defined oxidation potential of Pb component. The limit of detection for target DNA was as low as 0.26 fM.[49]

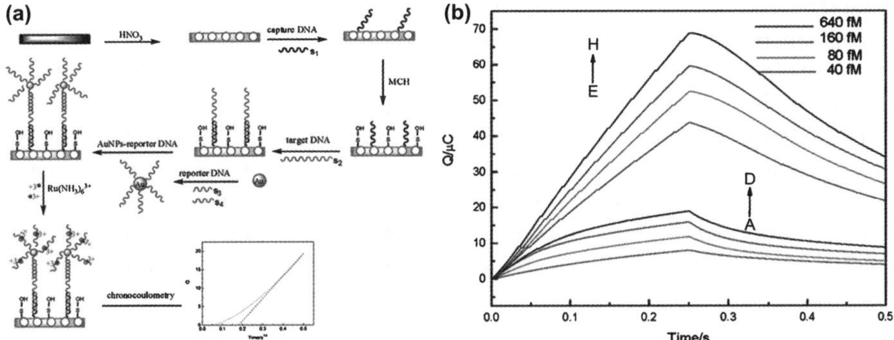

Figure 9.11 (a) Scheme of multilayer construction for chronocoulometry determination of DNA hybridization. (b) Chronocoulometry curves for the bare Au electrode (A–D) and the NPG electrodes (E–H) with capture probe hybridized with target DNA at a series of concentrations. The electrolyte was 10.0 mM Tris buffer (pH 7.0) containing 20 µM [Ru(NH$_3$)$_6$]$^{3+}$ (modified with permission from Hu *et al.*[48]).

NPG can act as both an electrode and an immobilization phase in an electrochemiluminescence (ECL)-based sensing device. For this application, the ultrathin NPG leaves are particularly useful, because they are nearly transparent, an essential character for ECL applications. Hu *et al.*[50] developed a new electrochemiluminescence (ECL) DNA assay using CdTe quantum dots (QDs) as labels and S$_2$O$_8^{2-}$ as co-reactant at the NPG leaf electrodes, on which the capture DNA was attached. The signal for successful DNA conjugation can then be converted to the ECL signal by monitoring the ECL emission from the QDs. With this method, a linear range of 5×10^{-15} to 1×10^{-11} mol L^{-1} was resolved for the detection of targeted DNA.[50] This ultrasensitive ECL assay was also tested in a model system to quantify the synthesized DNA, the key information regarding the specific gene expression levels in cells, and the results were fully comparable to those from a quantitative polymerase chain reaction.[50]

Ding *et al.*[51] described the combination of electrochemical immunoassay using NPG electrode with horseradish peroxidase (HRP) labeled secondary antibody-gold nanoparticles (AuNPs), bioconjugates for highly sensitive detection of hepatitis B surface antigen. The combination of the high active surface area of NPG electrode and the advantages of AuNPs, including biocompatibility and amplification, greatly enhanced the sensitivity of this method. The immunosensors were used to analyze hepatitis B surface antigen in human serum samples. Analytical results of clinical samples showed that the developed immunoassay was comparable with the enzyme-linked immunosorbent assay (ELISA) method.

9.3.4 Environmental Monitoring

Electroanalytical chemistry plays an important role in monitoring our environment. Many pollutants and toxic substances are electrochemically active on metal electrodes such as gold. Therefore, it is highly desirable to develop electrochemical sensor materials and technologies for fast and accurate detection of low-level environmentally concerned substances, using NPG, for example. Liu *et al.*[52] studied the application possibility of NPG as a novel electrochemical sensor for detection of *p*-nitrophenol, a common organic contaminant in waste water. The detection principle is based on the adsorption of *p*-nitrophenol on the NPG surface and the subsequent redox reaction of adsorbed electroactive species, which produces a pair of highly reversible redox peaks (Figure 9.12(a)). Especially, their peak areas are linear with the

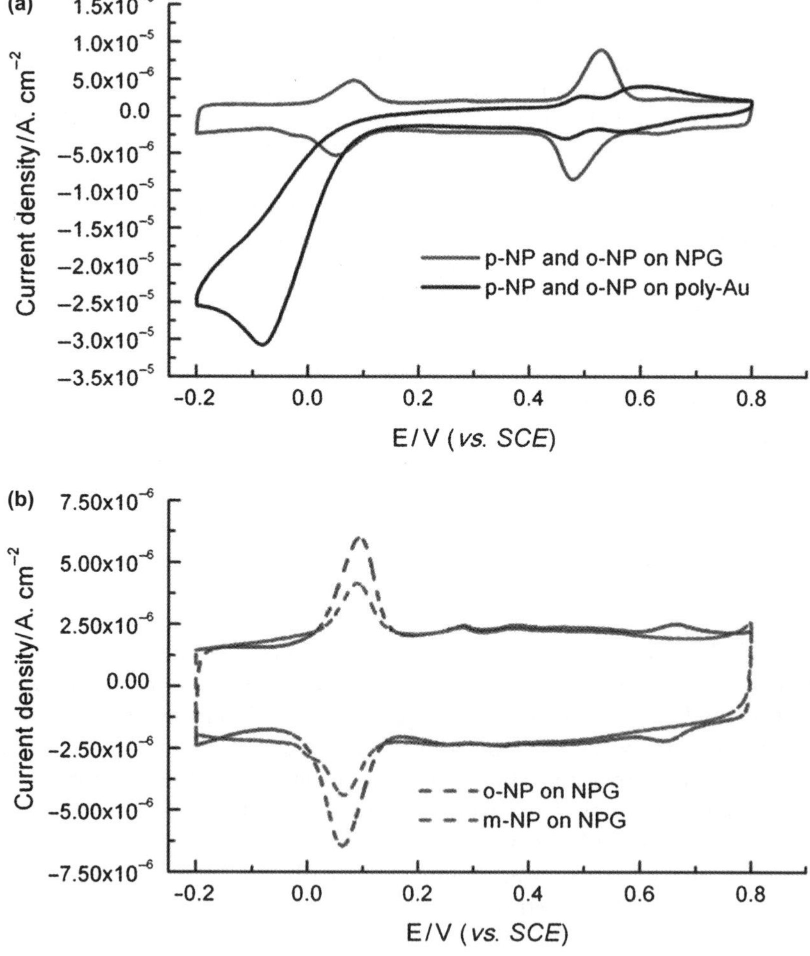

Figure 9.12 (a) Cyclic voltammogram curves for poly-Au and NPG electrodes in 0.1 M H$_2$SO$_4$ containing 10 mg L^{-1} *p*-NP and 10 mg L^{-1} *o*-NP. (b) Cyclic voltammograms for redox of 10 mg L^{-1} *o*-NP and 10 mg L^{-1} *m*-NP in 0.1 M H$_2$SO$_4$ on NPG (reproduced with permission from Liu *et al.*[48]).

p-nitrophenol concentration in the range of 0.25–10 mg L^{-1}.[52] Because this pair of redox waves is hardly affected by the isomers of *p*-nitrophenol, NPG can be used to accurately determine the *p*-nitrophenol in the presence of *o*-nitrophenol and *m*-nitrophenol (Figure 9.12(b)).[52]

Hydrazine is a toxic, easily vaporized substance. Even at a low concentration of 10 ppb, it may still cause serious liver and kidney damages. Therefore, the development of a sensitive method to detect hydrazine is critical in environmental monitoring. Yan *et al.*[53] studied the electrochemical behaviors of hydrazine oxidation on NPG, and found this electrode showed an onset potential at least 200 mV more negative than that of bulk Au. Hydrazine electro-oxidation on NPG also depended strongly on the pH value of the solution. This is because the hydrazine oxidation involves four electrons and four OH$^-$ species, and accordingly a higher concentration of hydroxide can promote the reaction. The detection of hydrazine was performed by amperometry using a rotating disk where NPG served as the electrode. The steady-state current of hydrazine oxidation was found to be proportional to the concentration in the range of 100 nM to 10.1 μM, resulting in a detection limit about 16 nM.[53]

Nitrite is ubiquitous within environmental, food, and physiological systems. It is also carcinogenic and teratogenic to human beings if their level in the blood exceeds the safety standard. Ge *et al.*[54] recently reported the application of NPG leaf in direct electrochemical detection of nitrite, which showed excellent sensitivity, stability, repeatability, and selectivity. Interestingly, the electrochemical responses of nitrite ions on this nanoelectrode were found not to be dependent on pH over a wide range from 4.5 to 8.0, which is markedly different from that of gold oxidation, a process known to be highly pH-sensitive (Figure 9.13). Amperometric study shows a linear relationship for nitrite determination in a concentration range from 1 μM to 1 mM. The presence of common species such as Na_2SO_4, KCl, NH_4NO_3, glucose, and C_2H_5OH was found not to interfere the detection, even at a concentration 100 times higher.[54]

Considering that the porous morphology of NPG-based materials is advantageous to trapping halide ions, Huang[55] fabricated an Ag UPD modified NPG electrode, and the resulting electrode possessed dual properties, including an intrinsically high surface area and the characteristics of the Ag UPD adlayer. It showed excellent sensitivity and selectivity in the electrochemical detection of chloride ions in the concentration range of 0.5–30 μM.

It is well known that the gold surface tends to form self-assembled monolayer (SAM) structures with molecules such as thiols, sulfides, and disulfides. The strong adsorption of these molecules onto the surface of NPG could change the physical properties (such as electrical conductivity) of the electrode, and this idea was demonstrated for the first time by Liu and Searson, who developed single NPG nanowire sensors that could monitor the adsorption of alkanethiol in real time.[5] Adsorption of a monolayer of octadecanethiol onto the NPG nanowire resulted in a change in resistance of about 3%. The sensitivity factor of 1.0×10^{-16} cm^2 was comparable to values reported for adsorption on ultrathin films.[5]

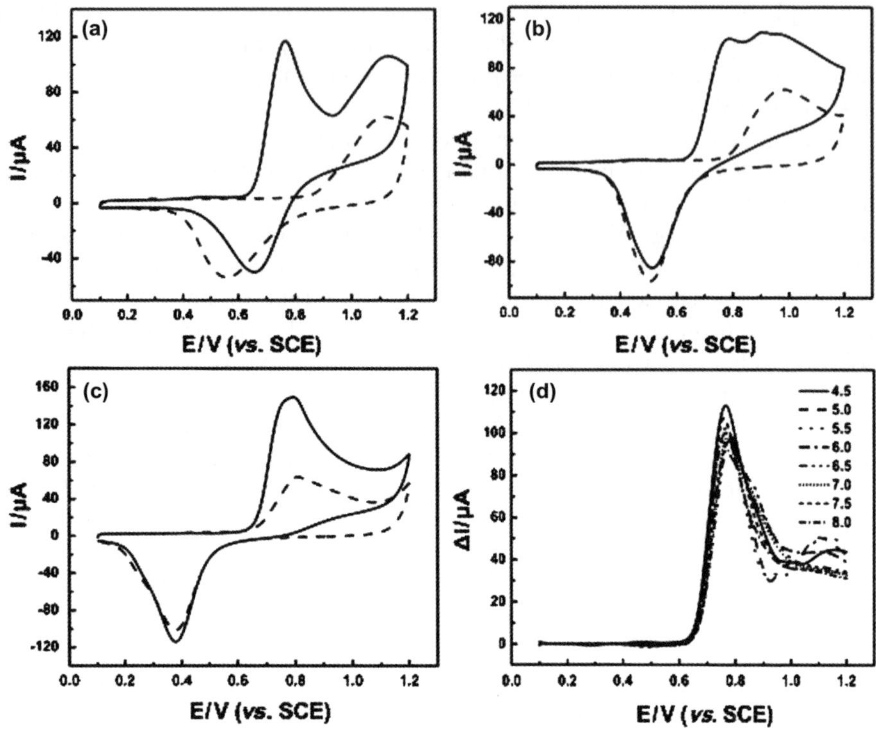

Figure 9.13   Cyclic voltammogram curves of NPGL/GCE in 0.1 M PBS with 3.0 mM NO_2^- at different pHs: (a) 4.5, (b) 6.0, and (c) 8.0. (d) Net contribution for anodic oxidation of nitrite on NPGL/GCE under different pH values. Voltammetric curves were plotted by subtracting the respective CV curves in the bare phosphate buffer solution without NO_2^- from the same electrode (reproduced with permission from Ge *et al.*[54]).

Huang and Sun also functionalized the surface of NPG by self-assembly of L-cysteine[56] or (3-mercaptopropyl)sulfonate (MPS)[57] molecules. These thiol-modified NPG electrodes not only significantly enhanced the sensitivity in detecting Cu^{2+} ions but also effectively prevented electrode surface fouling from surfactant adsorption, thus ensuring the high stability of the NPG electrode in selective detection of Cu^{2+} ions in industrial waste water.[57] In addition, Kim *et al.*[58] prepared the coral-like porous gold electrode modified by 1,6-hexanedithiol, appropriate for detecting Hg(II) ions at the ppb level. The above-mentioned studies indicate that NPG materials have great utility as sensing electrodes in the field of environmental monitoring.

9.4   Future Outlook

Different from bulk gold and gold nanoparticles, NPG is truly a nano-structured advanced material with excellent electrical conductivity, mechanical rigidity, permeability, chemical stability, biocompatibility, and other unusual properties, which are readily applicable in important areas, such as electrocatalysis, sensors, and environmental protection. NPG itself is an excellent catalyst or electrocatalyst. Moreover, this bulk nanomaterial provides a unique substrate with which to construct other binary or multifunctional materials with an even better catalytic performance, or to prepare electrochemical sensors with high sensitivity and selectivity. This chapter focuses on the recent advances regarding the electrocatalytical properties of NPG, in particular its applications in the above-mentioned areas. Despite a series of fascinating properties, the overall advancement in this area is still in its infancy. As a relatively new class of materials, most dealloyed nanoporous metals including NPG were investigated only within the last decade, and the majority of their structural properties and applications remain unexploited. In-depth research will be carried out in the near future, dealing with the continued exploration of new alloys and systems, the studies of their structural and interfacial properties, and their feasibility for real applications or device instrumentation. It is expected that more ground-breaking discoveries are yet to come for NPG-based or NPG-like nanomaterials, and a significant portion will be associated with important fields such as energy, environment, health, and biotechnology, where electrochemistry plays critical roles.

Acknowledgments

This work was sponsored by the National Basic Research Program of China (2012CB932801) and the National Science Foundation of China (90923011, 51171092). Y.D. is a Tai-Shan Scholar supported by the Independent Innovation Foundation of Shandong University (IIFSDU), the Research Fund for the Doctoral Program of Higher Education of China (20090131110019), and the Shandong Natural Science Fund for Distinguished Young Scholars. We thank Zhaona Liu, Cuicui Qiu, and other group members for their contributions to this chapter.

References

1. J. Erlebacher, M. J. Aziz, A. Karma, N. Dimitrov and K. Sieradzki, *Nature*, 2001, **410**, 450.
2. Y. Ding and M. Chen, *MRS Bull.*, 2009, **34**, 569.
3. V. Zielasek, B. Jurgens, C. Schulz, J. Biener, M.M. Biener, A.V. Hamza and M. Bäumer, *Angew. Chem. Int. Ed.*, 2006, **45**, 8241.
4. C. Xu, J. Su, X. Xu, P. Liu, H. Zhao, F. Tian and Y. Ding, *J. Am. Chem. Soc.*, 2007, **129**, 42.
5. Z. Liu and P. C. Searson, *J. Phys. Chem. B*, 2006, **110**, 4318.
6. H. Jin, X. Wang, S. Parida, K. Wang, M. Seo and J. Weissmüller, *Nano Lett.*, 2010, **10**, 187.
7. R. Zeis, A. Mathur, G. Fritz, J. Lee and J. Erlebacher, *J. Power Sources*, 2007, **165**, 65.

8. J. Lipkowski and P. N. Ross, *Electrocatalysis*, Wiley, New York, 1998.
9. M. Winter and R. J. Brodd, *Chem. Rev.*, 2004, **104**, 4245.
10. Y. Ding and J. Erlebacher, *J. Am. Chem. Soc.*, 2003, **125**, 7772.
11. Y. Ding, Y. J. Kim and J. Erlebacher, *Adv. Mater.*, 2004, **16**, 1897.
12. Y. Ding, M. Chen and J. Erlebacher, *J. Am. Chem. Soc.*, 2004, **126**, 6876.
13. X. Ge, R. Wang, P. Liu and Y. Ding, *Chem. Mater.*, 2007, **19**, 5827.
14. X. Lang, A. Hirata, T. Fujita and M. Chen, *Nature Nanotech.*, 2011, **6**, 23.
15. X. Yan, X. Ge and S. Cui, *Nanoscale Research Lett.*, 2011, **6**, 313.
16. L. Qian, Y. Ding, T. Fujita and M. Chen, *Langmuir.*, 2008, **24**, 4426.
17. R. Zeis, T. Lei, K. Sieradzki, J. Snyder and J. Erlebacher, *J. Catal.*, 2008, **253**, 132.
18. J. Zhang, P. Liu, H. Ma and Y. Ding, *J. Phys. Chem. C*, 2007, **111**, 10382.
19. Z. Borkowska, A. Tymosiak-Zielinska and G. Shul, *Electrochim. Acta*, 2004, **49**, 1209.
20. G. Tremiliosi-Filho, E. R. Gonzalez, A. J. Motheo, E. M. Belgsir, J.-M. Leger and C. J. Lamy, *Electroanal. Chem.*, 1998, **444**, 31.
21. K. A. Assiongbon and D. Roy, *Surf. Sci.*, 2005, **594**, 99.
22. C. Yu, F. Jia, Z. Ai and L. Zhang, *Chem. Mater.*, 2007, **19**, 6065.
23. J. Zhang, H. Ma, D. Zhang, P. Liu, F. Tian and Y. Ding, *Phys. Chem. Chem. Phys.*, 2008, **10**, 3250.
24. P. Liu, X. Ge, R. Wang, H. Ma and Y. Ding, *Langmuir*, 2009, **25**, 561.
25. S. H. Yoo and S. Park, *Adv. Mater.*, 2007, **19**, 1612.
26. T. Y. Shin, S. H. Yoo and S. Park, *Chem. Mater.*, 2008, **20**, 5682.
27. A. Kiani and E. N. Fard, *Electrochim. Acta*, 2009, **54**, 7254.
28. B. Hammer, Y. Morikawa and J. K. Nørskov, *Phys. Rev. Lett.*, 1996, **76**, 2141.
29. M. Ø. Pedersen, S. Helveg, A. Ruban, I. Stensgaard, E. Laegsgaard, J. K. Nørskov and F. Besenbacher, *Surf. Sci.*, 1999, **426**, 395.
30. Y. W. Rhee, S. Y. Ha and R. I. Masel, *J. Power Sources*, 2003, **117**, 35.
31. H. Okamoto, W. Kon and Y. Mukouyama, *J. Phys. Chem. B*, 2005, **109**, 15659.
32. X. Ge, R. Wang, S. Cui, F. Tian, L. Xu and Y. Ding, *Electrochem. Commun.*, 2008, **10**, 1494.
33. X. Ge, X. Yan, R. Wang, F. Tian and Y. Ding, *J. Phys. Chem. C*, 2009, **113**, 7379.
34. X. Gu, X. Cong and Y. Ding, *Chem. Phys. Chem.*, 2010, **11**, 841.
35. C. Xu, R. Wang, M. Chen, Y. Zhang and Y. Ding, *Phys. Chem. Chem. Phys.*, 2010, **12**, 239.
36. R. Wang, C. Wang, W. Cai and Y. Ding, *Adv. Mater.*, 2010, **22**, 1845.
37. F. Davis and S. P. J. Higson, *Biosens. Bioelectron.*, 2007, **22**, 1224.
38. Z. Liu, L. Huang, L. Zhang, H. Ma and Y. Ding, *Electrochim. Acta*, 2009, **54**, 7286.
39. H. Qiu and X. Huang, *J. Electroanal. Chem.*, 2010, **643**, 39.
40. H. Qiu, G. Zhou, G. Ji, Y. Zhang, X. Huang and Y. Ding, *Colloids Surf. B*, 2009, **69**, 105.

41. F. Jia, C. Yu, Z. Ai and L. Zhang, *Chem. Mater.*, 2007, **19**, 3648.
42. H. Qiu, L. Xue, G. Ji, G. Zhou, X. Huang, Y. Qu and P. Gao, *Biosen. Bioelectron.*, 2009, **24**, 3014.
43. H. Qiu, C. Xu, X. Huang, Y. Ding, Y. Qu and P. Gao, *J. Phys. Chem. C*, 2008, **112**, 14781.
44. H. Qiu, C. Xu, X. Huang, Y. Ding, Y. Qu and P. Gao, *J. Phys. Chem. C*, 2009, **113**, 2521.
45. A. A. Karyakin, *Electroanal.*, 2001, **13**, 813.
46. F. Ricci and G. Palleschi, *Biosens. Bioelectron.*, 2005, **21**, 389.
47. F. Jia, C. Yu, J. Gong and L. Zhang, *J. Solid State Electrochem.*, 2008, **12**, 1567.
48. K. Hu, D. Lan, X. Li and S. Zhang, *Anal. Chem.*, 2008, **80**, 9124.
49. K. Hu, P. Liu, S. Ye and S. Zhang, *Biosens. Bioelectron.*, 2009, **24**, 3113.
50. X. Hu, R. Wang, Y. Ding, X. Zhang and W. Jin, *Talanta*, 2010, **80**, 1737.
51. C. Ding, H. Li, K. Hu and J. Lin, *Talanta*, 2010, **80**, 1385.
52. Z. Liu, J. Du, C. Qiu, L. Huang, H. Ma, D. Shen and Y. Ding, *Electrochem. Commun.*, 2009, **11**, 1365.
53. X. Yan, F. Meng, S. Cui, J. Liu, J. Gu and Z. Zou, *J. Electroanal. Chem.*, 2011, **661**, 44.
54. X. Ge, L. Wang, Z. Liu and Y. Ding, *Electroanal.*, 2011, **23**, 381.
55. J. F. Huang, *Talanta*, 2009, **77**, 1694.
56. J. F. Huang and I. W. Sun, *Adv. Funct. Mater.*, 2005, **15**, 989.
57. J. F. Huang and B. T. Lin, *Analyst*, 2009, **134**, 2306.
58. H. Kim, Y. Kim, J. B. Joo, J. W. Ko and J. Yi, *Micropor. Mesopor. Mat.*, 2009, **122**, 283.

Nanoporous Gold in Sensor Applications

I-WEN SUN[*a] AND PO-YU CHEN[b]

[a] Department of Chemistry, National Cheng Kung University, Tainan, Taiwan; [b] Department of Medicinal and Applied Chemistry, Kaohsiung Medical University, Kaohsiung, Taiwan
*Email: iwsun@mail.ncku.edu.tw

10.1 Introduction

One important feature of a typical nanoporous metal is its high surface-to-volume ratio, *i.e.* a large fraction of surface atoms that has a lower coordination than that of bulk atoms. The low coordination of surface atoms normally improves the selectivity and activity of the nanoporous surface toward a variety of reactants.[1] The large surface and unique porous network microstructure give nanoporous metals enhanced heterogeneous catalytic, electrocatalytic, and optic properties, rendering nanoporous metals interesting for a wide range of applications including catalysis and sensing. Among the various porous metals, nanoporous gold (NPG) is of special interest because of its excellent stability and biocompatibility for use in sensors.[2] Due to the intrinsic hydrophilic nature of Au and the capillary force of nanopores, solutions and molecules can be readily pulled into the nanopores.

A nanoporous gold film (NPGF) can be prepared by the template-assisted approach in which gold is electrodeposited into the interspaces of a polystyrene sphere layer self-assembled on an electrode, and then the polystyrene template is subsequently removed.[3] The template-assisted technique has the advantage

RSC Nanoscience & Nanotechnology No. 22
Nanoporous Gold: From an Ancient Technology to a High-Tech Material
Edited by Arne Wittstock, Jürgen Biener, Jonah Erlebacher and Marcus Bäumer
© Royal Society of Chemistry 2012
Published by the Royal Society of Chemistry, www.rsc.org

of precise control over the size and microstructure of the final NPGF but is somewhat time-consuming. Alternatively, NPG films with a multimodal pore size distribution had been prepared by fast one-step direct electrodeposition from an aqueous solution using a hydrogen bubble template technique with NH_4Cl as the hydrogen bubble source.[4] In addition to deposition, a NPGF can also be prepared by repeated oxidizing and reducing a smooth gold electrode with square wave potential pulses[5,6] or by a rapid anodic potential step method with a smooth gold electrode in which NPGF formed through electrodissolution $(Au \rightarrow Au(I))$, disproportion $(Au(I) \rightarrow Au(II) + Au)$, and deposition (Au) processes during the anodic potential step in an HCl^7 or KCl solution.[8]

To date, the most popular method of generating NPGF is by dealloying gold alloy leaf containing a less noble component, such as silver or copper.[9–11] In this approach, the less noble silver or copper is selectively etched, either chemically or electrochemically, thus creating the nanoporous gold leaf, which is then attached to the surface of a substrate forming the NPGF material. Alternatively, NPGF can be fabricated by electrochemical formation of binary gold alloys such as $AuZn^{12}$ or $AuAg^{13}$ followed by dealloying of Zn or Ag.

For analytical sensing, the NPGF can be used either as prepared or after being functionalized. This was first illustrated by the following example.[12] First, an NPGF electrode was employed directly for electrochemical detection of the amino acid L-cysteine. L-Cysteine can be accumulated onto the NPGF *via* the thio group. Increasing the surface roughness (surface area) of the NPGF greatly increases the cyclic voltammetric peak currents, and thus the detection sensitivity of L-cysteine. Furthermore, the α-amino and α-carboxy groups on L-cysteine provide excellent coordination functionality; the complexation constant of the amino acid moiety with Cu(II) is four orders of magnitude larger than with other metal ions. Therefore, the L-cysteine-modified NPGF was used for sensitive and selective determination of Cu(II) ions in aqueous solution. To do this, the L-cysteine-modified NPGF electrode was immersed in the Cu(II) solution at open circuit potential to accumulate Cu(II) ions on the NPGF electrode through complexation with cysteine. After immersion, the Cu(II) ions that were accumulated at the L-cysteine-modified electrode were measured using the Osteryoung square wave stripping voltammetry technique. Due to the large surface area of the NPGF and the selectivity of L-cysteine, the sensitivity and selectivity for Cu(II) determination were significantly enhanced. Various NPGF-based sensors for the determination of species spanning from simple inorganic and organic compounds to biomolecules, especially DNA, have appeared after the above example.

10.2 Enzyme-Immobilized NPG Electrochemical Biosensors

It has been demonstrated that NPG can be a good carrier for biomolecules such as proteins and enzymes.[14–16] The successful immobilization of an enzyme was attributed to the spatial confinement of the enzyme to the microstructure of the NPG and the interaction between gold and the enzyme *via* gold–sulfur or gold–nitrogen bonds. The large pore size of the NPG provides the enzymes with sufficient degrees of freedom to take a preferred orientation so that the immobilized enzymes preserve good bioactivity. Moreover, a large signal and high stability are expected for an enzyme-immobilized NPG because the large surface of the NPG allows more enzymes to be immobilized, and the nanoscale porous structure prevents enzyme leaching. While porous silica or carbon-based materials may possess a similar stabilization effect for enzyme immobilization, they are not as good a conductor as NPG. Therefore, NPG may be a good material for constructing enzyme-immobilized electrochemical biosensors.

10.2.1 Enzyme-Modified NPG Glucose Sensor

The determination of glucose is an important issue.[17,18] Qiu *et al.*[19] demonstrated the development of enzyme-modified NPG electrochemical biosensors for detection of ethanol and glucose. The NPG leaf was prepared by dealloying an Au/Ag alloy leaf in concentrated nitric acid and fixed to a pre-polished glassy carbon electrode (GCE). As evidenced by great decrease in the required oxidation overpotential on the NPG/GCE compared with that on a gold sheet electrode, the resulting NPG/GCE showed a highly electrocatalytic activity towards the electrochemical oxidation of the enzyme-reaction products, nicotinamide adenine dinucleotide (NAD^+) and hydrogen peroxide (H_2O_2). The high electrocatalytic activity of the NPG/GCE was attributed to the high density of edge-plane-like defective site[20] and the large specific surface area of NPG. Based on the high electrocatalytic activity of NPG/GCE, sensors for the detection of ethanol and glucose, respectively, were prepared by loading alcohol dehydrogenase (ADH) and glucose oxidase (GOD), respectively, to the NPG/GCE at 4 °C. Each of the enzyme-loaded NPG/GCEs was further covered with a layer of Nafion, a cation-exchange polymer, to prevent leakage of enzyme. The amperometric current–time curves for detection of ethanol or glucose with the enzyme electrodes were recorded in stirred phosphate-buffer solution (PBS, pH 7.2) containing cofactor NAD^+ (for ethanol) or saturated with air (for glucose). The Nafion/ADH-NPG/GCE responded linearly to ethanol concentration over 1–8 mM with a sensitivity of 0.19 μA mM^{-1} and a detection limit of 120 μM (signal-to-noise ratio = 3). The Nafion/GOD-NPG/GCE showed a linear response over the glucose concentration range of 1–18 mM with a sensitivity of 0.049 A μM^{-1} and a detection limit of 196 μM (signal-to-noise ratio = 3). Due to the effective permselective barrier of the Nafion membrane, anionic ascorbic acid and uric acid at a physical level of 0.2 mM did not interfere significantly with the detection. The enzyme-immobilized NPG/GCEs lost less than 5.0% of the original current response after storage at 4 °C for a month, indicating their good stability.

10.2.2 Cytochrome C Encapsulated NPG Electrode for H_2O_2 Sensing

Zhu *et al.*[21] prepared an NPG film electrode to achieve direct electron transfer of cytochrome c (cyt. c) and subsequently constructed a H_2O_2 sensor by encapsulating cyt. c in the NPG electrode. The NPG film was prepared by a layer-by-layer route on transparent ITO substrates by alternatively assembling Au ($\sim$5 nm in diameter) and Ag ($\sim$10 nm in diameter) nanoparticles through 1,5-pentanedithiol as a cross-linker and then dissolving the Ag nanoparticles. Only a 2 nm difference was observed between the UV-visible absorption spectra of cyt. c in solution (411 nm) and cyt. c confined on the NPG surface (409 nm), indicating that no significant denaturation of the cyt. c protein occurred, and the NPG showed a good biocompatibility to retain the bioactivity of cyt. c molecules. In addition, analysis of the electrochemical data revealed that the electron transfer of cyt. c was greatly facilitated at the NPG with an electron-transfer rate constant (k_s) of 3.9 s^{-1}. Based on these results, an amperometric biosensor for H_2O_2 was constructed using the cyt. c encapsulated NPG electrode giving a linear concentration range of 1×10^{-5} to 1.2×10^{-2} M and a detection limit of 6.3×10^{-6} M. Interference of uric acid (UA) and ascorbic acid (AA) was minimized by adjusting the operating potential.

10.3 Non-enzymatic NPG-Based Sensors for Physiologic Important Species

Although enzyme-immobilized sensors have been intensively applied for monitoring various important species such as glucose, the long-term stability of the enzymatic sensors could be a problem because the activity of enzyme is usually affected by temperature, pH, humidity, toxic chemicals, *etc.*[22] Therefore, non-enzymatic sensors using noble metals are attracting considerable attention. Of these metals, gold has been shown to be a good candidate for developing non-enzymatic sensors due to its high catalytic activity towards, for example, glucose electro-oxidation in neutral and alkaline medium.[23] Compared with a smooth Au electrode, NPG gives a much better analytical performance due to its large specific surface area and high catalytic activity resulting from the larger number of low-coordinated Au atoms, *i.e.* the atoms on the corners and edges of the nano-ligaments.

10.3.1 Naked NPG Glucose Sensors

Xia *et al.*[8] investigated the glucose electro-oxidation on NPG in a 0.1 M PBS solution (pH 7.4) containing 0.1 M Na_2SO_4 or 0.1 M NaCl. As shown by the cyclic voltammograms in Figure 10.1, while no significant peak currents were observed on the smooth Au disk electrode, prominent peak currents were observed on the NPG electrode, indicating a substantially enhanced electro-catalytic oxidation of glucose on NPG. These voltammograms exhibit two anodic peaks during the positive potential scan. The first peak at a less positive potential was due to the electrosorption of glucose to form adsorbed intermediates. The intermediates accumulated and blocked the active sites of the NPG, leading to a decrease in current. As the potential was scanned to more positive values, Au-OH was formed, which catalytically oxidized and removed

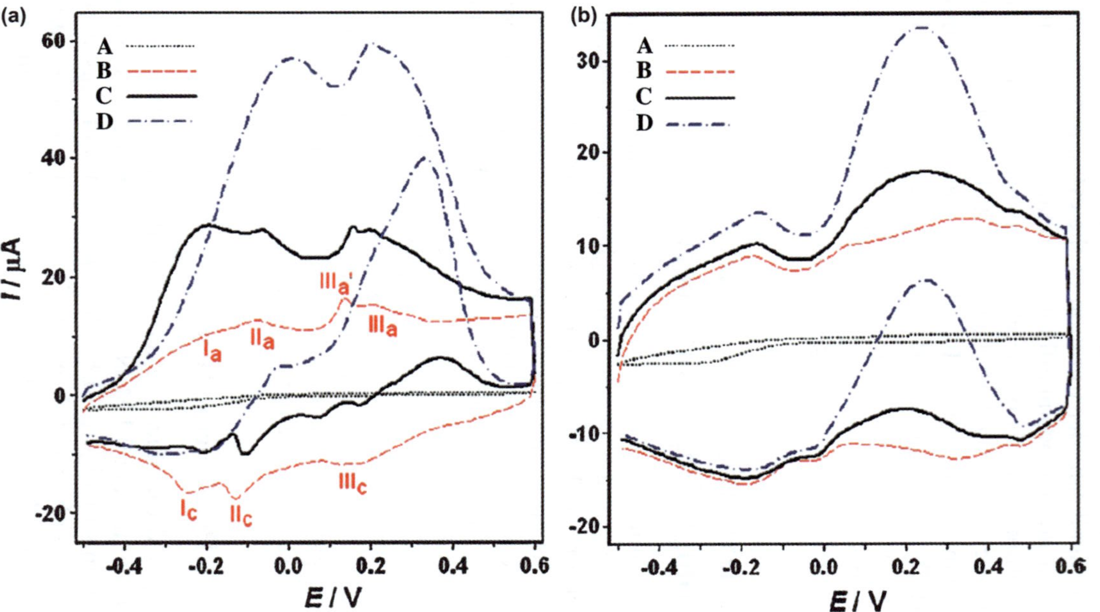

Figure 10.1 CV curves of (A) the polished gold electrode (the dotted lines) and (B–D) the NPGF electrode in the (B) absence and (A, C–D) presence of (A and C) 10 mM or (D) 50 mM glucose in (a) 0.1 M PBS + 0.1 M Na$_2$SO$_4$ or (b) 0.1 M PBS + 0.1 M NaCl. Scan rate: 20 mV s^{-1}.[8]

the intermediates to allow the direct oxidation of glucose. Therefore, the current rose until the potential was further shifted positively to the value where gold oxide formed, and the current diminished again. When the potential scan was reversed toward negative values, the gold oxide was reduced with the release of the active sites of NPG available for the direct oxidation of glucose, giving a net oxidation current until the potential was moved to more negative values where electrosorption of glucose occurred again, and the oxidation current dropped. Also revealed in Figure 10.1 is that chloride ions, which are usually present in physiological samples, suppressed the oxidation of glucose severely. In amperometric studies carried out in 0.1 M PBS (pH 7.4) containing 0.1 M NaCl, the NPG electrode exhibited a linear range from 10 μM to 11 mM with a sensitivity of 66.0 μA mM^{-1} cm^{-2} and a detection limit of 8.7 μM for glucose.

10.3.2 Pt-Decorated NPG Glucose Sensors

The property of the NPG can be modified by incorporating a second metal forming a bimetallic material. Special attention has been focused on Pt-decorated NPG (Pt–NPG) because of the excellent catalytic activity of Pt. Several methods have been developed for fabricating Pt–NPG materials. For example, Yoo and Park[24] prepared Pt–NPG nanorod arrays by coating bare NPG nanorods with a layer of copper through underpotential deposition (UPD) followed by spontaneous replacement of copper with Pt ions to obtain the ultrathin Pt-coated Pt–NPG nanorods. Alternatively, Pt–NPG was also prepared by electroless plating[25] and electroplating techniques.[26,27] The Pt–NPG materials display high electrocatalytic activities toward various reactions such as reduction of O_2, and oxidation of methanol, CO, and glucose.[24–27] Recently, Qui and Huang[28] prepared an NPG by square-wave potential pulse treatment of a Au substrate, coated the NPG with Pt by an immersion-deposition procedure, and compared the performance of the NPG and Pt–NPG electrodes as non-enzymatic glucose sensors in alkaline (0.2 M NaOH), neutral (0.1 M PBS), and acidic (0.1 M HClO$_4$) solutions. Voltammetry experiments indicated that the sensitivity of both the NPG and Pt–NPG decreased in the order: alkaline > neutral > acidic. An increase in electrocatalytic activity of Pt–NPG with increasing Pt coverage up to 60% was indicated by the negatively shifted onset potential and the enhanced current of glucose oxidation. Further increasing Pt coverage led to reduced electrocatalytic activity. This was attributed to the formation of the more active thin Pt film with low Pt coverge and the less active Pt islands with high Pt coverage. To simulate the physiological conditions, the amperometric response of the Pt–NPG sensor to glucose was recorded in neutral media. As shown in Figure 10.2, both NPG and Pt–NPG with a Pt coverage of 24% electrodes responded quickly ($\sim$5 s) to each addition of 1 mM glucose. The linear concentration range of glucose on both electrodes was 0.5–10 mM, but the sensitivity found on Pt–NPG (145.7 × 10^{-6} A cm^{-2} mM^{-1}) was higher than that on NPG (76.4 × 10^{-6} A cm^{-2} mM^{-1}). The detection limit was calculated to be 0.6 × 10^{-6} M on Pt–NPG and 1.0 × 10^{-6} M on NPG.

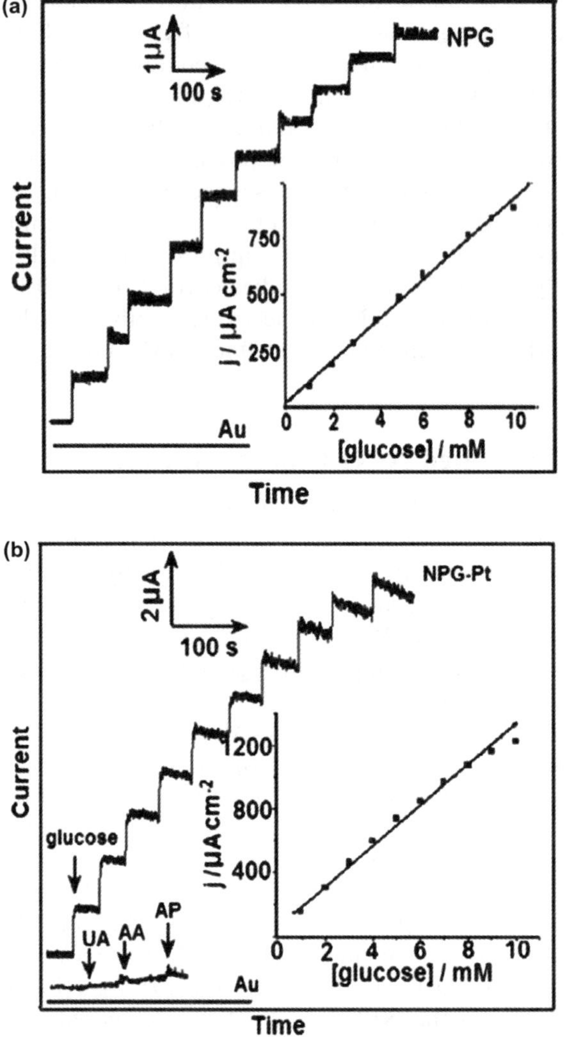

Figure 10.2   Current–time curves obtained at NPG (at $+0.35\,V$) (a) and NPG–Pt (at $+0.2\,V$) (b) electrodes with successive addition of glucose (1 mM) or UA (0.02 mM), AA (0.1 mM), AP (0.1 mM). Electrolyte: 0.1 M pH 7.4 PBS containing 0.1 M KCl. Insets show the calibration curves for glucose at NPG and NPG–Pt electrodes.[28]

10.3.3 Gold-Decorated Nanoporous Copper Core-Shell Composite Glucose Sensor

Considering that Au is a precious metal, and the high catalytic activity of NPG is mainly from nanoscale effects, it is more economical to construct a nanoporous gold material by coating a nanometer-thick Au film on the nanoporous structure of a non-noble metal such as Cu. Chen *et al.*[29] fabricated a gold-decorated nanoporous copper (Au@NPC) core-shell composite *via* a spontaneous displacement reaction at room temperature using NPC, which was prepared by dealloying $Cu_{30}Mn_{70}$ in a dilute HCl aqueous solution, as a reducing agent as well as a substrate according to the reaction in an $AuCl_4^-$ solution:

$$3Cu(NPC) + 2AuCl_4^- = 2Au@NPC + 3Cu^{2+} + 8Cl^-. \qquad (10.1)$$

SEM and TEM images confirmed that the pore size and ligament diameter of the Au@NPC composite remained similar to those of the as-prepared NPC. The electrocatalytic activity of the Au@NPC composite for the non-enzymatic detection of glucose under the physiology condition (pH 7.4) was evaluated. Shown in Figure 10.3[29] are the cyclic voltammograms (CVs) of glucose electro-oxidation on the NPG and Au@NPC electrodes in a PBS (pH 7.4) solution containing 0.05 M glucose at a scan rate of 20 mV s^{-1}. Compared with NPG, Au@NPC shows a better electrocatalytic activity toward glucose electro-oxidation, as indicated by a lower oxidation potential and higher peak current. As shown in Figure 10.3(b), the presence of chloride ions in the solution resulted in a positive potential shift and current decrease on both NPG and Au@NPC electrodes because chloride ions can occupy the active sites and retard the glucose reaction from proceeding. Nonetheless, the Au@NPC showed a better tolerance to chloride ions, as evidenced by less significant changes in the CVs. The glucose-sensing performance of the NPG and Au@NPC electrode was evaluated by amperometric experiments. The current on both electrodes responded quickly and sensitively to each addition of 1 mM glucose to the stirred solution with a linear range from 1 mM to 24 mM, but the Au@NPC exhibited a higher sensitivity in detecting glucose than that of NPG with a similar porosity.

It is important to note that the NPG-based sensors exhibit not only good sensitivity but also excellent selectivity for glucose sensing. It is well known that AA and UA at their normal physiological levels ($\sim$0.1 mM and $\sim$0.02 mM, respectively) interfered greatly with the amperometric glucose at the smooth, flat Au electrode. However, these species show negligible interference in the detection of glucose on the NPG-based electrodes. The high selectivity of NPG-based electrode can be attributed to the following. The electro-oxidation of glucose is a sluggish kinetic-controlled process that is favored by the large active surface area provided by NPG. The oxidations of AA and UA, however, are diffusion-controlled process, which depends on the apparent geometric area and is much less sensitive to the enlarged active surface area of the NPG. Therefore, the selectivity of the NPG-based glucose sensors increases with increasing surface roughness. Various methods for the determination of glucose are summarized in Table 10.1.

10.3.4 Pt–NPG Sensor for *Escherichia coli* (*E. coli*)

Routine monitoring *E. coli* in food or water is important in health issues.[30–32] Recently, a Pt–NPG electrode was developed by Yang *et al.*[33] for amperometric

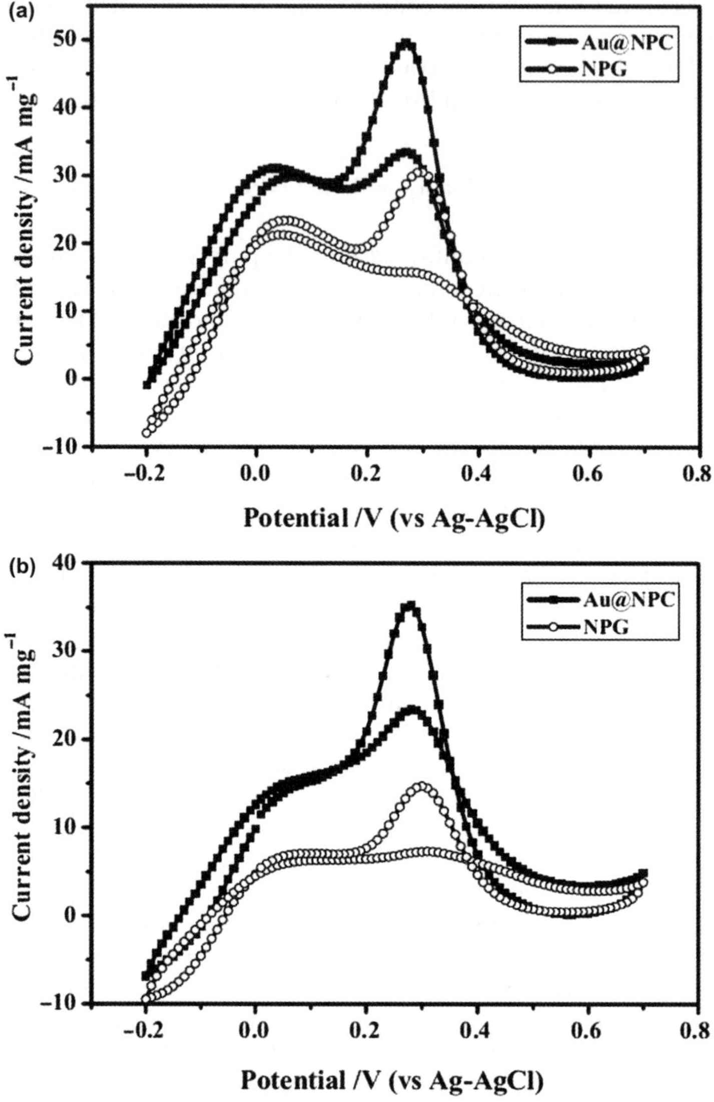

Figure 10.3   (a) CV curves of Au@NPC and NPG electrodes in 0.1 M PBS (pH 7.4) containing 0.05 M glucose and (b) in the presence of 0.1 M KCl. Scan rate: 20 mV s^{-1}.[29]

detection of *E. coli*. The Pt–NPG was prepared by electrochemical deposition of Pt nanoparticles on a pre-made NPG in a solution of 0.02 M H_2PtCl_6 and 0.05 M H_2SO_4 with a scan rate of 10 mV s^{-1}. An enzyme-galactosidase was released from *E. coli* to catalyze the hydrolysis of *p*-aminophenyl-β-D-galac-topyranoside producing electroactive *p*-aminophenol. The *E. coli* was then determined by measuring the electrochemical signal produced by oxidation of

Table 10.1 Comparison of the various assays for the determination of glucose.

Sensor type	Solution pH	Linear range	Detection limit	Reference
Au nanoparticles	PBS (pH 9.2)	0–8 mM	50 nM	17
Pt nanotubule	PBS (pH 7.4)	2–14 mM	1 μM	18
Glucose oxidase-modified NPG	PBS (pH 7.2)	1–18 mM	196 μM	19
NPG	PBS (pH 7.4)	10 μM to 11 mM	8.7 μM	8
Pt-decorated NPG	PBS (pH 7.4)	0.5–10 mM	0.6 μM	28
Au-decorated nanoporous Cu core-shell composite	PBS (pH 7.4)	1–24 mM	–	29

p-aminophenol on the anode. Bacterial cultures were measured by bare Au, NPG, and Pt–NPG electrodes. The current responses at NPG and Pt–NPG electrodes were found to be about two and five times, respectively, higher than that at the bare Au electrode, indicating that the Pt–NPG is more sensitive than NPG. The current increased linearly with *E. coli* concentration from 2×10^1 to 1×10^6 cfu mL^{-1} with a detection limit of 10 cfu mL^{-1}. The concentration of *E. coli* in a river-water sample determined with Pt–NPG agreed with the result obtained by the plate count method. The Pt–NPG electrode could be regenerated by immersing in nitric acid solution.

10.3.5 NPG Sensor for Dopamine in the Presence of Ascorbic Acid

Dopamine (DA) is an important neurotransmitter related to several serious diseases such as Parkinsonism and schizophrenia.[34,35] The major problem of electrochemical detection of DA is the interference due to the coexisting AA, which is about 1000 times higher than DA in biological samples. Many efforts therefore have focused on the sensitive and selective detection of DA using various modified electrodes.[36–39] Recently, Qiu *et al.*[40] reported the use of an NPG-coated glassy carbon electrode (NPG/GCE) for sensitive detection of DA in the presence of AA. It was found that a linear scan voltammetric peak current of the AA electro-oxidation varied with the square root of scan rate, indicating that the oxidation of AA was a diffusion-controlled process. On the other hand, the peak current of DA varied linearly with the scan rate, indicating that the oxidation of DA was a surface (adsorption) controlled process. When differential pulse voltammetry was used, well-resolved oxidation peaks of DA and AA were observed (Figure 10.4), making selective determination of DA in the presence of AA possible. It was also found that DA could be accumulated on the NPG/GCE through the interaction of the amino group of DA with the NPG surface, while AA could not. Using differential pulse voltammetry, a submicrolevel of DA could be detected in the millimolar level of AA with a detection limit of 17 nM. A similar conclusion was also reported by Choi and coworkers[41] using an electrochemically fabricated NPG film electrode with pores approximately 30 nm in diameter and 150 nm thick.

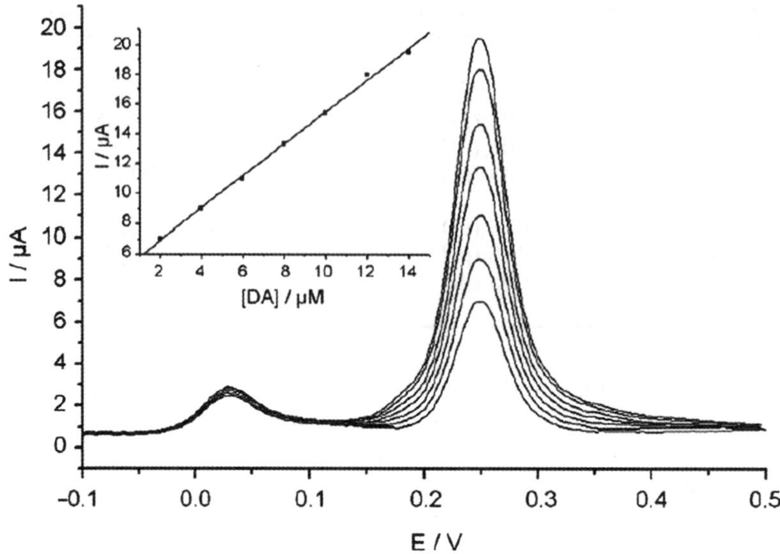

Figure 10.4 DPVs of different concentrations of DA (2, 4, 6, 8, 10, 12, and 14 μM) at NPG/GCE in 0.1 M phosphate buffer (pH 7.0) containing 10 μM AA. Scan rate: 50 mV s^{-1}; pulse amplitude: 20 mV; pulse width: 60 ms. The inset shows the calibration curve.[40]

10.3.6 NPG Immunosensor for Detection of Cancer Biomarker

The use of label-free immunosensors has received considerable attention, as this is an ideal analytical technique for the simple and rapid detection of biomolecules. Among the various methods for label-free immunosensors, electrochemical methods have the advantages of being highly sensitive and cost-effective. Due to its excellent biocompatibility and large specific surface area, NPG is an ideal material for immobilizing biomolecules in constructing immunosensors. Recently, Wei *et al.*[42] demonstrated the preparation of a NPG-based immunosensor for label-free detection of prostate specific antigen (PSA), which is a biomarker for the diagnosis and prognosis of prostate cancer. An NPG leaf was carefully coated onto a GCE, and then anti-PSA antibody (Ab) was adsorbed onto the NPG. Unbound Ab was removed by washing. Next, bovine serum albumin was used to eliminate the nonspecific binding between the antigen and the electrode surface. Subsequently, PSA was captured by incubating PSA solution on the electrode at room temperature followed by washing, and the electrode was ready for measurement in a solution of $Fe(CN)_6^{3-}$. The sensing was based on monitoring the redox current of $Fe(CN)_6^{3-}$ on the electrode. The amperometric current of $Fe(CN)_6^{3-}$ decreased linearly with PSA concentration from 0.05 to 26 ng mL^{-1} and gave a detection limit of 3 pg mL^{-1}. The sensor showed good reproducibility and selectivity, and was stable for more than two weeks when stored at 4 °C, implying that the antibody Ab was firmly bond to the NPG. The practical application of the sensor was tested by detection PSA in human serum samples, and the recovery was found to be 98.9% and 102.9%, thus indicating the feasibility of NPG immunosensors for clinical determination. The excellent performance of the sensor could be attributed to the large specific surface area and biocompatibility of NPG that enabled the adsorption of a large amount of antibody.

10.4   NPG-Based DNA Sensors

Highly sensitive detection of DNA is important for various applications including clinical diagnosis and therapy of genetic diseases, forensic analysis, and environmental control.[43–48] Specific detection of DNA relies on the 'sandwich hybridization' reaction, which involves a dual hybridization event in which the target DNA hybridizes with the capture DNA at one end and with the reporter DNA at the other end. A signal transducer is linked to the reporter DNA to produce the desired analytical signal after the sandwich hybridization reaction is completed. The sensitivity of the detection can be improved by amplification strategies by attaching a larger number of signal elements to each reporter DNA.

The fact that thiolated DNA can bound strongly at the gold surface *via* Au-thiol binding makes gold a promising material as a DNA sensor. Due to its high surface-to-volume ratio, good stability, and biocompatibility, NPG appears to be an ideal material for immobilizing capture DNA to fabricate target DNA sensors. This section provides a summary of the NPG-based DNA sensors that have been proposed from 2008 to date.

10.4.1   NPG-Based DNA Sensors with $[Ru(NH_3)_6]^{3+}$ Transducer

Zhang and coworkers[49] reported the first NPG-based electrochemical DNA sensor. This system involved four different DNA sequences including a thiolated capture DNA S_1 as probe, target DNA S_2, thiolated reporter DNA S_3, and thiolated signal DNA S_4 (Table 10.2). While the reporter DNA S_3 was complementary to target DNA S_2, the signal DNA S_4 was noncomplementary. Before fabricating the sensor, DNA S_3 and DNA S_4 were labeled onto gold nanoparticles (Au-NP) by adding a mixture of the two DNAs to freshly prepared Au-NP solution (Scheme 10.1). It was noted that there were about 27 S_3 and 100 S_4 on each Au-NP. The fabrication of the target DNA sensor is shown in Scheme 10.1. The NPG electrode was prepared by selective dissolution of silver from silver/gold alloy leaf in nitric acid. The produced NPG foil, which has a 9.2-fold enhancement in effective surface area with respect to a flat gold electrode with the same geometric area, was then coated onto a glassy carbon electrode to obtain the NPG/GCE. The single-strand thiolated capture probe DNA S_1 was immobilized onto the NPG/GCE electrode surface by a thiol–Au interaction by immersing the NPG/GCE in a solution of S_1. To obtain a well-aligned DNA monolayer, the DNA-modified NPG/GCE was further

Table 10.2 DNA sequences used in a DNA biosensor based on NPG.

Name	Sequence	Description
Capture DNA S_1	5′-SH-$(CH_2)_6$-TCG TAC GAT CGA TCC-3′	Thiolated probe immobilized on NPG
Target DNA S_2	5′-GCC GCT CAC ACG ATA TTT TTT TTG GAT CGA TCG TAC GA-3′	Complementary to S_1 and S_3
Reporter DNA S_3	5′-TAT CGT GTG AGC GGC TTT TTT TT$(CH_2)_6$-SH-3′	Thiolated and complementary to S_2
Signal DNA S_4	5′-GCT CAT ATG GAC CTC TTT TTT TT$(CH_2)_6$-SH-3′	Thiolated and noncomplementary to S_2

immersed in a 6-mercapto-1-hexanol solution. After removing the unspecific adsorbed S_1, the target DNA S_2 was hybridized with S_1 by immersing the DNA S_1/NPG/GCE into the DNA S_2 solution at 37 °C. The DNA S_2/DNA S_1/NPG/ GCE was further immersed in the DNA S_3/DNA S_4-loaded Au-NPs solution so that the target DNA S_2 was secondly hybridized with the reporter DNA S_3 loaded on An-NPs. After hybridization, the electrode was rinsed with washing buffer and dried under nitrogen prior to electrochemical measurement. When the electrode was immersed in a Tris-HCl solution containing $[Ru(NH_3)_6]^{3+}$, the electrochemical signal of the $[Ru(NH_3)_6]^{3+}$ that bound to the signal DNA S_4 *via* electrostatic interaction was measured by chronocoulometry. The signal was proportional to the amount of DNA S_4 attached on Au-NP and further related to the concentration of the target DNA S_2. It was found that the signal detected on the NPG electrode was 30 times higher than that on the bare gold electrode with the same target DNA. Analytical results indicated that this DNA-modified NPG/Au-NP biosensor was able to detect the target DNA with a linear range from 8.0×10^{-17} to 8.0×10^{-16} M and a limit of detection of 2.8×10^{-17} M. This strategy is more sensitive than many other reported assays. Because only the DNA matched perfectly with the target DNA produced prominent signal, this strategy gave excellent selectivity for detecting the target DNA. Since the hybridized DNA could be removed by thermal denaturation, this NPG electrode could be regenerated by keeping it in hot water (90 °C) for 1 min. The superior performance of this NPG-based biosensor could be attributed to the larger effective surface area and high electrical conductivity provided by the NPG. The higher effective surface area allowed the loading of a larger number of capture DNA S_1 at the NPG electrode than at the flat gold electrode. Moreover, since about 100 signal DNA S_4 were loaded on each single Au-NP that was linked to the target DNA through the reporter DNA S_3, a significant amplification effect for the detection of the target DNA S_2 was realized.

The same strategy was adopted by Qiu *et al.* in the fabrication of an aptasensor for thrombin.[50] First, a layer of thiolated aptamer was assembled on the NPG; then, one binding site of thrombin was bound to the aptamer by the specific affinity, and the other side of thrombin was then bound to aptamer

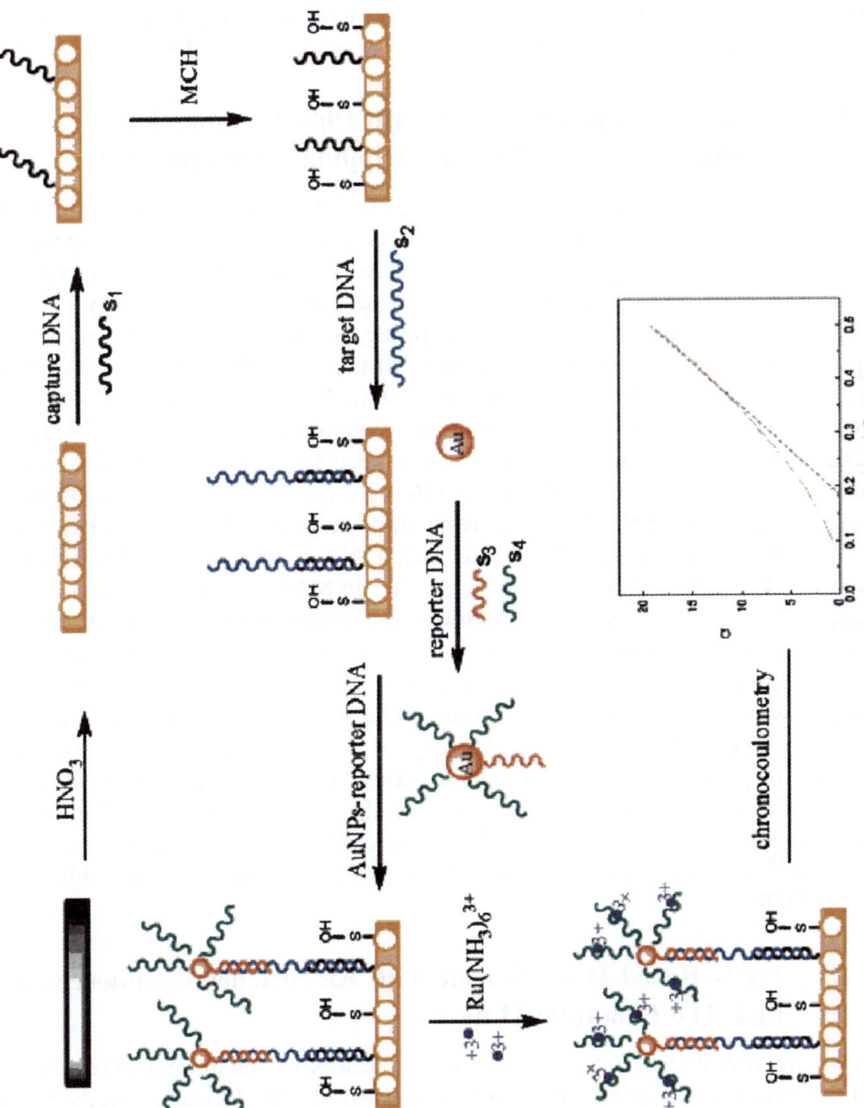

Scheme 10.1 Chronocoulometry determination of DNA hybridization through two steps of amplification.[49]

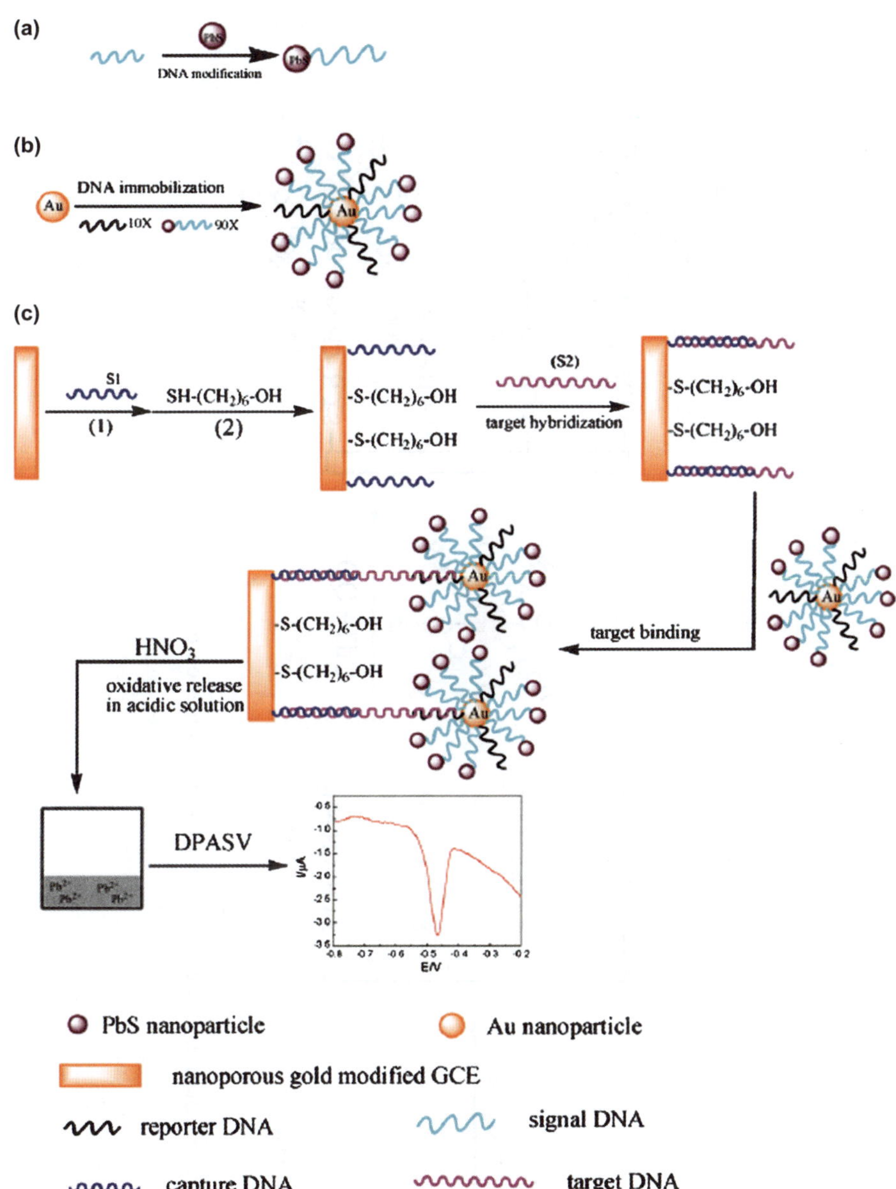

Scheme 10.2 Schematic illustration of the detection for DNA based on sandwich-type DPASV.[52]

Using this approach, Jin and coworkers[53] demonstrated a DNA-assay procedure using CdTe QDs as DNA labels at NPG electrode. Mercaptopropionic acid (MPA) capped CdTe QDs with size of 2.5 nm were prepared by reacting NaHTe with $CdCl_2$ and MPA at a pH of 9–10. The steps of this assay are shown in Scheme 10.3. The NPG surface was first modified with thioglycolic

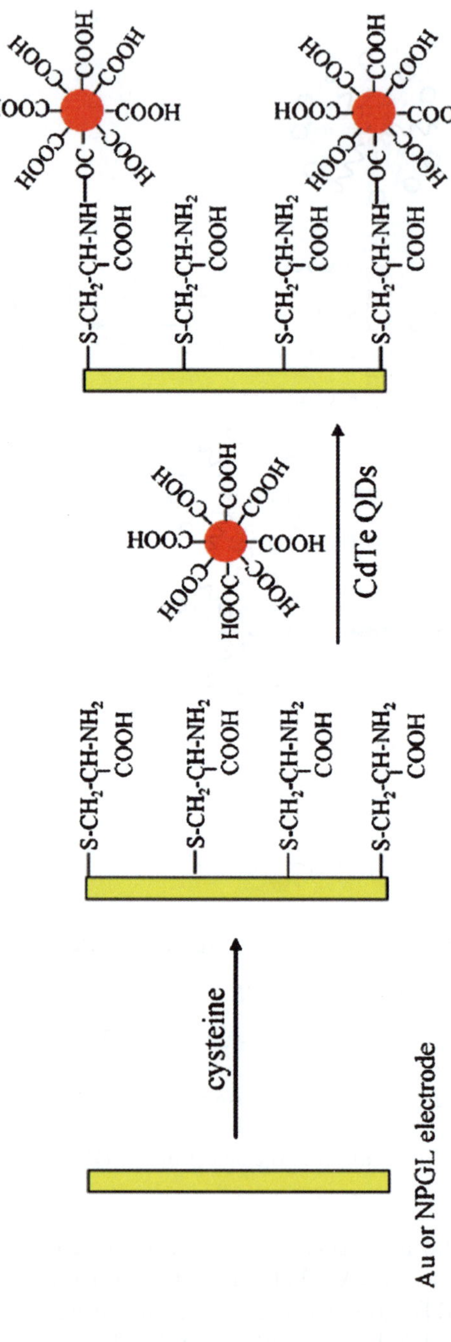

Scheme 10.3 Schematic representation of the process of ECL determination of DNA.[53]

acid (TGA), and amino-modified capture DNA S_1 was bound to TGA. The unbound TGA was blocked with bovine serum albumin to prevent nonspecific binding. Following this, the target DNA S_2, which contains complementary sequences to both capture DNA S_1 and probe DNA S_3, was hybridized with capture DNA S_1. Then, the target DNA S_2 was further hybridized with the amino-modified probe DNA S_3, forming the sandwich hybrid on the NPG electrode. Subsequently, MPA-capped CdTe QDs were labeled to the amino group end of the DNA S_3 of the sandwich hybrids. Finally, the ECL emission of the QDs-labeled DNA hybrids on the NPG was measured in the presence of co-reactant. The efficiency of various co-reactants including H_2O_2, O_2, and $S_2O_8^{2-}$ was compared, and $S_2O_8^{2-}$ was found to give the best result. In the absence of $S_2O_8^{2-}$, the CdTe QDs displayed negligible ECL, whereas prominent ECL signal was observed in the presence of $S_2O_8^{2-}$. The possible mechanism of this ECL process can be expressed by reactions (10.2)–(10.5). $S_2O_8^{2-}$ and CdTe QDs are electrochemically reduced to SO_4^{*-} and $CdTe^{*-}$ radicals, respectively. The SO_4^{*-} reacts with $CdTe^{*-}$ at the electrode producing SO_4^{2-} and excited $CdTe^*$, which emits light.

$$S_2O_8^{2-} + e^- \rightarrow SO_4^{2-} + SO_4^{*-} \tag{10.2}$$

$$CdTe + e^- \rightarrow CdTe^{*-} \tag{10.3}$$

$$SO_4^{*-} + CdTe^* \rightarrow CdTe^* + SO_4^{2-} \tag{10.4}$$

$$CdTe^* \rightarrow CdTe + h\nu. \tag{10.5}$$

The maximum ECL intensity of the QD-labeled DNA hybrids on the NPG electrode was proportional to target DNA S_2 concentration with a linear range of 5×10^{-15} to 1×10^{-11} M and a limit of detection of 4.7×10^{-15} M. Unlike the previously described procedures in which a large number of the transducer elements ($[Ru(NH_3)_6]^{3+}$ or PbS-NPs) were associated with each target DNA to achieve an amplification effect, this procedure linked only one CdTe QD to each target DNA. Thus, the performance of the ECL detection might be further improved if the amplification effect was incorporated.

A comparison of the various assays for the detection of DNA is summarized in Table 10.3. In summary, the above assays have shown that NPG is an excellent material for fabrication of DNA sensors. Although the above assays are single-target protocols, sensors for simultaneous detection of multiple DNA targets can be developed in the near future by immobilizing simultaneously various capture DNAs on the same NPG substrate to capture corresponding target DNAs followed by hybridization with reporter DNAs. Furthermore, sensors other than electrochemical techniques should also be investigated.

Table 10.3 Comparison of the various assays for the determination of DNA.

Label or indicator	Analytical technique	Detection limit of target DNA	Reference
Au-nanoparticles	Surface plasmon resonance	10 pM	43
Au-nanoparticles	Laser diffraction	50 fM	44
Liposome	Liposome-amplified electrochemical detection	50 fM	45
ZnS and CdSe quantum dots	Fluorescence	2 nM	46
Au-nanoparticles with Ag amplification	Bio-barcode-amplified scanometric detection	500 aM	47
PbS nanoparticles	Anodic stripping voltammetry	4.38 pM	48
NPG with bio-barcode amplification	Chronocoulometry	28 aM	49
NPG with PbS nanoparticles tags	Bio-barcode-amplified differential pulse anodic stripping voltammetry	0.26 fM	52
NPG with CdTe quantum dots	Electrochemiluminescence	4.7×10^{-15} M	53

10.5 NPG Sensors for Nitrogen-Containing Compounds

10.5.1 NPG Sensor for Detection of *p*-Nitrophenol

The *p*-nitrophenol (*p*-NP) is a toxic compound widely found in the environment. Therefore, it is important to develop simple and sensitive methods for trace analysis of *p*-NP. Recently, the sensitive detection of *p*-NP using an NPG electrode was evaluated by Liu *et al.*[54] In H_2SO_4 solution, the irreversible electro-reduction of *p*-NP produces 4-(hydroxyamino)phenol, which can be oxidized to 4-nitrosophenol at a more positive potential. The 4-nitorsophenol can be electrochemically reduced back to 4-(hydroxyamino)phenol. It was found that 4-nitrosophenol adsorbed more easily on the rough surface of NPG than on the smooth surface of polycrystalline gold. As a result, the redox currents of the 4-(hydroxyamino)phenol/4-nitrosophenol couple on the NPG were sensitive to the variation of *p*-NP concentration from 0.25 to 10 mg dm^{-3}. In addition, cyclic voltammograms showed that the current and peak potentials for the 4-(hydroxyamino)phenol/4-nitrosophenol couple were unaffected by other NP isomers, such as *ortho*-nitrophenol (*o*-NP) and *meta*-nitrophenol (*m*-NP) because the 2-(hydroxyamino)phenol/2-nitrosophenol and 3-(hydroxyamino)phenol/3-nitrosophenol couples occurred at potentials well separated from that of the 4-(hydroxyamino)phenol/4-nitrosophenol couple. This allowed the selective detection of *p*-NP in the presence of *o*-NP and *m*-NP. It should be noted that UV-visible spectrophotometry is unable to distinguish *p*-NP from *o*-NP or *m*-NP.

10.5.2 NPG Sensor for Amperometric Determination of Nitrite

Nitrite, NO_2^-, is a carcinogenic and teratogenic compound, and the determination of nitrite is important for environmental monitoring and health protection. Recently, Ge *et al.*[55] developed a method using NPG leaf-coated GCE (NPGL/GCE) as the working electrode for amperometric determination of nitrite by electrocatalytic oxidation of nitrite. As evidenced by dramatically reduced overvoltage required for NO_2^- oxidation, NPGL/GCE, compared with bare GCE, exhibited an outstanding electrocatalytic activity toward the oxidation of NO_2^- to NO_3^-. Interestingly, it was found that within a pH range from 4.5 to 8.0, the NO_2^- oxidation at NPGL/GCE is fairly pH-insensitive. The linear scan voltammetric anodic peak current of NO_2^- was proportional to the scan rate, indicating that the electrocatayltic oxidation of NO_2^- was a surface electron-transfer process. The amperometric response of NO_2^- on NPGL/GCE was at least 10 times higher than that at GCE. The amperometric signal varied linearly with the NO_2^- concentration from 1 μM to 1 mM with a detection limit of 1 μM. Species such as Na_2SO_4, KCl, NH_4NO_3, glucose, and C_2H_5OH at a concentration level higher than that of NO_2^- did not show any apparent interference with the determination of NO_2^- on NPGL/GCE.

10.6 NPG as Promising Substrates for Surface-Enhanced Raman Scattering

Nanostructured metals exhibit considerable differences from their bulk counterparts in optical properties because of the surface plasmon resonance (SPR) phenomenon. There are two types of SPR, propagating SPR and localized SPR, the properties of which are associated with the nanostructures of metals. NPG layers have simultaneous excitation of propagating SPR in planar metal films and localized SPR.[56,57] The SPR of NPG films shows wavelength dependence on the characteristics of the gold ligaments. In the other words, the size of the pore and the ligament on the NPG films determine the SPR wavelength. The SPR excitation of NPG provides the possibility of potential applications such as the detection of biomolecules by taking advantage of surface-enhanced Raman scattering (SERS).[57] Recently, NPG has been considered to be a promising substrate for SERS. SERS results from the enhanced inelastic scattering of the molecules adsorbed on nanostructured metals and alloys.[58,59] The enhancement of SERS is significantly determined by the nanoscale characteristics of the metallic substrates such as the surface morphology and size of the particles or ligaments. Although a dramatically high enhancement of SERS has been achieved at 'hot spots' in which strong electromagnetic fields are believed to produce at interparticle, it is still essential to develop a method for the fabrication of reliable SERS substrates that can provide ultrahigh enhancement of SERS and good reproducibility. A variety of attempts have been made to prepare high-quality SERS substrates, and NPG has recently attracted much attention because NPG has a very large surface area and high porosity. Qian *et al.* have studied the dependence of SERS intensities on the nanopore size of NPG.[60–62] They used rhodamine 6 G (R6G) as a probe molecule adsorbed on NPG films to study the dependence of SERS intensity on the nanopore size of NPG films. They found that the Raman-scattering intensities of R6G decreased significantly with increasing nanopore size from ~5 nm to ~700 nm, thus indicating that the SERS enhancement of the NPG films is easily affected by the nanopore size. The most apparent SERS enhancement of NPG was observed at the NPG film with an ultrafine nanopore size of ~5 nm. The integrated intensity of the Raman band 1650 cm^{-1} for the R6G molecule is almost twice as strong as that observed at the NPG film with a pore size of ~700 nm. This size dependence for SERS enhancement is normal for many other organic molecules rather than a unique behavior for the R6G molecule.[60,63] The SERS enhancement of NPG films is attributed to two reasons: the chemical enhancement from the interaction between the adsorbed molecules and the NPG films, and the electromagnetic enhancement from the combination of localized SPR and the incident laser. Many studies have demonstrated that the electromagnetic-field enhancement mainly contributes to the SERS enhancement of the NPG films, although the nanoporous structure really provides a huge surface area for the adsorption of analytes, resulting in strong Raman signals.

Although SERS is still not employed in chemical sensors, the SERS effect has been utilized for detecting organic and inorganic species including single molecule detection. A gold nanoporous structure has been made from gold nanoparticles, and this structure shows an excellent SERS reproducibility, physical stability, and strong Raman enhancement; the magnitude of the SERS enhancement factor was found to be twice as high as that from traditional dealloy nanoporous films. The structure obtained in this study may be a universal-type SERS substrate.[64] Hybrid gold nanopore-gold nanoparticle films with embedded hot spots have been prepared as an SERS substrate. A detection limit close to 10^{-11} mol L^{-1} and a giant enhancement factor of 1.5×10^{13} were observed. This substrate provides an opportunity for detecting the spatial orientation of single molecule within hot spots.[65] NPG has also been used for the detection of divalent metal ions at trace levels. A ligand was adsorbed on the NPG film to form complex ions with divalent metal ions, and the complexation-induced shift of a characteristic Raman peak of the metal ion receptor was detected.[66]

10.7　Concluding Remarks

A field in the development of sensors and diagnostic devices centered around NPG as alternatives to smooth polycrystalline gold and other semiconductor nanoporous materials is rapidly emerging. In this chapter, we have reviewed and discussed several examples of recent progress in various sensor formats that gain significant advantage from the incorporation of NPG, *i.e.* a high surface area favoring the surface-confined phenomena, high activity resulting

Table 10.4 Examples of NPG sensors.

Species determined	Sensor type	Reference
Ethanol and glucose	Enzyme-modified NPG	19
H_2O_2	Cytochrome c-encapsulated NPG	21
Glucose	Nonenzymatic NPGF/GCE	8
Glucose	Nonenzymatic monolayer Pt-coated NPGF	28
Glucose	Au-decorated nanoporous copper core-shell composite electrode	29
Escherichia coli	Pt-coated NPG	33
Dopamine	NPG/GCE	40, 41
Cancer biomarker: prostate-specific antigen	Antibody-adsorbed NPG	42
DNA	Sandwich hybridization on NPG with $[Ru(NH_3)_6]^{3+}$ transducer	49
Thrombin	Sandwich hybridization on NPG with $[Ru(NH_3)_6]^{3+}$ transducer	50
DNA	Sandwich hybridization on NPG with PbS nanoparticles transducer	52
DNA	Sandwich hybridization on NPG with CdTe nanoparticles electrochemiluminescene	53
p-Nitrophenol	NPG	54
Nitrite	NPG	55

from their unique nano-structural environment, and easily modifiable for selective interaction with various species. Examples include the as-prepared NPG, enzyme-modified NPG, catalytic metal-modified NPG, and DNA-immobilized NPG. NPG materials are just opening the door to novel analytical methods that are not possible using classical smooth gold. As summarized in Table 10.4, to date, work has focused mainly on NPG as electrochemical sensor platforms in liquid phase. However, in view of the fact that NPG exhibit excellent catalytic effects toward many gas compounds and extraordinary optical enhancement effects, novel applications may emerge in areas that have not even been envisaged, and we certainly believe the best is yet to come.

References

1. A. Wittstock, J. Biener and M. Baumer, *Phys. Chem. Chem. Phys.*, 2010, **12**, 12919–12930.
2. J. L. West and N. J. Halas, *Annu. Rev. Biomed. Eeng.*, 2003, **5**, 285–292.
3. Y. Bai, W. Yang, Y. Sun and C. Sun, *Sens. Actuat. B Chem.*, 2008, **134**, 471–476.
4. S. Cherevko and C.-H. Chung, *Electrochem. Commun.*, 2011, **13**, 16–19.
5. W. Huang, M. Wang, J. Zheng and Z. Li, *J. Phys. Chem. C*, 2009, **113**, 1800–1805.
6. M. Wooten, J. H. Shim and W. Gorski, *Electroanalysis*, 2010, **22**, 1275–1277.

7. Y. Deng, W. Huang, X. Chen and Z. Li, *Electrochem. Commun.*, 2008, **10**, 810–813.

8. Y. Xia, W. Huang, J. Zheng, Z. Niu and Z. Li, *Biosens. Bioelectron.*, 2011, **26**, 3555–3561.

9. R. C. Newman and K. Sieradzki, *MRS Bull.*, 1999, **24**, 12.

10. Y. Ding and J. Erlebacher, *J. Am. Chem. Soc.*, 2003, **125**, 7772.

11. D. V. Pugh, A. Dursun and S. G. Corcoran, *J. Mater. Res.*, 2003, **18**, 216–221.

12. J.-F. Huang and I.-W. Sun, *Adv. Funct. Mater.*, 2005, **15**, 989–994.

13. B. Seo and J. Kim, *Electroanalysis*, 2010, **22**, 939–945.

14. O. V. Shulga, K. Jefferson, A. R. Khan, V. T. D'Souza, J. Liu, A. V. Demchenko and K. J. Stine, *Chem. Mater.*, 2007, **19**, 3902.

15. H. Qiu, C. Xu, X. Huang, Y. Ding, Y. Qu and P. Gao, *J. Phys. Chem. C*, 2008, **112**, 14781.

16. H. Qiu, C. Xu, X. Huang, Y. Ding, Y. Qu and P. Gao, *J. Phys. Chem. C*, 2009, **113**, 2521–2525.

17. B. K. Jena and C. R. Raj, *Chem. Eur. J.*, 2006, **12**, 2702–2708.

18. J. Yuan, K. Wang and X. Xia, *Adv. Funct. Mater.*, 2005, **15**, 803.

19. H. Qiu, L. Xue, G. Ji, G. Zhou, X. Huang, Y. Qu and P. Gao, *Biosens. Bioelectron.*, 2009, **24**, 3014–3018.

20. C. E. Banks and R. G. Compton, *Analyst*, 2005, **130**, 1232–1239.

21. A. Zhu, Y. Tian, H. Liu and Y. Luo, *Biomaterials*, 2009, **30**, 3183–3188.

22. R. Wilson and A. P. F. Turner, *Biosens. Bioelectron*, 1992, **7**, 165.

23. M. Tominaga, T. Shimazoe, M. Nagashima and I. Taniguchi, *Electrochem. Commun.*, 2005, **7**, 189–193.

24. S. H. Yoo and S. Park, *Adv. Mater.*, 2007, **19**, 1612–1615.

25. Y. Ding, M. Chen and J. Erlebacher, *J. Am. Chem. Soc.*, 2004, **126**, 6876–6877.

26. P. Liu, X. Ge, R. Wang, H. Ma and Y. Ding, *Langmuir*, 2009, **25**, 561–567.

27. J. Jia, L. Cao and Z. Wang, *Langmuir*, 2008, **24**, 5932–5936.

28. H. Qiu and X. Huang, *J. Electroanal. Chem.*, 2010, **643**, 39–45.

29. L. Y. Chen, T. Fujita, Y. Ding and M. W. Chen, *Adv. Funct. Mater.*, 2010, **20**, 2279–2285.

30. A. S. Mittelmann, E. Z. Ron and J. Rishpon, *Anal. Chem.*, 2002, **74**, 903.

31. B. Serra, M. D. Morales, J. Zhang, A. J. Reviejo, E. H. Hall and J. M. Pingarron, *Anal. Chem.*, 2005, **77**, 8115–8121.

32. P. Geng, J. Zheng, X. Zhang, Q. Wang, W. Zhang, L. Jin, Z. Feng and Z. Wu, *Electrochem. Commun.*, 2007, **9**, 2157–2162.

33. Q. Yang, Y. Liang, T. Zhou, G. Shi and L. Jin, *Electrochem. Commun.*, 2009, **11**, 893–896.

34. R. M. Wightman, L. J. May and A. C. Michael, *Anal. Chem.*, 1988, **60**, 769.

35. J.-W. Mo and B. Ogorevc, *Anal. Chem.*, 2001, **73**, 1196–1202.

36. G. Z. Hu, D. P. Zhang, W. L. Wu and Z. S. Yang, *Colloid Surf. B: Biointerf.*, 2008, **62**, 199–205.

37. C. R. Raj, K. Tokuda and T. Ohsaka, *Bioelectrochemistry*, 2001, **53**, 183–191.

38. J. Huang, Y. Liu, H. Hou and T. You, *Biosens. Bioelectron.*, 2008, **24**, 632–637.

39. S. Zhu, H. Li, W. Niu and G. Xu, *Biosens. Bioelectron.*, 2009, **25**, 940–943.
40. H. J. Qiu, G. P. Zhou, G. L. Ji, Y. Zhang, X. R. Huang and Y. Ding, *Colloid Surf. B: Biointerf.*, 2009, **69**, 105–108.
41. W. A. El-Said, J.-H. Lee, B.-K. Oh and J.-W. Choi, *Electrochem. Commun.*, 2010, **12**, 1756–1759.
42. Q. Wei, Y. Zhao, C. Xu, D. Wu, Y. Cai, J. He, H. Li, B. Du and M. Yang, *Biosens. Bioelectron.*, 2011, **26**, 3714–3718.
43. L. He, M. D. Musick, S. R. Nicewarner, F. G. Salinas, S. J. Benkovic, M. J. Natan and C. D. Keating, *J. Am. Chem. Soc.*, 2000, **122**, 9071–9077.
44. T. A. Taton, C. A. Mirkin and R. L. Letsinger, *Science*, 2000, **289**, 1757–1760.
45. F. Patolsky, A. Lichtenstein and I. Willner, *Angew. Chem. Int. Ed.*, 2000, **39**, 940–943.
46. D. Gerion, F. Chen, B. Kannan, A. Fu, W. J. Parak, D. J. Chen, A. Majumdar and A. P. Alivisatos, *Anal. Chem.*, 2003, **75**, 4766–4772.
47. J.-M. Nam, S. I. Stoeva and C. A. Mirkin, *J. Am. Chem. Soc.*, 2004, **126**, 5932–5933.
48. W. Sun, J. Zhong, P. Qin and K. Jiao, *Anal. Biochem.*, 2008, **377**, 115–119.
49. K. Hu, D. Lan, X. Li and S. Zhang, *Anal. Chem.*, 2008, **80**, 9124–9130.
50. H. Qiu, Y. Sun, X. Huang and Y. Qu, *Colloid Surf. B: Biointerf.*, 2010, **79**, 304–308.
51. J. Wang, G. Liu and A. Merkoci, *J. Am. Chem. Soc.*, 2003, **125**, 3214–3215.
52. K. Hu, P. Liu, S. Ye and S. Zhang, *Biosens. Bioelectron.*, 2009, **24**, 3113–3119.
53. X. Hu, R. Wang, Y. Ding, X. Zhang and W. Jin, *Talanta*, 2010, **80**, 1737–1743.
54. Z. Liu, J. Du, C. Qiu, L. Huang, H. Ma, D. Shen and Y. Ding, *Electrochem. Commun.*, 2009, **11**, 1365–1368.
55. X. Ge, L. Wang, Z. Liu and Y. Ding, *Electroanalysis*, 2011, **23**, 381–386.
56. M. C. Dixon, T. A. Daniel, M. Hieda, D. M. Smilgies, M. H. W. Chan and D. L. Allara, *Langmuir*, 2007, **23**, 2414–2422.
57. F. Yu, S. Ahl, A.-M. Caminade, J.-P. Majoral, W. Knoll and J. Erlebacher, *Anal. Chem.*, 2006, **78**, 7346–7350.
58. M. Moskovist, *Rev. Mod. Phys.*, 1985, **57**, 783.
59. A. Campion and P. Kambhampati, *Chem. Soc. Rev.*, 1998, **27**, 241.
60. L. H. Qian, X. Q. Yan, T. Fujita, A. Inoue and M. W. Chen, *Appl. Phys. Lett.*, 2007, **90**, 153120.
61. L. H. Qian, A. Inoue and M. W. Chen, *Appl. Phys. Lett.*, 2008, **92**, 093113.
62. L.-H. Qian, Y. Ding, T. Fujita and M.-W. Chen, *Langmuir*, 2008, **24**, 4426–4429.
63. L.-Y. Chen, J.-S. Yu, T. Fujita and M.-W. Chen, *Adv. Funct. Mater.*, 2009, **19**, 1221–1226.
64. N.-J. Kim and M. Lin, *J. Nanosci. Nanotechnol.*, 2010, **10**, 5077.
65. L. Qian, B. Das, Y. Li and Z. Yang, *J. Mater. Chem.*, 2010, **20**, 6891.
66. Y. Zhao, J. N. Newton, J. Liu and A. Wei, *Langmuir*, 2009, **25**, 13833–13839.

Subject Index